|  |  |  | IIIA | IVA | VA | VIA | VIIA | O |
|---|---|---|---|---|---|---|---|---|
|  |  |  |  |  |  |  | 1.00797 ₁H Hydrogen | 4.0026 ₂He Helium |
|  |  |  | 10.811 ₅B Boron | 12.01 ₆C Carbon | 14.01 ₇N Nitrogen | 15.999 ₈O Oxygen | 18.998 ₉F Fluorine | 20.182 ₁₀Ne Neon |
|  |  |  | 26.98 ₁₃Al Aluminum | 28.09 ₁₄Si Silicon | 30.97 ₁₅P Phosphorus | 32.06 ₁₆S Sulfur | 35.45 ₁₇Cl Chlorine | 39.95 ₁₈Ar Argon |
| VIII | IB | IIB |  |  |  |  |  |  |
| 58.71 ₂₈Ni Nickel | 63.54 ₂₉Cu Copper | 65.37 ₃₀Zn Zinc | 69.72 ₃₁Ga Gallium | 72.59 ₃₂Ge Germanium | 74.92 ₃₃As Arsenic | 78.96 ₃₄Se Selenium | 79.91 ₃₅Br Bromine | 83.80 ₃₆Kr Krypton |
| 106.4 ₄₆Pd Palladium | 107.9 ₄₇Ag Silver | 112.4 ₄₈Cd Cadmium | 114.82 ₄₉In Indium | 118.69 ₅₀Sn Tin | 121.75 ₅₁Sb Antimony | 127.6 ₅₂Te Tellurium | 126.90 ₅₃I Iodine | 131.30 ₅₄Xe Xenon |
| 195.09 ₇₈Pt Platinum | 196.97 ₇₉Au Gold | 200.59 ₈₀Hg Mercury | 204.37 ₈₁Tl Thallium | 207.12 ₈₂Pb Lead | 208.98 ₈₃Bi Bismuth | (209) ₈₄Po Polonium | (210) ₈₅At Astatine | (222) ₈₆Rn Radon |

| 157.25 ₆₄Gd Gadolinium | 158.9 ₆₅Tb Terbium | 162.5 ₆₆Dy Dysprosium | 164.9 ₆₇Ho Holmium | 167.3 ₆₈Er Erbium | 168.9 ₆₉Tm Thulium | 173.0 ₇₀Yb Ytterbium | 175.0 ₇₁Lu Lutetium |
|---|---|---|---|---|---|---|---|
| (247) ₉₆Cm Curium | (247) ₉₇Bk Berkelium | (251) ₉₈Cf Californium | (254) ₉₉Es Einsteinium | (253) ₁₀₀Fm Fermium | (256) ₁₀₁Md Mendelevium | (253) ₁₀₂No Nobelium | (257) ₁₀₃Lr Lawrencium |

# PHYSICAL SCIENCE

a dynamic approach

# ROBERT T. DIXON

*Associate Professor of Astronomy*
*Physical Science Department*
*Riverside City College*

PRENTICE-HALL, INC.
Englewood Cliffs, New Jersey 07632

# PHYSICAL SCIENCE

## a dynamic approach

*Library of Congress Cataloging in Publication Data*

Dixon, Robert T
  Physical Science.

  Includes index.
  1. Science. I. Title.
Q161.2.D59    500.2    78-23209
ISBN 0-13-669820-4

Physical Science: A Dynamic Approach
Robert T. Dixon

Editorial/production supervision: Eleanor Henshaw Hiatt
Interior design and cover: Linda Conway
Manufacturing buyer: John Hall

© 1979 by Prentice-Hall, Inc., Englewood Cliffs, N.J. 07632

All rights reserved. No part of this book
may be reproduced in any form or
by any means without permission in writing
from the publisher.

Printed in the United States of America

10 9 8 7 6 5 4 3 2 1

PRENTICE-HALL INTERNATIONAL, INC., *London*
PRENTICE-HALL OF AUSTRALIA PTY. LIMITED, *Sydney*
PRENTICE-HALL OF CANADA, LTD., *Toronto*
PRENTICE-HALL OF INDIA PRIVATE LIMITED, *New Delhi*
PRENTICE-HALL OF JAPAN, INC., *Tokyo*
PRENTICE-HALL OF SOUTHEAST ASIA PTE. LTD., *Singapore*
WHITEHALL BOOKS LIMITED, *Wellington, New Zealand*

# summary contents

**complete contents** vii
**preface** xiii
**chapter one** early concepts of nature 1
**chapter two** the rebirth of science 19
**chapter three** motion 31
**chapter four** gravity 45
**chapter five** energy 59
**chapter six** waves and sound 77
**chapter seven** electricity and magnetism 97
**chapter eight** electromagnetic spectrum 125
**chapter nine** relativity 149
**chapter ten** the atom 159
**chapter eleven** the periodic nature of elements 183
**chapter twelve** states of matter 205
**chapter thirteen** chemical energy 221
**chapter fourteen** chemistry of living organisms 231
**chapter fifteen** the dynamic earth 245
**chapter sixteen** an ocean of air and water 271
**chapter seventeen** earth–moon, a binary system 301
**chapter eighteen** the solar system 321
**chapter nineteen** stars and nebulae 355
**chapter twenty** the cosmos 373
**chapter twenty-one** extraterrestrial life 385
**appendixes** 391
**index** 395

# complete contents

summary contents v

preface xiii

## early concepts of nature 1
EARLY HUMANS 2    NOMAD TURNS TO FARMING 3
EGYPTIAN CIVILIZATION 4    EGYPTIAN CALENDAR 5
BABYLONIA, CRADLE OF ASTROLOGY 6
SUPERSTITION VERSUS SCIENTIFIC EXPLANATION 7    STONEHENGE 8
GREECE—BIRTHPLACE OF NEW IDEAS 9    THE ATOM ENVISIONED 10
A REAL UNIVERSE 11    ALEXANDRIAN ACADEMY 13
A BROAD SCIENTIFIC INTEREST 16    THE PTOLEMAIC MODEL 17
QUESTIONS 18

## the rebirth of science 19
ARABIC INFLUENCE 20    SPAIN PLAYS AN IMPORTANT ROLE 22
A GREEK STRONGHOLD 22    REBIRTH OF LEARNING IN EUROPE 23
CHURCH DOGMA 23    CONTRIBUTIONS BY MANY 24
ART, SCIENCE, AND BOOKMAKING 24    THE COPERNICAN REVOLUTION 25
GALILEO, FATHER OF EXPERIMENTAL SCIENCE 26
THE WEIGHT OF CHURCH DOGMA 28    THE INDUCTIVE METHOD IS BORN 29
QUESTIONS 29

## motion 31
METRIC SYSTEM 32    LENGTH, AREA, AND VOLUME 33
MASS AND DENSITY 34    TIME 35    VELOCITY 35    AVERAGE VELOCITY 36
SIR ISAAC NEWTON 37    MOMENTUM 40    IMPULSE 41
ANGULAR MOMENTUM 42    QUESTIONS 42

## gravity 45
THE CONCEPT OF GRAVITATION 46    THE UNIVERSAL NATURE OF GRAVITY 49
ACCELERATION BECAUSE OF GRAVITY 51    CIRCULAR MOTION 53
ORBITING SATELLITES 56    QUESTIONS 57

## 5 energy 59

WORK *60*     EFFICIENCY AND MECHANICAL ADVANTAGE *61*     POWER *64*
ENERGY—ITS MANY FORMS *65*     POTENTIAL ENERGY *65*
KINETIC ENERGY *67*     POTENTIAL ENERGY VERSUS KINETIC ENERGY *68*
TEMPERATURE VERSUS HEAT  *69*     HEAT, A FORM OF ENERGY *71*
HEAT, MECHANICAL EQUIVALENT *72*     STEAM ENGINE *73*
LAWS OF THERMODYNAMICS *74*     ENTROPY—A MEASURE OF DISORDER *74*
QUESTIONS *75*

## 6 waves and sound 77

THE GRAPH OF A VIBRATING OBJECT *79*     PROPERTIES OF A WAVE *81*
REFLECTION *83*     REFRACTION *83*     INTERFERENCE *84*     DIFFRACTION *85*
SOUND *86*     SPEED OF SOUND *88*     THE HUMAN EAR *90*
SOURCES OF MUSICAL SOUNDS *90*     QUALITY OF SOUND *94*
HIGH FIDELITY REPRODUCTION OF SOUND *94*
SOURCE OF VIBRATION IN WIND INSTRUMENTS *95*     QUESTIONS *95*

## 7 electricity and magnetism 97

ELECTRICAL FORCES *98*     THE ATOMIC MODEL OF A CHARGED BODY *100*
THE ELECTROSCOPE *100*     CHARGING AN ELECTROSCOPE *101*
CHARGING BY INDUCTION *101*     COULOMB'S LAW OF CHARGES *103*
ELECTRICAL AND GRAVITATIONAL FORCES *104*
ELECTRICAL FORCE FIELD *104*     THE VOLTAIC CELL *106*
ELECTROMOTIVE FORCE *106*     ELECTRIC CIRCUITS *107*
SERIES CIRCUITS *108*     PARALLEL CIRCUITS *108*
A PRACTICAL CONSIDERATION *109*     MAGNETISM *109*
THE ATOMIC MODEL OF MAGNETISM *111*     THE SOLENOID *112*
VAN ALLEN BELTS *114*     TORQUE—IN A MAGNETIC FIELD *115*
ALTERNATING CURRENT GENERATOR *116*
TRANSMISSION OF ELECTRICAL ENERGY *120*
DIRECT-CURRENT GENERATORS *122*     QUESTIONS *123*

## 8 electromagnetic spectrum 125

AN OSCILLATING CHARGED PARTICLE *126*     THE SPEED OF LIGHT *127*
WAVELENGTHS *128*     LIGHT TRAVELS IN A STRAIGHT LINE *129*
THE LAW OF REFLECTION *131*     REFRACTION *131*
THE REFRACTOR TELESCOPE *131*     DISPERSION OF LIGHT *133*
THE SPECTRUM *133*     THE BOHR MODEL OF THE ATOM *134*
THE SUN'S SPECTRUM *136*     DOPPLER EFFECT *137*
TOOLS FOR OBSERVATION—THE HUMAN EYE *140*     THE CAMERA *141*
THE TELESCOPE *141*     THE BINOCULAR *142*     THE MICROSCOPE *144*
ELECTRON MICROSCOPE *145*     QUESTIONS *146*

viii

## 5 energy — 59

WORK 60     EFFICIENCY AND MECHANICAL ADVANTAGE 61     POWER 64
ENERGY—ITS MANY FORMS 65     POTENTIAL ENERGY 65
KINETIC ENERGY 67     POTENTIAL ENERGY VERSUS KINETIC ENERGY 68
TEMPERATURE VERSUS HEAT 69     HEAT, A FORM OF ENERGY 71
HEAT, MECHANICAL EQUIVALENT 72     STEAM ENGINE 73
LAWS OF THERMODYNAMICS 74     ENTROPY—A MEASURE OF DISORDER 74
QUESTIONS 75

## 6 waves and sound — 77

THE GRAPH OF A VIBRATING OBJECT 79     PROPERTIES OF A WAVE 81
REFLECTION 83     REFRACTION 83     INTERFERENCE 84     DIFFRACTION 85
SOUND 86     SPEED OF SOUND 88     THE HUMAN EAR 90
SOURCES OF MUSICAL SOUNDS 90     QUALITY OF SOUND 94
HIGH FIDELITY REPRODUCTION OF SOUND 94
SOURCE OF VIBRATION IN WIND INSTRUMENTS 95     QUESTIONS 95

## 7 electricity and magnetism — 97

ELECTRICAL FORCES 98     THE ATOMIC MODEL OF A CHARGED BODY 100
THE ELECTROSCOPE 100     CHARGING AN ELECTROSCOPE 101
CHARGING BY INDUCTION 101     COULOMB'S LAW OF CHARGES 103
ELECTRICAL AND GRAVITATIONAL FORCES 104
ELECTRICAL FORCE FIELD 104     THE VOLTAIC CELL 106
ELECTROMOTIVE FORCE 106     ELECTRIC CIRCUITS 107
SERIES CIRCUITS 108     PARALLEL CIRCUITS 108
A PRACTICAL CONSIDERATION 109     MAGNETISM 109
THE ATOMIC MODEL OF MAGNETISM 111     THE SOLENOID 112
VAN ALLEN BELTS 114     TORQUE—IN A MAGNETIC FIELD 115
ALTERNATING CURRENT GENERATOR 116
TRANSMISSION OF ELECTRICAL ENERGY 120
DIRECT-CURRENT GENERATORS 122     QUESTIONS 123

## 8 electromagnetic spectrum — 125

AN OSCILLATING CHARGED PARTICLE 126     THE SPEED OF LIGHT 127
WAVELENGTHS 128     LIGHT TRAVELS IN A STRAIGHT LINE 129
THE LAW OF REFLECTION 131     REFRACTION 131
THE REFRACTOR TELESCOPE 131     DISPERSION OF LIGHT 133
THE SPECTRUM 133     THE BOHR MODEL OF THE ATOM 134
THE SUN'S SPECTRUM 136     DOPPLER EFFECT 137
TOOLS FOR OBSERVATION—THE HUMAN EYE 140     THE CAMERA 141
THE TELESCOPE 141     THE BINOCULAR 142     THE MICROSCOPE 144
ELECTRON MICROSCOPE 145     QUESTIONS 146

# complete contents

summary contents v

preface xiii

## early concepts of nature   1

EARLY HUMANS 2    NOMAD TURNS TO FARMING 3
EGYPTIAN CIVILIZATION 4    EGYPTIAN CALENDAR 5
BABYLONIA, CRADLE OF ASTROLOGY 6
SUPERSTITION VERSUS SCIENTIFIC EXPLANATION 7    STONEHENGE 8
GREECE—BIRTHPLACE OF NEW IDEAS 9    THE ATOM ENVISIONED 10
A REAL UNIVERSE 11    ALEXANDRIAN ACADEMY 13
A BROAD SCIENTIFIC INTEREST 16    THE PTOLEMAIC MODEL 17
QUESTIONS 18

## the rebirth of science   19

ARABIC INFLUENCE 20    SPAIN PLAYS AN IMPORTANT ROLE 22
A GREEK STRONGHOLD 22    REBIRTH OF LEARNING IN EUROPE 23
CHURCH DOGMA 23    CONTRIBUTIONS BY MANY 24
ART, SCIENCE, AND BOOKMAKING 24    THE COPERNICAN REVOLUTION 25
GALILEO, FATHER OF EXPERIMENTAL SCIENCE 26
THE WEIGHT OF CHURCH DOGMA 28    THE INDUCTIVE METHOD IS BORN 29
QUESTIONS 29

## motion   31

METRIC SYSTEM 32    LENGTH, AREA, AND VOLUME 33
MASS AND DENSITY 34    TIME 35    VELOCITY 35    AVERAGE VELOCITY 36
SIR ISAAC NEWTON 37    MOMENTUM 40    IMPULSE 41
ANGULAR MOMENTUM 42    QUESTIONS 42

## gravity   45

THE CONCEPT OF GRAVITATION 46    THE UNIVERSAL NATURE OF GRAVITY 49
ACCELERATION BECAUSE OF GRAVITY 51    CIRCULAR MOTION 53
ORBITING SATELLITES 56    QUESTIONS 57

## 9 relativity — 149

RELATIVE MOTION *150*    MICHELSON-MORLEY EXPERIMENT *152*
THE LORENTZ CONTRACTION *154*    ALBERT EINSTEIN *154*
A THOUGHT EXPERIMENT *155*
THE EQUIVALENCY OF MASS AND ENERGY *156*    QUESTIONS *157*

## 10 the atom — 159

MODERN ATOMIC THEORY *160*    RADIOACTIVE ATOMS *161*
GAMMA RAYS *162*    THE ELECTRON *163*    THE NUCLEUS *165*
THE QUANTUM THEORY OF RADIATION *166*
FORCES WITHIN THE NUCLEUS *170*    RADIOACTIVITY *171*
METHOD OF DECAY *174*    ATOMIC ENERGY *176*
THERMONUCLEAR FUSION *179*    QUESTIONS *180*

## 11 the periodic nature of elements — 183

UNITS OF MATTER *184*    ATOMIC WEIGHTS *185*
FAMILY GROUPS OF ELEMENTS *185*    METALS VERSUS NONMETALS *189*
ATOMIC ENERGY LEVELS *190*    QUANTUM MECHANICAL MODEL *191*
VALENCE *193*    CHEMICAL BONDS *194*    COVALENT BONDING *196*
POLAR COVALENT BONDS *197*    ELECTRONEGATIVITY *201*    QUESTIONS *202*

## 12 states of matter — 205

GASES—THEIR PROPERTIES *206*    KINETIC-MOLECULAR THEORY *207*
GAS LAWS *207*    THE LIQUID STATE *209*    THE SOLID STATE *211*
OTHER CRYSTALLINE FORMS *215*
MOLECULAR MOTION IN SOLIDS *216*    CHANGE OF STATE *217*
PLASMAS *218*    QUESTIONS *220*

## 13 chemical energy — 221

EXOTHERMIC OR ENDOTHERMIC REACTIONS *222*
OXIDATION-REDUCTION REACTIONS *224*    BALANCING EQUATIONS *225*
GRAM-FORMULA-WEIGHT *226*    ACIDS AND BASES *227*    BASES *228*
ACID-BASE REACTIONS *229*    QUESTIONS *230*

## 14 chemistry of living organisms — 231

CARBON, THE CENTRAL ELEMENT *232*    CARBON RINGS *235*    FOODS *236*
CARBOHYDRATES *236*    LIPIDS (FATTY ACIDS) *237*
SOAPS AND DETERGENTS *239*    STEROIDS *241*    PROTEINS *241*
DNA—THE CODE OF LIFE *242*    QUESTIONS *243*

## 15 the dynamic earth — 245

EARTHQUAKES 246   SEISMOGRAPH 248   VOLCANOS 250
PLATE TECTONICS 251   COLLIDING PLATES AND SUBDUCTION 253
EROSION 257   READING THE HISTORY OF THE EARTH IN THE ROCKS 258
MINERALS AND CRYSTAL STRUCTURE 261   RADIOACTIVE ELEMENTS 263
AGE OF ROCKS 265   DATING ORGANIC MATERIAL 266
GEOLOGICAL ERAS 267   QUESTIONS 269

## 16 an ocean of air and water — 271

EARTH'S ATMOSPHERE 272   A FUNCTION OF TEMPERATURE 274
ATMOSPHERIC PRESSURE 274   A UNIQUE ATMOSPHERE 276
HUMIDITY 276   SOLAR HEATING 277
CURRENTS OF AIR (PREVAILING WINDS) 278   THE EFFECT OF TIDES 285
GEOLOGY OF THE OCEAN FLOOR 287   WEATHER 287
CLOUD SEEDING 289   CLOUD FORMS 291
WEATHER FRONTS (SEASONAL VARIATION) 292   WEATHER MAPS 293
POLLUTION OF THE ATMOSPHERE 295   CARBON MONOXIDE 295
HYDROCARBONS 295   NITRIC OXIDES 296   PHOTOCHEMICAL SMOG 296
PAN 297   LONDON-TYPE SMOG 297   PARTICULATE MATTER 297
SOLUTION TO POLLUTION 298   QUESTIONS 298

## 17 earth-moon, a binary system — 301

ROTATION OF THE EARTH 302   REVOLUTION OF THE EARTH 302
SIGNS OF THE ZODIAC 305   PRECESSION OF THE EQUINOXES 308
AGE OF AQUARIUS 308   EARTH-MOON, A BINARY SYSTEM 310
PHASES OF THE MOON 311   ECLIPSES 313   LUNAR ECLIPSE 316
A TRIP TO THE MOON 317   AGE AND HISTORY OF THE MOON 318
QUESTIONS 319

## 18 the solar system — 321

WANDERERS IN THE SKY 322   SIXTEENTH CENTURY OBSERVATIONS 323
SEVENTEENTH CENTURY OBSERVATIONS 323
PHYSICAL PROPERTIES OF THE PLANETS 326   MERCURY 328   VENUS 330
MARS 331   JUPITER 339   SATURN 342   URANUS 343   NEPTUNE 345
PLUTO 346   ASTEROIDS 348   COMETS 349   METEORS 351
ORIGIN OF THE SOLAR SYSTEM 353   QUESTIONS 353

## 19 stars and nebulae — 355

STAR ENERGY 356   STARS IN GENERAL 358   BINARY STARS 360
THE LIFE CYCLE OF A STAR 361   BLACK HOLES 365
INTERSTELLAR MATERIAL 366   INTERSTELLAR MOLECULES 368
DUST NEBULAE 368   COSMIC RAYS 369   QUESTIONS 370

## the cosmos 373
VARIETY IN GALAXIES *374*   UNUSUAL GALAXIES *378*   QUASARS *380*
RED SHIFT LAW *380*   COSMOLOGIES *382*   QUESTIONS *382*

## extraterrestrial life 385
WHAT IS LIFE? *386*   THE SEARCH FOR LIFE *387*
DIALOGUE OR MONOLOGUE? *387*   OUR COMMITMENT *389*   QUESTIONS *389*

## appendixes 391
## index 395

# preface

This book has grown out of a series of lectures that have been presented, with great success, to students who typically have little or no background in science or mathematics. My aim is to present, in a rather broad spectrum, the spirit of scientific investigation and to develop in the student an appreciation for the physical universe. To accomplish this aim, concepts are developed with reference to everyday experiences.

The history of science is used as a tool whereby the student is made aware of the flow of ideas that have issued from the minds of humans. Of even greater significance, however, the book shows the interaction and interdependence of one person's work upon that of another.

Each chapter is developed from the most obvious relationships (based on experiences that are common to almost everyone or at least those that can be easily visualized) to the more obscure relationships. Technical terms are not used without definition; in fact, the order of introduction of technical terms is the key to this readable, understandable text.

Certainly this book will explain many of the phenomena students have wondered about and will aid in a new appreciation of the physical universe. I hope it will convey some of the thought processes that brought human beings to their present understanding of the universe and will create an interest for deeper study in the physical sciences.

The book is designed for a one-semester or two-quarter course for liberal arts students. Basically it utilizes a descriptive approach, but the quantitative flavor of science is also introduced and developed in a number of sections. Where this occurs, concepts are always verbalized to provide understanding in cases where mathematical background is lacking.

I would like to acknowledge the fine cooperation of the numerous professional and technical organizations who supplied illustrations for this text; the very generous assistance of my students who tested many of the approaches I have taken, and of my colleagues who have criticized portions of the work—Dr. Virginia Holten and Professors John Elliott, Harold Nemer, John Georgakakos, Ray Hawley, and William Wood. I would also like to thank the staff at Prentice-Hall, especially Logan Campbell, college editor; Eleanor Henshaw Hiatt, production editor; and Linda Conway, designer, for their conscientious transformation of the manuscript into a book.

And thanks especially to my wife, Marian, for her constant encouragement, and to my children for their patience, for without them this book could not have been written.

*Riverside, California* R.T.D.

# 1

# early concepts of nature

Humans are forever trying to relate to the universe in which they live—largely out of necessity, particularly in the early stages of their development. They must live harmoniously with nature or continually lock horns with it. The very fact that humans are intelligent, thinking beings suggests that they have observed things about themselves, have seen relationships between various factors, and have proceeded to turn these natural phenomena to their own advantage. These actions comprise the three fundamental aspects of early "science":

1. Making observations.
2. Seeing relationships.
3. Making use of these relationships.

## EARLY HUMANS

Humans' most basic needs—food, clothing, and shelter—motivated them to continually seek better ways to provide these necessities. Imagine people of the Stone Age who had only used a club to kill animals coming upon an animal that had been accidentally impaled on a broken branch of a tree. Making such an observation, they may have recognized the advantage of impaling the animal (rather than clubbing it), and thus the concept of the spear was born. The stone (flint) tools shown in Figure 1.1 illustrate significant advances beyond the first spearlike objects. By repeated attempts to chip stone, people of the Stone Age learned to produce such tools rather consistently. The arrow-shaped stones provided a more efficient point for the spear, piercing the hide of the animal better than a sharpened stick; and the scraper [Figure 1.1(b)] tells the story of skins scraped to provide clothing.

Perhaps the next most significant discovery was fire, for it was to change the approach to making tools, and it permitted the development of crafts. We would hesitate to say that humans "invented" fire; rather they simply observed a natural phenomenon that sometimes resulted when lightening struck a forest, burning that forest, and leaving behind material that was different from the growing trees—namely, charred, blackened wood. Today we would say that they witnessed a chemical change. They also recognized the value of the fire in cooking food and

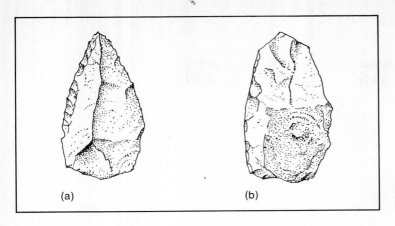

**Figure 1.1.** Flint tools of the Stone Age: (a) spear head; (b) scraper for working hides.

warming a cave for their own comfort, and here we see the beginning of an aspect of science that produces creature comforts beyond the bare necessities of life.

By 4000 B.C., near the end of the Stone Age, humans had moved into what is called the Neolithic period—the age of working with metals. Perhaps this period began with the discovery of a shiny material—gold—which sometimes appears as almost a pure nugget. In an attempt to shape such material, these later Stone Age people found that they could beat the gold into a flat, sharp-edged instrument (or ornament) without it breaking; however, such an instrument would not hold an edge but would blunt easily.

Eventually other metals such as copper and tin were found; but these were not in nugget form and had to be separated from the rock (ore) by heating in a fire. Thus smelting was born, and by 3000 B.C. humans had learned to produce metal instruments that ranged from hunting spears to cooking utensils. Copper and tin were blended by heating to form bronze; hence the designation of this period as the Bronze Age. The fact that bronze was capable of maintaining a sharp edge was discovered only after many trials, some of which surely led to useless products. In a similar fashion the path of modern science is not always a smooth one but often leads into blind alleys from which the researcher must turn back and take another road.

## NOMAD TURNS TO FARMING

Because early humans were basically hunters, they were also nomads, always on the move in search of a new food supply. By about 5000 B.C., however, a significant change took place. Humans learned to domesticate animals and grow crops, which allowed them to remain in one place. What would such persons look for if they were settling in one place to farm? They would certainly desire fertile soil, a water supply, and warmth of the sun for at least a part of the year. The earth provides all these elements in certain naturally fertile river valleys.

Although nomadic people would certainly have noticed cycles of nature such

**4**

early
concepts
of nature

as day and night and perhaps the monthly cycle of moon phases, as farmers their success or failure would depend on recognizing the cycle of seasons. If they planted seed in the wrong season, they could not expect a harvest. At first this cycle of sowing and reaping must have gone through a period of experimentation (trial and error), but within a few years certainly these farmers must have recognized a correlation between the height of the noonday sun and the proper planting time.

Let us suppose that these early people had decided that the springtime was the best time to plant seed. How would they recognize such a season when no calendar existed? They might have simply placed a stick in the ground pointing directly to the noonday sun on the day that they had found best for planting. On this day the stick would have no shadow, but as the sun appeared more and more to the north (in summer), the stick would have a shadow even at noon (see Figure 1.2). Then as the sun worked its way southward in the wintertime, the stick would also cast a shadow at noon. And when the sun returned to the intermediate position and the stick cast no shadow, it would be time to plant seed again. A year had passed.

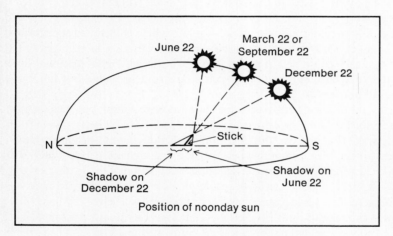

**Figure 1.2.** A stick pointing to the noonday sun on March 22 casts no shadow, but as the sun moves northward it casts a shadow even at noon, and likewise when it moves southward. By this simple experiment, one may tell when a year has passed.

The fact that men and women could remain in a single location, producing their own food, led to another significant step—that is, family groups banding together for their mutual benefit—beginning of civilizations. Because of the agricultural nature of this life, it was only natural that early civilizations should arise in fertile river valleys, such as along the Tigris and Euphrates rivers of Mesopotamia (today called Syria) and in the Nile Valley of Egypt.

## EGYPTIAN CIVILIZATION

The very existence of life in Egypt was based on the flooding of the Nile River, for this phenomenon brought natural irrigation and fertile silt to the broad farm lands on the river banks. Characteristic of any civilization is the division of labor and

**5**

early concepts of nature

specialization, and it soon became the prerogative of the priestly order to predict the overflow of the Nile. This might be possible if the cycle of its overflow happened to match one of the other recognizable cycles of nature. Sure enough, the Nile was observed to begin to rise when the sun appeared at its highest point at noon (to us, June 22); so then it became a matter of predicting the longest day in the year. As a testimony to the significance of this day, we find numerous monuments oriented toward the rising point of the sun on that longest day of the year—the summer solstice. These monuments take the form of temples with corridors oriented toward that point, and pyramids with edges or corridors likewise oriented. Some of the temples were built with corridors aligned through several buildings so that only on the day of the summer solstice would the sunlight flood the entire length and seem to "flash" upon the "holy of holies." It is easy to see how the sun could become a godlike symbol for worship. An expression of this worship is the Egyptian concept of the universe. The sun god Ra was believed to move in his barge up over the back of the sky goddess Nut and into the hands of the god of the underworld at night, then to be conducted under the flat earth to reappear the next day (see Figure 1.3).

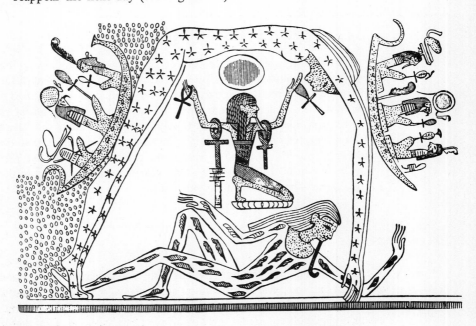

**Figure 1.3.** An Egyptian concept of the universe and the motion of the sun across the sky. The sun god Ra travels in his barge up over the sky goddess Nut. (Yerkes Observatory)

## EGYPTIAN CALENDAR

Probably the first people who came to the Nile Valley brought with them a lunar (moon) calendar. The moon takes about 30 days to go through all its phases, and 12 such 30-day months produce a year of 360 days. As soon as the early

Egyptians began to observe successive summer solstices, it became apparent that their 360-day year was in error by about five days, and if left uncorrected, the summer solstice would appear about five days earlier each year. Such a calendar did not reflect seasonal changes, however; by adding a mini-month of five days each year, the Egyptians brought their calendar into approximate synchronization with the seasons. Did they not know that the average year is $365\frac{1}{4}$ days? We believe that the priests realized this, for in only eight years of counting days between the summer solstices they would have recorded the following:

| | |
|---|---|
| First year | 365 days |
| Second year | 365 days |
| Third year | 366 days |
| Fourth year | 365 days |
| Fifth year | 365 days |
| Sixth year | 365 days |
| Seventh year | 366 days |
| Eighth year | 365 days |
| Average | $2{,}922 \div 8 = 365\frac{1}{4}$ days per year |

This illustrates a very important principle that has always been valid in scientific investigations: The result of making many independent observations of a phenomenon and then averaging those observations can be assumed to produce a result that is more accurate than any single observation chosen at random.

We may speculate that even though the priests knew that there were $365\frac{1}{4}$ days in a year, they may have retained 365 days in their calendar so that only they would know exactly where they were in the cycle. This may have been a move to retain their power. (The idea of a leap year was not instigated until the time of Julius Caesar—hence the Julian calendar.)

## BABYLONIA, CRADLE OF ASTROLOGY

We find evidence that the Babylonians were also concerned with the cycle of day and night and the phases of the moon and seasons, but they exhibited still another interest—the study of astrology. The astrologer charted the motion of the sun, moon, and planets among the stars on a daily basis, with the belief that these objects exerted a direct influence on the affairs of humans and allowed them to predict future events. There was virtually no distinction between astrology and astronomy at this time, as both disciplines were based on similar types of observations, and, the dual nature of the combined disciplines persisted well into the sixteenth and seventeenth centuries. Although astronomers and astrologers have parted company in modern times, astronomers are still indebted to early astrologers for the records they kept on planetary motion.

Both the Egyptians and Babylonians utilized the 12 divisions of the sky called

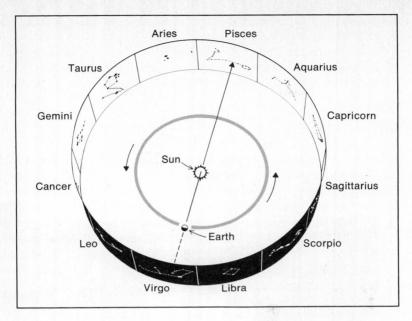

**Figure 1.4.** As the earth revolves about the sun, the sun appears to align itself with each sign of the zodiac in succession. The sun is shown aligned with Pisces, but in one month's time it will align with Aries, and so forth. In one year, the sun will again align with Pisces.

the *signs of the zodiac*. The sun, moon, and planets appear to move through these 12 constellations—the moon in a period of approximately one month and the sun in a period of one year (see Figure 1.4). The entire sky is divided into 88 constellations, each one covering a specific area like states within a country. Most of these constellations were named by the Babylonians and Egyptians.

## SUPERSTITION VERSUS SCIENTIFIC EXPLANATION

It was traditionally the responsibility of the astrologer to predict eclipses, for if an eclipse had been predicted, then the terror and stigma attached to such events were mollified; but woe be to the astrologer who failed to predict an eclipse. Usually the astrologer was in the employ of a king or ruler, and failure to predict correctly might cost the astrologer his life.

For early humans the only explanation for an eclipse lay in superstitions that had grown over the years. For example, tradition suggests that certain cultures thought that a dragon had swallowed the sun, and that by banging pans the people could frighten the dragon who would then regurgitate the sun. Only after such an event is explained in terms of the moon passing between the earth and the sun, and only after such an event becomes predictable, can the superstition be dispelled.

**8**

early
concepts
of nature

You will see this pattern repeated many times as humans' understanding of the universe broadens and deepens.

We find evidence of early civilizations in India and China; however, the evidence is not as clearly defined as in Babylon and Egypt. What is a bit surprising is that we also have evidence of an early civilization in the British Isles.

## STONEHENGE

There are numerous arrangements of stones dotting the plains of England and Brittany (France), but the most spectacular is found in the Salisbury Plain of England, known as Stonehenge. Scientists have dated this structure to 4000 B.C., yet it shows a sophistication that seems impossible for that time. First, the stones are arranged so precisely that the midwinter sunrise aligns perfectly with two principal stones (see Figure 1.5), as does the midwinter sunset and the midsummer sunrise and sunset—in this way measuring the seasons as precisely as did the Egyptians. But Stonehenge does much more. It also records rising and setting points of the moon that vary significantly in one month and also in one year. As a result, it is possible to predict the alignment of the moon, sun, earth, and, hence, the occurrence of eclipses. The unannounced and unexplained occurrence of an eclipse has always been a fearsome thing. Imagine the sky going dark in the middle of the day and then mysteriously becoming light again!

The superstition that surrounds such events gradually gave way to understanding as we move forward in time to the civilization of Greece.

**Figure 1.5.** Stonehenge, on Salisbury Plain, England. (British Tourist Authority)

By 1000 B.C. we see the birth of civilization that was to touch the intellectual pursuits of humans for thousands of years. Fundamentally different than the earlier cultures, the Greeks were not farmers but rather manufacturers, traders, and philosophers. This is to say that they had turned their crafts to production beyond their own individual needs, and, furthermore, their socio-economic structure permitted certain members to devote their time to pursuits that were motivated by curiosity—to philosophize on the nature of things about them. For example, craftsmen for years had taken a bluish stone, heated that stone, and then obtained a red metal (copper) from it, perhaps without thinking too much about what they were doing—they were satisfied with the results. Now in ca. 600 B.C. a Greek named Thales (thought of by many as the founder of Greek science) was not only thinking about that particular phenomenon but was generalizing to the point of thinking that everything was made of the same substance—simply having different forms. If you were looking for a single substance still very evident about you today, what would you choose? Thales chose water as his universal substance, for he knew of the Mediterranean, rivers, lakes, wells, and moisture in the air, as well as the necessity of water to life itself. It is little wonder that his model of the universe included a flat earth floating on water with a canopy of stars overhead.

Soon after, another Greek, Anaximenes (ca. 550 B.C.), suggested that air was the basic element and that it manifested itself in different ways depending on the degree to which it was compressed. When air was uncompressed, it was invisible; but a slight compression would be revealed as a wind. Still higher levels of compression would turn the air into clouds, water, earth, and rock. Consequently, his universe consisted of a flat earth floating in air.

Heraclitus (ca. 500 B.C.) had still another idea; he said fire was the essential element on which everything is made and which brings about change. Change was so much a part of the thinking of Heraclitus that he visualized the sun (a ball of fire) as being created anew each day. By ca. 450 B.C., another Greek philosopher, Empedocles had united these three elements (water, air, and fire) with his own element, earth, to provide a theory of his own—that all objects are simply a mixture of varying amounts of these elements. Thus, he inferred that everything is solid, liquid, gaseous, or fiery, or may represent combinations of these states. We shall see how closely these states represent modern views in Chapter 12. The important aspect of these early intellectual models is the fact that they prompted further thinking as to the fundamental units of matter—an exercise that still persists today.

By this time (ca. 550 B.C.) Pythagoras had carried the Greek influence to southern Italy, and it was there that he (or his students) noticed that boats appeared to "sink" into the sea as they sailed away and seemed to "rise" from the sea as they approached; thus concluding that the earth was more like a ball rather than flat. If the earth were spherical, then perhaps the entire universe would likewise be more like a ball, and certainly this would simplify our model of it. We remember Pythagoras for his geometric theorem that states: "The square of the hypotenuse of a right triangle is equal to the sum of the squares of the other two sides." Indeed, Pythagoras was motivated to describe the universe in numbers. We will see how nearly correct he was as we consider the essential role of numbers in describing the universe.

## THE ATOM ENVISIONED

Humans, in their search for understanding, look in two directions from the immediate realm of objects about them: (1) outward to the macrocosmos (a view of the universe at large) and (2) inward to the microcosmos (a view of the exceedingly small). You may be surprised to find that the original concept of the atom as a fundamental unit of matter had its origin in thought about 450 B.C. Without any observational evidence for such, Leucippus, a Greek philosopher, asked a simple thought question: "If one could continually subdivide a substance into smaller and smaller parts, would it be possible to find a particle that could not be subdivided further?" By ca. 440 B.C., Democritus, a student of Leucippus, continued this line of reasoning, concluding that such a particle might be obtained. He called it *atomos*, meaning "indivisible." His idea was that different kinds of atoms, having different sizes and shapes, made up various substances.

Several hundred years later the Roman poet Lucretius wrote about Democritus' ideas about the atom as follows:

> Lest you yet
> Should tend in any way to doubt my words
> Because the primal particles of things
> Can never be distinguished by the eye,
> Consider now these further instances
> Of Bodies which you must yourself admit
> Are real things, and yet cannot be seen.
> First the wind's violent force scourges the sea,
> Whelming huge ships and scattering the clouds. . .
>
> Winds therefore must be invisible substances
> Beyond all doubt, since in their works and ways
> We find that they resemble mighty rivers
> Which are visible substance. Then again
> We can perceive the various scents of things,
> Yet never see them coming to our nostrils:
> Heats too we see not, nor can we observe
> Cold with our eyes nor ever behold sounds:
> Yet must all these be of a bodily nature,
> Since they are able to act upon our senses.
> For naught can touch or be touched except body.
>
> An image illustrating what I tell you
> Is constantly at hand and taking place
> Before our very eyes. Do but observe:

## 11
early concepts of nature

>Whenever beams make their way in and pour
>The sunlight through the dark rooms of a house,
>You will see many tiny bodies mingling
>In many ways within those beams of light
>All through the empty space, and as it were
>In never-ending conflict waging war,
>Combating and contending troop with troop
>Without pause, kept in motion by perpetual
>Meetings and separations; so that this
>May help you to imagine what it means
>That the primordial particles of things
>Are always tossing about in the great void.
>
>               Such wavering indicate
>That underneath appearance there must be
>Motions of matter secret and unseen.
>For many bodies you will here observe
>Changing their course, urged by invisible blows,
>Driven backward and returning whence they came,
>Now this way and now that, on all sides round.

Basic to early Greek thought was their unquestioning belief in certain eternal principles. From these principles the philosopher would reason in a logical manner and, thereby, reach valid conclusions—at least they were valid provided the original assumptions were valid.

Plato (ca. 400 B.C.) felt that he could build a mathematical model of the physical universe, reason logically about that model, and thereby be certain of his conclusions. He attempted to unify previously stated theories by showing that only five regular convex solids could be constructed (see Figure 1.6), and to these he assigned the properties of fire, earth, air, water, and ether. In effect, he attempted to prove that there could be no other elements. Plato believed that the universe must be treated abstractly because the observer saw only an illusion of the real world.

## A REAL UNIVERSE

Aristotle (ca. 350 B.C.), Plato's pupil, digressed significantly from his point of view by treating the observed universe as the real universe. This brought the role of careful observation more nearly to its rightful place, but deductive reasoning from basic assumptions still dominated his work. For instance, his basic assumption that the perfect geometric solid is a sphere was very clearly reflected in his model of the universe. He represented a spherical earth, standing still in the center of the universe, and surrounded by a crystal sphere of stars. Because the sun and moon each moved in relation to the background of stars, each needed its own circular sphere on which to move. Likewise, the planets appeared to move in

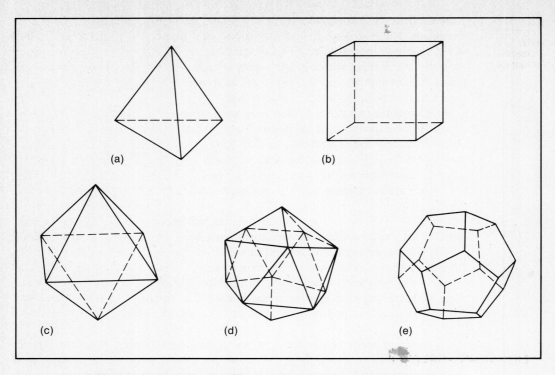

**Figure 1.6.** The Platonic solids: (a) tetrahedron; (b) cube; (c) octahedron; (d) icosahedron; (e) dodecahedron.

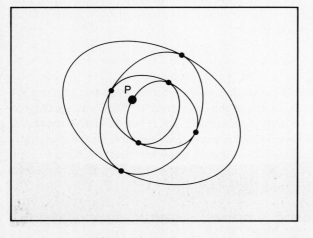

**Figure 1.7.** Aristotle's modification of the model of Eudoxis.

relation to the star background; however, their motion appeared substantially more complicated, sometimes moving slower than at other times and periodically stopping only to move backward for a time. In order to account for these motions, Aristotle modified an earlier model, which resulted in a total of 55 interconnected crystalline spheres. Figure 1.7 shows the interconnected spheres for only a single planet.

This revised model illustrates the extent to which the Greeks would go to preserve a basic assumption, one that would seem to have many counterexamples today. By contrast, however, it is interesting to note that Heraclides (ca. 350 B.C.), also a student of Plato, was bold enough to suggest that the earth rotates, thus producing the apparent daily motion of all celestial objects. The Greek astronomer Aristarchus (ca. 230 B.C.) carried the idea of a nonstationary earth a step further when he suggested that the earth revolves around the sun as a center. However, both of their ideas violated basic assumptions of Aristotle; and because Aristotle's work was so highly regarded in many fields, these so-called radical ideas were not accepted. Perhaps their denial was prompted by a lack of positive observational evidence. What evidence would you have given to verify the rotation and revolution of the earth at that time? It should be noted that at least these men were free to suggest new ideas, and that eventually someone would read of their thoughts and, thereby, reopen the question at a more favorable time.

## ALEXANDRIAN ACADEMY

By the third century B.C., the Greek influence had spread over much of the Mediterranean world, and Alexander the Great had established a center for learning in Alexandria, Egypt. One of the finest libraries in the world was assembled, and an atmosphere of study and research was provided. It is here that we see a blending of cultures—the Greek, the Egyptian, and the Babylonian—with an emphasis on the development of philosophy, mathematics, astronomy, and medicine.

Eratosthenes (ca. 220 B.C.), one of the first directors of the Alexandrian Academy, made a very significant observation concerning the earth. When in Syene (Aswan), Egypt, he noted that the sun illuminated the bottom of a vertical well on the longest day of the year, indicating that it was directly overhead then. At the same instant a vertical post cast a shadow at noon in Alexandria, which is about 500 miles north of Syene. In fact, the angle of the noon sun was about $7\frac{1}{5}°$ from the vertical. Now he reasoned that if the earth is like a sphere, the extension of both a vertical well and a vertical post would meet at the center of the earth (see Figure 1.8). The angle measured at the center should also be $7\frac{1}{5}°$ because of the parallel nature of the sun's rays. Since $7\frac{1}{5}°$ is contained in an entire circle 50 times ($50 \times 7\frac{1}{5} = 360°$), there must be 50 sectors of 500 miles each, giving the circumference of the earth as 25,000 miles, a figure very close to that accepted today. Then, utilizing a relationship known to the Greeks, namely, $C = \pi d$, it is possible to obtain a diameter just under 8000 miles. Although the unit of length we call a *mile* had not been defined as such in that day, a comparable diameter was deter-

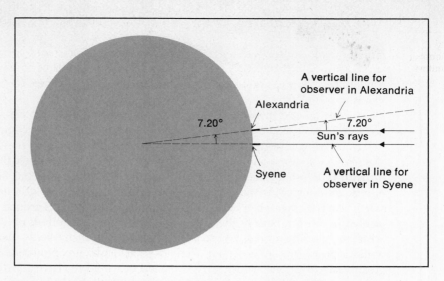

**Figure 1.8.** The method by which Eratosthenes measured the size of the earth.

mined in a unit called a *stadia*. This method of Eratosthenes clearly illustrates an early use of a method of indirect measure.

The school of Alexandria also produced the great mathematician-philosopher Archimedes (ca. 250 B.C.), who reasoned out the relationship we now call the Law of Levers. This principle can be illustrated easily by a teeter-totter on which a father who weighs 150 lb wishes to balance his daughter who weighs only 50 lb; so he adjusts his distance from the pivot (fulcrum) so as to be just one-third of the daughter's distance from the fulcrum (see Figure 1.9).

**Figure 1.9.** A father and daughter balance a teeter-totter to illustrate the principle of the lever.

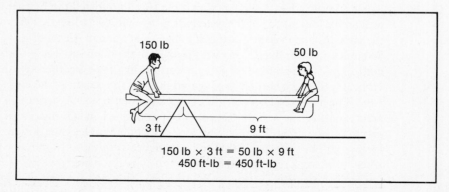

We see the relationship that Archimedes discovered:

Weight$_1$ × distance to fulcrum = weight$_2$ × distance to fulcrum

In Chapter 5 we will see how this principle is still being applied to many simple tools and devices today.

Archimedes returned to his homeland of Sicily, where he served as the king's advisor. The king presented Archimedes with a rather perplexing problem. The king had given a certain weight of gold to a goldsmith from which a crown was to be fashioned. When the smith had completed the crown and presented it to the king, the king suspected the smith of substituting a metal of lesser value for part of the gold. It was Archimedes' task to determine whether the king's suspicions were well founded; however, purity of the crown was not easily tested. One day while at the public bath, Archimedes noticed the water rise and overflow as he entered the tub. It was obvious to him that the water and his body could not occupy the same space. From this observation he proceeded to measure the volume of the king's crown by immersing it in a full container of water and collecting the water that overflowed. The volume of water collected equaled the volume of the crown (see Figure 1.10). However, when Archimedes placed in the container a lump of gold of equal weight to that which had been given to the smith, a lesser amount of water overflowed. Thus Archimedes concluded that the smith had substituted a metal of lesser density, perhaps silver, for a portion of the gold.

Archimedes also experimented with floating objects and generalized his discoveries by recognizing that any body is buoyed up (lifted) by a force equal to the weight of liquid displaced by the object. This certainly represents an understanding of one aspect of the physical universe that seems ahead of its time.

Also a credit to the Alexandrian Academy was the Greek "engineer", Hero (ca. A.D. 50), who created a device that may be viewed as the forerunner of the steam turbine, the jet engine, and the rocket. Figure 1.11 illustrates this device. When heated, the water is converted to steam, which, due to expansion, issues forth from the two nozzles. The action of the steam leaving the system produces a reaction on the device itself, causing it to turn as indicated. Hero's understanding of air was quite advanced for his time.

**Figure 1.10.** Archimedes measured the volume of the crown and of an equal weight of gold by measuring their displacement in water.

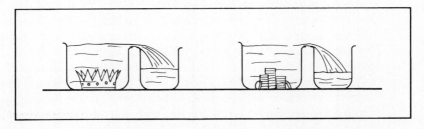

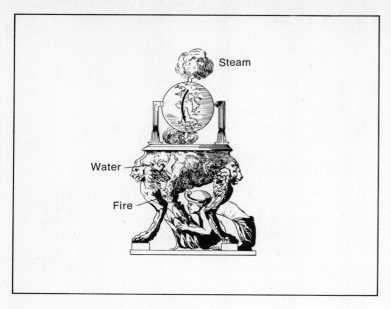

**Figure 1.11.** Hero's steam turbine.

## A BROAD SCIENTIFIC INTEREST

Egyptian science was also expanding in many fields. The Egyptians had already recognized certain chemical reactions, primarily those used in embalming the dead and those that could produce magical potions. However, their methods were couched in secrecy to the point that one worker did not even know how another functioned. On the other hand, within the atmosphere of the Alexandrian Academy we can see the beginnings of open communications among observers and investigators—an essential part of science today.

Motivated by the scarcity and ever-increasing value of gold, early "chemists" gave their very life to the proposition that they should be able to change base metals, such as lead or tin, into gold. After all, they believed that metals only differed because they contained differing mixtures of basic elements. This search, together with the search for an elixir that would produce long life, was to dominate much of that phase of science called *alchemy*, for we will find sixteenth- and seventeenth-century chemists still enamored by the idea.

The Greek astronomer, Hipparchus (ca. 150 B.C.), often called the "father of positional astronomy," made observations in Rhodes and in Alexandria that led him to a variety of scientific "discoveries," including (1) a more exact measure of the year, namely, $365\frac{1}{4}$ days less $\frac{1}{300}$ day; (2) a knowledge of the subtle gyrating of the earth, which requires 26,000 years; (3) a method whereby he calculated the distance to the moon; (4) a very accurate cataloging of stars according to position and apparent brightness; and (5) a realization that the sun varies in its distance from the earth. It was in response to this last observation that he suggested the idea of the epicycle (see Figure 1.12). He placed the sun on a smaller circle that turned as it moved along the larger orbit; thus it did not destroy the Aristotelian assumption

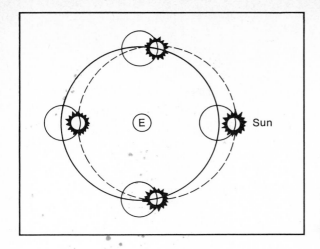

**Figure 1.12.** Hipparchus invented the epicycle as a device for explaining that the sun appears nearer the earth at certain times and farther away at others.

of the perfectness of circles. Hipparchus had achieved an orbit that in effect was not a circle.

## THE PTOLEMAIC MODEL

This idea reached fruition in the last significant period of the Alexandrian Academy, under the leadership of Claudius Ptolemaeus (Ptolemy) (ca. 150 A.D.). We might say that Ptolemy "put it all together." Without changing the basic assumptions of Aristotle, Ptolemy started with a spherical earth, standing still in the center of the universe, a sphere for the moon's orbit, one for the sun, and an outer sphere for the stars. From this point his model is quite different—each planet turning on its own epicycle, and each epicycle traveling on a larger orbit that he called the *deferent* (see Figure 1.13). You can see that by assigning the proper motion to each

**Figure 1.13.** The Ptolemaic system.

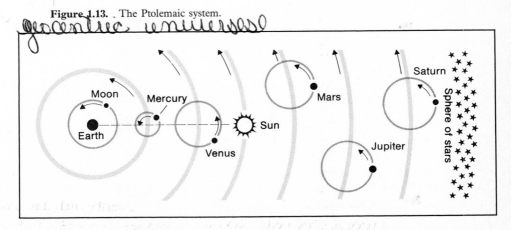

epicycle, a planet appears to back up periodically. This model was greatly simplified as compared to that of Aristotle's with its 55 spheres.

The real test, however, for any model is threefold. First, a model must be consistent with observed phenomena—this is to say that the planets must move in the model as they appear to move in the real sky. Second, a model must not violate known physical laws. Had it been known that the earth rotates on its axis and revolves around the sun, obviously Ptolemy would not have constructed a model in violation of those facts. Third, a model should allow one to predict the position of a planet in the future. Ptolemy's model seemed to meet all three conditions, at least for a time; however, small errors forced minor modification from time to time.

In spite of the error in Aristotle's basic assumption, his work (with modifications such as Ptolemy brought to it) stood almost unchallenged until the seventeenth century.

## QUESTIONS

1. What advantage did instruments made of bronze have over those made of gold?
2. What basic advantages prompted nomadic man to turn to farming and to become civilized?
3. In what sense did the Egyptian calendar of 360 days per year fail?
4. Why did early civilization arise in Babylon and Egypt? What did these two regions have in common?
5. In what way does scientific explanation dispel superstition?
6. The alignment of stones in Stonehenge was evidently used to record and predict certain astronomical events. What were these events?
7. What kind of thought process led Democritus, in 450 B.C., to the idea of the atom?
8. Aristotle assumed that the earth stood still in space. What evidence can you cite that shows its motion (both rotation and revolution)?
9. In the Alexandrian Academy we see a blending of what cultures? What positive elements existed, within the Academy, to prompt a form of scientific thought?
10. Was it necessary for Eratosthenes to assume that the earth had a spherical shape in order to complete his calculation of its circumference? What difference would it have made if the earth were "egg-shaped"?
11. Explain how Archimedes was able to determine the volume of the king's crown in spite of its irregular shape.
12. Why did Ptolemy utilize the epicycle in his model of the planets?
13. In Ptolemy's model, the moon is not given an epicycle. Can you tell why?
14. In many cultures, the priests were the observers of the "heavens." What power could they derive from this activity?

# 2

## the rebirth of science

In this chapter we want to search out the numerous factors that prepared the way for a dramatic rebirth of science in the sixteenth and seventeenth centuries. The Greeks had produced a very valuable foundation for future development—if only their thoughts and culture could be preserved through the tumultuous plundering by barbarian hoards who overran much of the Mediterranean world. The Greek concept of moral order, as codified by Roman law, did tend to stabilize life in the Dark Ages. In addition to the contributions of the Greeks were those of Indian and Chinese scholars. By what agent were the works of these cultures brought into contact, interacted, preserved, and advanced—in spite of the political and economic turmoil that characterized so much of Europe from the fifth to tenth centuries? By the seventh century the Arabs had united many nations under the Islam religion; and as we trace their influence, you will see that they provided the unifying force whereby Europe would one day experience the rebirth of science.

ARABIC INFLUENCE

One of the essential roles of the Arabs was that of translation. The only way that the works of China, India, Egypt, Greece, Syria, and Mesopotamia could experience a blending or at least an opportunity to interact was through a common language; that language for at least several centuries was Arabic. Furthermore, in the early portion of the Arabic influence (eighth century), the city of Bagdad provided a center for scholars from all these diverse cultures to share their ideas. Thus we schematically show in Figure 2.1 the flow of ideas from many cultures into this city. Although much of the work could be characterized as *encyclopedic* (a compilation of facts), advances were made in the areas of mathematics (number concepts and algebra) and astronomy (observations of the motions of planets).

The Chinese had already learned how to harness water power to turn huge circular devices by which stars and planets could be followed in the sky. Early clocks had also been devised by the Chinese. Arabic medicine emphasized the distillation of spirits for medicinal purposes. Although the Arabs also developed

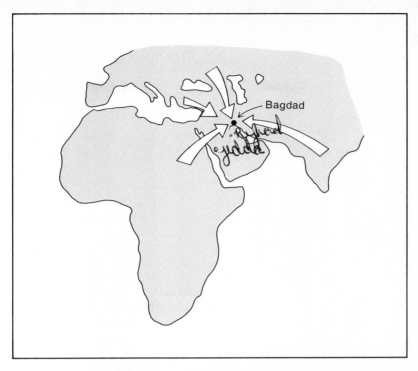

**Figure 2.1.** The Arabs conquered a large segment of the Middle East during the second to the seventh century A.D. Bagdad became a center for the cataloging and preservation of knowledge from many cultures.

the study of optics, utilizing lenses and mirrors, they did not turn this technology skyward in the form of a telescope. Chemistry as we know it today did not exist yet, for the emphasis was still on alchemy and transmutation of base elements (like lead) into gold. One significant contribution was the realization that properties such as heat, cold, dryness, and moistness could be separated (abstracted) from matter itself; however, it was erroneous to think that gold could be created simply by finding the right combination of these elements. Some of the motivation for an early type of chemistry came from the practical need for dies (for cloth), pigments (for paints), and glazes (for pottery). A fundamental motivation for Arabic science, in general, stems from the fact that those who governed (the caliphs) expressed a personal interest in its advancement. They sponsored the search for manuscripts worthy of translation. Moslems considered astronomy to be one of the highest forms of science, for it provided guidance for the observation of religious holidays. Figure 2.2 reveals the spread of Arabic influence during the tenth, eleventh, and twelfth centuries. The fact that Spain came under this influence was to play a major role in preparing Europe for the Renaissance. We still see the Moorish (Arabic) influence in the architecture of cities like Toledo and Cordova today.

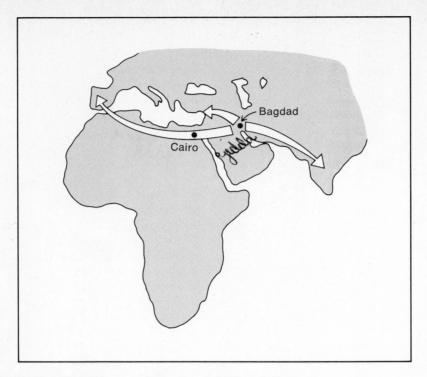

**Figure 2.2.** The Arabic influence spread as far west as Spain and helped set the stage for the rebirth of science in Europe.

## SPAIN PLAYS AN IMPORTANT ROLE

In Spain, science also experienced the impetus of royal interest. King Alphonso X, in 1252, commissioned observers to prepare tables of planetary motion: correcting errors in the Ptolemaic tables and accounting for the very slow precession of the earth's axis—a 26,000-year cycle. The Spanish city of Cordova became a center of learning that attracted scholars from all over Europe. At this time the Jewish and Christian scholars played a further role in translation, for they worked side by side with the Arabs in producing parallel translations in Latin—a crucial step in making the great works of the world available to Europeans in this universal language. Thus, Spain could be accurately described as the bridge between the East and the West.

## A GREEK STRONGHOLD

One of the few exceptions to the flow of Greek ideas through the Arabic path into Europe is represented by the fact that the Arabs were not able to conquer Constantinople in several attempts between the seventh and fifteenth centuries. This city was able to remain a Hellenic stronghold largely because the Byzantine "scientists" had learned to manufacture a sort of "napalm bomb," which, if fired into Arab ships still offshore, burned those ships with a fire that could not be extinguished by water. Thus, both Greek and Oriental works were preserved in Constantinople for many centuries. In the fifteenth century when the city was finally conquered, the Greek scholars fled westward into Europe with their cultural treasures. This resulted in a direct infusion of Greek traditions into Europe just at a time when Europe was ready for a rebirth of learning.

## REBIRTH OF LEARNING IN EUROPE

We have traced the happenings in the countries surrounding Europe, so let us now look to factors that prepared Europe internally for a rebirth of science. As early as the sixth century the Benedictine Order had promulgated a new concept of the value of work—a work ethic. In direct contrast to the Greek attitude that manual labor was not meant for the intelligentsia, the monks of this order dedicated themselves to diligent work, copying manuscripts and performing experiments.

By the eleventh and twelfth centuries, Europe had begun a reawakening of interest in learning; the economic picture had improved; and a semblance of order was evident in the development of cities. The thirteenth century produced universities and what could be called the "golden age" of scholasticism. A product of such scholasticism was the Italian theologian St. Thomas Aquinas (1225-1274) who was trained in the beliefs of the Christian Church, yet stood at the crossroads of intellectual thought as the philosophy of Plato and Aristotle entered Europe (specifically Italy). Aquinas was confronted by: (1) Plato's assertion that the universe existed only in man's thoughts; (2) Aristotle's contradiction of Plato asserting the universe of reality with the earth as its center, the planets being moved about by gods; and (3) the fatalistic ideas of Mohammed that "everything possible will be and that which will never be is impossible."

## CHURCH DOGMA

It was inevitable that the Church must meet these ideas head-on at some point in time. Aquinas in his thorough study of Aristotelianism recognized certain aspects that complemented the Church's stand, namely, that the earth was the center of the universe and stood still in space. It was largely through his efforts that the Church incorporated this concept in its dogma, placing it on a par with the Holy Scriptures. This one fact was to hinder the advancement of man's understanding of the universe for many centuries, as you will see later in this chapter. On the other hand, Aquinas also called attention to "errors" in Aristotle's work, and the Church encouraged scientific investigations that would identify such errors. Thus, the Church played both positive and negative roles in the advancement of science between the thirteenth and seventeenth centuries. Christian philosophers who had been steeped in the thought that the only source of knowledge was through divine

## CONTRIBUTIONS BY MANY

The great English scientist Roger Bacon (1220-1292) expressed this new spirit. He studied the tenth century work of the Arabian Al-Hazen on optics and experimented with simple curved lenses, predicting their eventual use in microscopes and telescopes. Bacon was also far ahead of his time in speculating on the mechanical propulsion of ships and submarines and flying machines. His ideas were so revolutionary that he was criticized severely by his own Franciscan Order.

Bacon experimented with the use of gun powder that had been discovered much earlier by Eastern alchemists. In the fourteenth century the fact that gun powder could be used to fire bullets changed the whole social and political structure of Europe, for the feudal system had been based on the strength of horsemen and spears. Now guns were more effective in battle.

Consider for a moment the flight of a bullet through the air. By what force does it continue its flight after leaving the barrel of the gun? Essentially this is the same question that had confronted early planetary observers; that is, What force keeps the planets moving in their orbits? The almost unanimous answer had been belief in some intelligent, godlike beings pushing the planets, or in the case of the bullet, air rushing in at the back pushing it forward. Neither explanation would stand the tests of experimentation, and it was in the early fourteenth century that the English scholar William of Ockham (1280-1349) challenged Aquinas's belief that only if the moving force is in contact with an object can it move that object. Ockham made use of a theory suggested much earlier, that of impetus. Ockham said that if an object is thrust forth from a gun or a hand (perhaps even the hand of God), impetus is imparted to the object, which sustains its motion; thus, even the planets need no continuing force to move them, only an initial impetus. The French mathematician Nicolas Oresme applied this idea to the earth in suggesting that the apparent motion of all objects across the sky during one day (and/or night) was really due to the earth's rotation. After all, if the earth had been given a rotation at the time of formation, then it would retain that rotation without further force—the result of impetus.

The speculations of these fourteenth-century observers led the way for the work usually credited to Copernicus, Galileo, and Newton several centuries later.

## ART, SCIENCE, AND BOOKMAKING

In the fifteenth century, the life and work of Leonardo da Vinci beautifully illustrates the complementary aspects of art and science, of architecture and engineering, of manual labor and intelligent thought. For a person to be a natural artist or scientist, careful observation is required. It has been said that the real genius of an artist lies in where he sits (choosing a proper point from which to view nature). This is also true for the scientist whose genius often lies with his perspective (seeing the universe from a new point of view) and then being able to ask questions that lead to understanding. In this period there was an increasing sense that what happened in nature was the direct (and related) result of forces and/or actions that preceded the event. This is to suggest that natural causal relationships exist and may be discovered by careful research. Certainly this intuitive feeling motivates scientists to this day.

By the sixteenth century printing and bookmaking were an established art, and their value to science in the dissemination of information cannot be overestimated. This fact was fundamental to the Reformation movement of Luther, placing Bibles in the hands of ordinary citizens. The Church was threatened by such vulgarization of the Scriptures. In fact, it was threatened on numerous fronts, and in an attempt to thwart any further erosion of its authority, the Church instituted the Inquisition—a form of trial that placed the ends of the Church before justice.

## THE COPERNICAN REVOLUTION

For fear of the Inquisitors, the Polish astronomer Nicholaus Copernicus (1473-1543) declined to publish his thoughts on the nature of the solar system. However, in the year of his death his work was published—*De Revolution bus Orbium Celestium* [The revolution of the heavenly orbs]. The essence of his scheme is expressed in his own words:

> At rest in the middle of everything is the sun. For in this most beautiful temple, who would put this lamp in another or better position than from which it can illuminate the whole thing at the same time? Thus, indeed, as though seated on a royal throne, the sun governs the family of planets revolving around it.

What a revolutionary idea! Copernicus would certainly have been tried and found guilty of heresy, for the dogma of the Church stated: "The earth is the center of the universe and does not move."

This marked the beginning of the Copernican Revolution. Note the double meaning of the word *revolution*. Perhaps this is the derivation of its present-day usage. Copernicus certainly turned things upside down, even though he was merely verbalizing once again concepts some of which had been voiced as early as the second century B.C. Copernicus had no proof that the sun was the center of the solar system, the earth being displaced from its favored position and spinning on its own axis to create the daily motion of objects in the sky; but certainly his system was more beautiful in its simplicity than that of Ptolemy.

The Church, however, was not convinced of the virtue of simplicity but only in tightening its stand against such heretical depravity. But even the Church could not stop the thoughts of humans; it could only stop outright enunciation of contrary principles. Perhaps even this restriction might be escaped by journeying outside the domination of Rome, and so it was in the republic of Venice that Italian scientist Galileo Galilei (1564–1642) was able to work without hindrance. An imaginative young man, he had already noticed the swinging of a chandelier in the cathedral of Pisa (the place of his birth), and from it he caught the idea that a clockwork might be regulated by the length of its pendulum (a fact still in use today).

## GALILEO, FATHER OF EXPERIMENTAL SCIENCE

Galileo is sometimes called the "father of experimental science" because instead of just thinking about what might happen under certain conditions, he devised an experiment to discover what actually did happen under those conditions. Epitomizing this attitude, Galileo climbed a tower and dropped two round stones (of differing mass) to see which would reach the ground first. Philosophers had thought the result to be so obvious they had not tried such an experiment—surely the more massive stone would hit the ground first (for the earth obviously pulled on it harder). To everyone's surprise the two rocks fell side by side to the ground, neither moving ahead of the other at any time. The results of this experiment may seem more difficult to explain than if the more massive stone had hit first. We will see the solution under Newton's work in Chapters 3 and 4. Galileo also studied the effect of gravity by allowing balls to roll down inclining planes (ramps), and his study was to serve as the foundation for Sir Isaac Newton's work. Newton was born in England the year Galileo died.

Galileo's interest included magnetism, architecture, engineering, acoustics, the measurement of temperature, hydrostatics, and the nature of light. Certainly Galileo was not the first to work with lenses, nor to make a simple telescope, but the genius of this man lay in his use of such an instrument. Having demonstrated its value in the identification of ships while they were still far off from the port of Venice, he turned his instruments heavenward to discover facts never known before: that the moon was rough with mountains and valleys and pockmarked; that the sun was imperfect for it had dark spots; and that the moons of Jupiter revolved as regularly as the hands of a clock, the planet itself serving as their center of revolution. Thus Galileo demonstrated that the earth was not the only center of revolution in the universe.

The discovery that was most threatening to the model of Ptolemy and to the dogma of the Church was the fact that Venus appeared in the telescope to experience changing phases like the moon (see Figure 2.3). If Ptolemy's model had been correct, you can see that Venus would never appear in any phase greater than that of a crescent (see Figure 2.4); however, it actually appeared more than half full at times. This observation represented a counterexample to an existing theory—an outright proof that Ptolemy's model was wrong. Furthermore, the phases of Venus could be explained as observed by the Copernican model (see Figure 2.5). This did not in itself constitute a proof that the sun held a central position in the system of planets, but it certainly was consistent with that model. Furthermore, the apparent

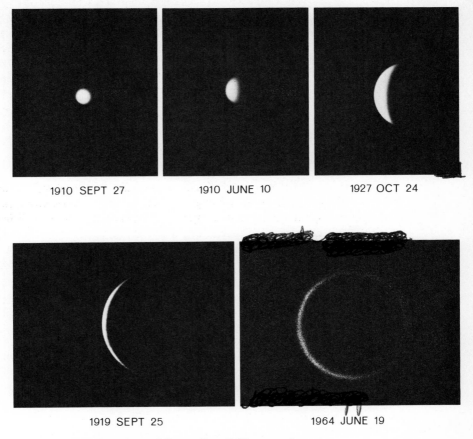

**Figure 2.3.** The five phases of Venus. (Lowell Observatory)

**Figure 2.4.** The phases of Venus: Ptolemaic system.

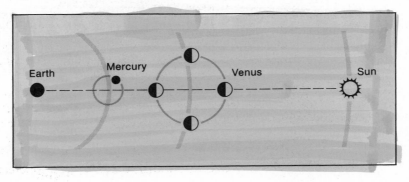

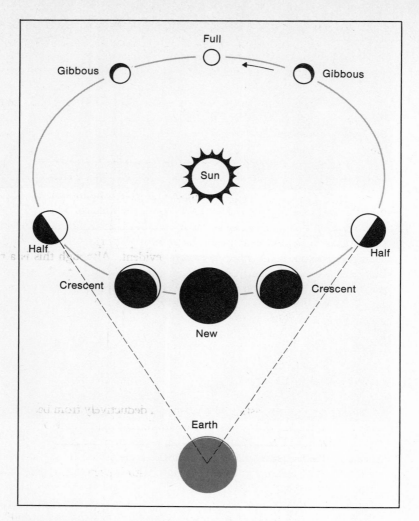

**Figure 2.5.** The phases of Venus: Copernican system.

*retrograde* (backward) motion of the planets could also be explained using the Copernican model. When Venus (or Mercury) passes the earth, that planet appears to back up among the stars. Likewise, when the earth passes Mars (or Jupiter or Saturn), that planet appears to retrograde as when two trains leave a station together. Picture yourself riding on the faster train—the slower train will appear to back up.

## THE WEIGHT OF CHURCH DOGMA

Thinking that his reputation was strong enough and that the weight of observational evidence was sufficient, Galileo set out to try to convince the Church fathers (the Pope and Bishops) that their dogmatic stand in support of the Aristotelian model was ill-founded. Galileo failed in this attempt. The Church authority was so threatened on every hand that not even this man nor this evidence could be given a fair hearing; and Galileo was convicted of heresy, was forced to recant his own beliefs, and was imprisoned in his own home for the remaining years of his life.

The Italian philosopher Giordono Bruno (1548-1600) had been burned at the stake in 1600 for his imaginative speculations regarding the infinite nature of the universe and the possibility that many worlds surround other stars, perhaps inhabited like the earth. Such actions on the part of the Church were to stifle astronomical speculations among Italian scientists for at least another century; however, the same kind of oppression was not being experienced in England.

Let us see how the work of one of Galileo's contemporaries, English philosopher Francis Bacon (1561-1626), was to set the tone for science even to modern times. Bacon believed that if he made careful observations of natural happenings (and collected enough examples), the general principle (natural law) that governs such phenomena would become clearly evident. Although this is a rather naive description of the inductive method, it points the way to the modern scientific method that involves the following steps as guidelines: observation of a given phenomenon, classification, experimentation, formulation of a theory to explain the phenomenon, further experimentation suggested by the theory, and refinement of the theory.

## THE INDUCTIVE METHOD IS BORN

In contrast to the scholastics who reasoned deductively from basic assumptions to specific conclusions, Bacon's writings extolled the merits of looking to the specifics first (observation of facts) and then trying to discover the general law that explained those facts, an inductive method. Herein "fact" refers to any observable event. Hence, facts of nature, in this sense, are neither proven nor unproven; they are simply events that can be observed.

The sixteenth and seventeenth centuries produced other greats in science like Robert Boyle, who studied the nature of gases; Evangelista Torricelli, who measured the pressure due to the earth's atmosphere utilizing a barometer; Otto von Geuricke, who invented the vacuum pump and made one of the earliest electrical machines; and William Gilbert, whose work in magnetism and electricity helped found modern science. The work of these persons, together with that of eighteenth-, nineteenth-, and twentieth-century scientists will be treated in the appropriate chapters, which follow.

## QUESTIONS

1. The Arabs were instrumental in preserving and unifying the knowledge of many different cultures during the Dark Ages of Europe. How did they accomplish this? 20, 21

the rebirth of science

2. Where was the center of Arabic influence?
3. What were the primary goals of chemists (alchemists) between the second and twelfth centuries A.D.?
4. The work of the modern scientist is often supported by federal or private grants. What parallel for this practice was found in early times?
5. In what sense was Spain the bridge between East and West?
6. How did Constantinople survive, as a Greek stronghold, when the Arabs conquered the remainder of Greece?
7. In what sense was Constantinople another bridge between East and West?
8. What were several of the conflicts between the Roman Catholic Church and the early philosophers?
9. Express your own thoughts as to why the Church adopted the Aristotelian model of the universe—with the earth standing still at its center.
10. How did the use of gunpowder change the social and political structure of Europe? Can you think of other scientific discoveries that have changed the social and political structures of countries up to the present time?
11. How would you react to the concept of impetus when you find that your car soon stops moving when you turn off the engine?
12. Why did the Church feel so threatened during the sixteenth and seventeenth century? Can you recall additional reasons that were not stated in your text?
13. When Copernicus called the sun a "lamp," what did this reveal about his understanding of the nature of the sun?
14. What made Galileo's approach toward science different from many other inventors or thinkers of his time?
15. Explain how Galileo's discovery that Venus goes through phases, somewhat like the moon, disproves the Ptolemaic theory.
16. Can you explain why the planets appear to back up among the stars periodically?
17. Contrast the inductive method of the modern scientist with the deductive method of the Greek philosopher.

# 3

motion

We live in the most mobile age of all time—people going here and there at an ever-quickening pace. If we are to describe the motion of people or objects accurately, we must first develop basic concepts of length and time. For example, to say that the distance between two cities is "100" has no meaning; for we must immediately ask, "100 what? Miles, kilometers, degrees, or . . .?" During medieval times, in an effort to give "length" a meaning so that the concept could be communicated, the distance from a king's nose to the tip of his outstretched arm was defined as a *yard*. The length of his thumb joint was defined as an *inch*, and the average length of the feet of 12 men, measured as they marched from church, was defined as a *foot*. Obviously such standards do not meet the basic requirement for international units—reproducibility and constancy.

## METRIC SYSTEM

In the late eighteenth century an attempt was made to define the *meter* as 1/10,000,000th (1 ten-millionth) of the distance from the earth's pole to the equator along a meridian through Paris, a method that was not easy to reproduce. In order to provide an international standard that was accessible, a standard meter was ruled on a platinum-iridium alloy bar that was maintained at 0°C in the International Bureau of Weights and Measures near Paris. From this standard meter, secondary standards could be copied and maintained throughout the world. From these secondary standards, meter sticks were manufactured for everyday use. However, it would seem to be more advantageous to define a meter in terms of natural phenomena that could be reproduced in any well-equipped laboratory. Hence, the standard meter was compared with the wavelength of the orange light produced by krypton-86 when in an excited state. It was found that the meter is 1,650,763.73 times that wavelength. Since krypton-86 atoms always produce the same wavelength, this wavelength presents an unchanging and reproducible standard from which the meter can be determined. The concepts of an excited atomic state and the resultant production of light are more fully explained in Chapter 8.

The meter is subdivided into centimeters (1 cm = 1/100th of a meter), and millimeters (1 mm = 1/1000th of a meter); and the meter is multiplied by 1000 to produce the kilometer (km). Although other subdivisions and multiples will be used later, these are the most common units. This system based on the meter as the fundamental unit of length is called the *metric system*. However, this basic system has now been extended to include basic units of length, mass, time, and temperature. It is now called the International System of Units (Système Internationale d'Unites) or SI system.

## LENGTH, AREA, AND VOLUME

As the United States moves from the British units of inches, yards, miles, ounces, pounds, quarts, and gallons to corresponding units of the SI system, the following table of length conversions (Table 3.1) may be useful:

**Table 3.1**

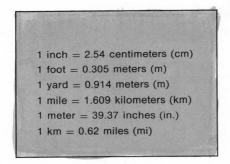

From the basic unit of length we can define the units whereby a surface area or volume may be measured. The metric unit of area is defined as a *square meter*, equivalent to the area of a flat square surface one meter long and one meter wide. This basic unit may be subdivided into square centimeters (1 cm$^2$ = 1/10,000 m$^2$) or may be multiplied to form square kilometers (1 km$^2$ = 1,000,000 m$^2$), which would be useful in land measurement.

**Figure 3.1.** Metric units of volume.

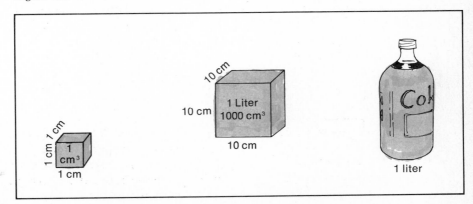

Volume measures the amount of space an object occupies, and the basic unit is a cubic meter (m³). If we were measuring solutions in the laboratory or buying milk or other liquid goods, we would use as our unit of measure a subdivision of the cubic meter, namely, the cubic centimeter (cm³)—a cube one centimeter on each side. One thousand (1000) cm³ are equivalent to one liter, which is just a little more than one quart (1 liter = 1.06 quart). Soon we will be buying gasoline in such units (1 gallon = 3.785 liters) (see Figure 3.1).

## MASS AND DENSITY

Although the concept of mass will have a broader definition as our understanding of energy develops in Chapter 5, let us note that the *rest mass* of an object is a measure of the amount of material in that object. As a unit of mass, the amount of material in one liter (1000 cm³) of fresh water, just above freezing, is defined to be one kilogram (1 kg = 1000 grams); hence the mass of 1 cm of fresh water is 1 gram (g). Does this definition of the unit of mass meet the conditions of reproducibility and constancy? Yes, for what could be more accessible than fresh water; and under the conditions specified, its density remains constant. But what is the meaning of the term *density*? If you held a steel marble in one hand and a glass marble of the same size in the other, could you tell why the steely feels heavier than the glassy? One might answer, "There is more material packed into the steely than in the glassy, even though they are the same size and have the same volume." The analysis provides the correct idea—that density is a measure of how much material is packed into a given space. Written in more usable terms: "For any given object, its density equals its mass divided by its volume ($D = M/V$), and the unit of density is grams per cubic centimeter ($g/cm^3$)."

The density of some common materials is given in Table 3.2.

We have developed the units of area, volume, mass, and density from the basic definition of length, together with the natural density of water; but before we

**Table 3.2**

| MATERIAL | DENSITY g/cm³ |
|---|---|
| Water | 1 |
| Ice | 0.9 |
| Sea water | 1.03 |
| Cork | 0.25 (approximately) |
| Iron | 7.6 (approximately) |
| Mercury | 13.6 |
| Silver | 10.5 |
| Gold | 19.3 |
| Platinum | 21.4 |

can proceed to the concepts of velocity, acceleration, and momentum, we need the concept of time.

## TIME

The fundamental unit of time is the second, which for centuries was based on the rotation of the earth. The *second* was defined as 1/86,400th of a mean solar day, where 86,400 is the number of seconds in an average day. However, the average solar day is not constant, and so a much better basis was found, namely, the vibrations in a cesium-133 atom. These vibrations are believed to be so nearly constant that a cesium clock will not vary more than one second in 6000 years.

In 1967 at the General Conference on Weights and Measures, the second was defined to be an interval of time equivalent to that of 9,192,631,770 vibrations of cesium-133. Accuracy characteristic of a cesium clock is necessary in verifying some of Einstein's predictions regarding time changes in moving systems (see Chapter 9). You are familiar with the basic units of one minute (min) (60 sec = 1 min) and one hour (hr) (60 min = 1 hr). These units are the same in the British and the SI systems.

## VELOCITY

Whereas the measurement of speed usually involves only two factors—the distance covered in a given time—we add the further condition of "direction" when specifying velocity. Direction requires a frame of reference, and the one we are most familiar with is that of north, south, east, or west. Let us suppose that a car is driven from one city to another for a distance of 100 km in 2 hours and that the car is traveling due north. We define velocity as distance divided by time ($v = d/t$), and specify the average velocity of this car as 50 kr/hr (100 km/2 hr = 50 km/hr) due north.

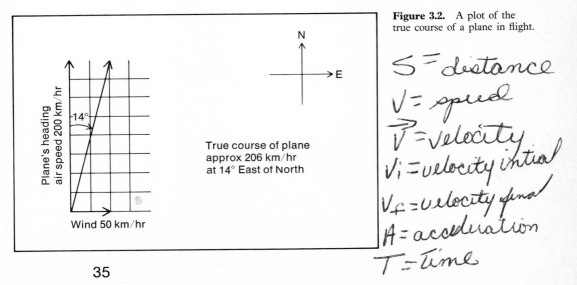

**Figure 3.2.** A plot of the true course of a plane in flight.

$S =$ distance
$V =$ speed
$\vec{V} =$ velocity
$V_i =$ velocity initial
$V_f =$ velocity final
$A =$ acceleration
$T =$ time

**36**

**motion**

A plane flies (heading due north) at the rate of 200 km/hr, but a wind is blowing from the west at 50 km/hr. What is the plane's velocity, including direction over the ground? (See Figure 3.2.) By representing each velocity as a vector pointing in the correct direction with a length proportional to speed, it is possible to predict the true course of the plane.

## AVERAGE VELOCITY

If you drove from one city to another, say 300 km in 6 hr, it would be easy to compute your average velocity to be 50 km/hr (300 km/6 hr = 50 km/hr). However, this does not mean that you traveled at exactly that velocity throughout the trip. At one point you may have been traveling at 70 km/hr, while at another point you may have been doing only 40 km/hr. These velocities could be called *instantaneous velocities* and could have been noted at any given moment by glancing at the speedometer. If you were to keep track of how far you had traveled at the end of each half-hour interval as in Table 3.3, you could plot time and distance on a graph as in Figure 3.3.

Can you read the instantaneous velocity at any given time during the trip? No; but you could compute the average velocity during any half-hour interval, say from $A$ to $B$ as follows: distance at $B$ (140 km) minus distance at $A$ (110 km) gives 30 km, covered in one-half hr. The average velocity equals distance divided by

**Table 3.3**

| T | 0 | $\frac{1}{2}$ | 1 | $1\frac{1}{2}$ | 2 | $2\frac{1}{2}$ | 3 | $3\frac{1}{2}$ | 4 | $4\frac{1}{2}$ | 5 | $5\frac{1}{2}$ | 6 |
|---|---|---|---|---|---|---|---|---|---|---|---|---|---|
| D | 0 | 30 | 55 | 80 | 110 | 140 | 160 | 180 | 190 | 210 | 240 | 275 | 300 |

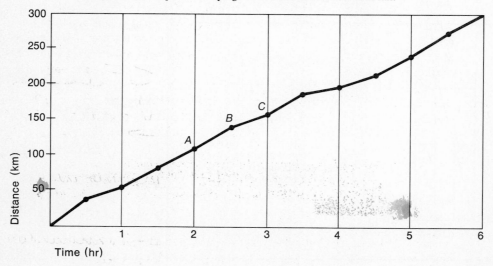

**Figure 3.3.** A plot of the progress of a car in half-hour intervals.

time (30 ÷ 1/2), yielding 60 km/hr for that interval. See if you can compute the average velocity during the interval from B to C. (*Answer:* 40 km/hr)

Suppose you had kept track of the distance covered every 15 min, or every 5 min, or every minute. Could you come any closer to knowing the instantaneous velocity at a given moment? Yes; in fact if you could record the distance covered every second or every hundredth of a second, you could come still closer to reading off an instantaneous velocity. This idea of shortening the intervals of time is the basis for what Newton called *fluxions*—a branch of mathematics eventually called the *calculus*.

## SIR ISAAC NEWTON

Sir Isaac Newton was born in England the year Galileo died (1642), and by age 23 he had demonstrated his unusual insight into relationships of force, mass, velocity, changes in velocity (acceleration), gravity, the nature of light, and mathematics. In fact, he invented fluxions to aid in the analysis of a body moving under a given force, that of gravity. One might say that Newton picked up where Galileo left off; and surely Newton referred to Galileo (and others) when he said, "If I have been able to see farther than others, it is because I was standing on the shoulders of giants."

We have mentioned two terms that are yet undefined, namely, *force* and *acceleration*. Newton discovered that the concepts of force and acceleration are closely related, as we will see. Every person has undoubtedly pushed a wagon in his childhood, and as a result of pushing (exerting a force), the wagon moved. Hence, we might define a force as that which moves an object. Is this an adequate definition? No, for one might exert a force on a telephone pole all day and still not move it. The individual would tire even though the pole did not move. The above definition is inadequate in another respect. Suppose you were driving along a level road at 30 km/hr and you turned off the engine of your car, coasting in neutral. What would happen? Would you continue coasting at 30 km/hr, or would you gradually come to rest? All experience shows that you would come to rest; and, therefore, the only way to maintain 30 km/hr is to apply a force by keeping the engine running and the car in gear. Newton transcended the limitation of such experience and realized that if the opposing force of friction could be removed, then a body in motion would tend to remain in motion (at the same speed) in a straight line forever. Thus, a force is not necessary to maintain a constant velocity in a frictionless environment. He had hit upon the concept of inertia.

*Newton's First Law—Inertia: An object at rest tends to remain at rest unless acted upon by an outside force and an object in motion tends to remain in motion in a straight line and at the same speed unless acted upon by an outside force.*

Before Newton's time, observers had thought that the natural tendency of any object was to come to rest, and in the world of friction this appeared to be true.

Newton's genius is shown by his ability to sense what might happen in the frictionless world, even though he could not experiment directly in such a world. Can you see that his first law not only suggests that a force is needed to get an object going but also to get a moving object stopped or turned in another direction? Thus, a force is capable of speeding up the motion of an object or of slowing it down or of changing its direction. Any of these kinds of changes can be called *acceleration*. Hence, acceleration is defined as the change in velocity or direction per unit of time. For instance, if you stepped on the accelerator of a car and changed its velocity from 10 m/sec (36 km/hr) to 20 m/sec (72 km/hr) in 5 sec, then you may compute the acceleration as

$$\frac{20 \text{ m/sec} - 10 \text{ m/sec}}{5 \text{ sec}} = \frac{10 \text{ m/sec}}{5 \text{ sec}} = 2 \text{ m/sec/sec}$$
$$= 2 \text{ m/sec}^2$$

The final result is read as "2 meters per second per second," or "2 meters per second squared." Assuming that your acceleration was uniform, you may interpret such a calculation as saying that you started out at 10 m/sec, but one second later you were doing 12 m/sec; after two seconds, 14 m/sec; after 3 seconds, 16 m/sec; after 4 seconds, 18 m/sec; and after 5 seconds, 20 m/sec. Thus you increased your velocity 2 meters per second every second.

Had you applied your brake (also a force, but in the negative sense), you might have slowed from 20 m/sec to 14 m/sec in 2 seconds. Your negative acceleration (deceleration) would have been

$$\frac{14 \text{ m/sec} - 20 \text{ m/sec}}{2 \text{ sec}} = \frac{-6 \text{ m/sec}}{2 \text{ sec}} = -3 \text{ m/sec/sec}$$
$$= -3 \text{ m/sec}^2$$

Is there a relationship between the amount of force applied and the amount of acceleration resulting? Yes; we would say the harder you apply the brakes, the faster you slow down; or the harder you push the wagon, the faster it speeds up. Newton found a simple relationship between force and acceleration. When he doubled the force on an object, he noticed that the acceleration also doubled. When he tripled the force, the acceleration likewise tripled. He stated this relationship in his second law:

> *Newton's Second Law—$F = ma$: If a force produces a motion, a double force will produce a double motion and a triple force will produce a triple motion.*

This concept can be expressed algebraically as $F = m \times a$ (force = mass × acceleration), and without regard to units can be illustrated simply. A force of 24 acting on a mass of 4 produces an acceleration of 6 ($24 = 4 \times 6$); whereas a force of 48 acting on the same mass (4) produces an acceleration of 12.

But suppose we experiment with two objects of differing masses. Let object $A$ have a mass of 4 (as above) and object $B$ have a mass of 8. The force of 24 acting an object $A$ (mass of 4) produces an acceleration of 6 ($24 = 4 \times 6$); but the same force of 24 acting on object $B$ (mass of 8) produces an acceleration of only 3 ($24 = 8 \times 3$). What force would be needed to produce an acceleration of 6 in object $B$? (*Answer:* 48 [$48 = 8 \times 6$].)

At this point the concept of mass really comes into better focus, for the mass of an object could be defined by simply solving the equation $F = m \times a$ for $m$ itself, namely, $m = F/a$. Thus, the mass of an object may be determined by simply applying a known force and observing the resulting acceleration. For example, if a force of 8 is continuously applied to an object, resulting in an acceleration of 2, then the mass of the object is 4 ($8/2 = 4$). We have neglected the units of measure in order to focus on the concept.

Now let's define at least one possible set of units. If we measure the acceleration in meters per second squared (m/sec$^2$) and the mass in kilograms (kg), then force will be measured in newtons (to honor Sir Isaac Newton). A *newton* (nt) is defined to be the force needed to accelerate a 1 kg mass 1 m/sec$^2$.

This discussion exemplifies a real change in the spirit of science during the sixteenth and seventeenth centuries—a spirit of quantification. Earlier observers had been content with statements like, "The harder you push, the faster an object will go," whereas quantification is far more specific. It assigns quantities (numbers and units) to concepts like force, mass, acceleration, etc., and shows specific relationships as opposed to vague generalities. So fundamental is Newton's work in the physical realm that we speak of his formalization of concepts as being the beginning of a new era in science. He certainly predicted the modern rocket when he stated:

> *Newton's Third Law—Action and Reaction: For every action there is an equal and opposite reaction.*

You may have experienced this principle if you have ever dived from the stern of a small unanchored boat. When you came up, the boat had moved in the opposite direction from its original position (see Figure 3.4).

**Figure 3.4.** A swimmer who dives from an unanchored boat finds, upon surfacing, that the boat has also moved away from him—an illustration of the law of action (on the swimmer) and reaction (on the boat).

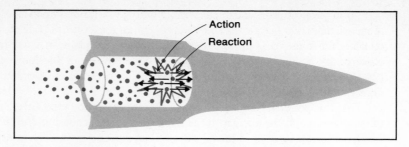

**Figure 3.5.** The acceleration of a rocket depends upon the action on its propellant and the reaction on the rocket itself.

If we call the motion of the diver the *action*, then the motion of the boat must be called the *reaction*. Action and reaction never occur on the same object. In the firing of a gun, the action may refer to the motion of the bullet, the reaction being the motion of the gun itself. You feel this reaction as a "kick" against the shoulder.

Airplanes fly in the atmosphere largely through action-reaction phenomena. The turning propeller is designed to produce a backward motion on air molecules that it encounters. As a consequence, the propeller (and the plane as a whole) experiences a reaction force in a forward direction. However, because a propeller must react with air molecules, such a conventional type aircraft cannot fly above the atmosphere. On the other hand, a rocket carries its own supply of mass particles (fuel); and for each unit of force exerted to expel these mass particles out of the rear of the rocket, an equal (forward) force is experienced by the rocket itself (see Figure 3.5). The direction a rocket is traveling may be changed by small rockets directed to either side. For instance, if one desires to change the rocket's motion to the right, then a small rocket must be discharged to the left.

## MOMENTUM

It may seem unreasonable to expect particles as small as molecules to propel a massive rocket. Is it necessary for the total mass of the molecules ejected to be equal to the mass of the rocket? No, for then the mass of the rocket plus the mass of the fuel would have to be less than or equal to the mass of the rocket itself. The quantity that must be equal is called *momentum*. To understand this quantity, consider the hypothetical situation in which you take the place of the ten pins at the end of the bowling alley. First, a volley ball is rolled down the alley at 10 m/sec. You would not hesitate to catch the ball. On the other hand, if a bowling ball were rolled down the alley at the same speed, you would probably step aside for fear of injury. Why? Perhaps it is because you sense that the bowling ball is much harder to stop than the volley ball in spite of the fact that their velocity is the same. The mass of the bowling ball is so much greater, and it is this fact together with its velocity that makes it harder to stop. We say that the bowling ball has greater momentum, where this quantity is defined as the product of mass times velocity ($M = m \times v$).

The concept of momentum is particularly useful in considering collisions, for the physicist has observed that the total momentum of a system before a collision is equal to the total momentum afterward—that is, momentum is conserved. The momentum of a body is a vector quantity, like velocity, in that direction is always part of the quantity. Suppose a truck that has a mass of 10,000 kg is traveling due north at 75 km/hr and a car with a mass of 2000 kg is traveling due south at 100 km/hr. We may arbitrarily assign a plus (+) to the truck's momentum and a minus (−) to the car's momentum. Computing the total momentum:

$$\text{Truck: } M = 10,000 \times 75 = +750,000 \text{ kg km/hr}$$

$$\text{Car: } M = 2000 \times 100 = -200,000 \text{ kg km/hr}$$

$$\text{Total: } M = 750,000 - 200,000 = 550,000 \text{ kg km/hr}$$

Suppose that the truck and the car had a head-on collision, and their bumpers became interlocked so that they moved as one object with a total mass of 12,000 kg. What does the conservation of momentum predict for their direction and speed? Solving the equation $M = mv$ for $v$, we get:

$$v = \frac{M}{m} = \frac{+550,000 \text{ kg km/sec}}{12,000 \text{ kg}} = +46 \text{ km/sec}$$

The total wreckage would tend to move due north at approximately 46 km/sec. The resultant momentum of the car alone would be:

$$M = mv = 2000 \text{ kg} \times (+46) \text{ km/hr} = +92,000 \text{ kg km/hr}$$

## IMPULSE

One can imagine the force necessary to convert the momentum of the car from $-200,000$ kg km/hr to $+92,000$ kg km/hr. In fact, we can define such a change in momentum by the term impulse.

$$\text{Impulse} = mv_1 - mv_2$$

where

$mv_1$ = momentum of object before collision

$mv_2$ = momentum of object after collision

In order to see the true nature of the concept of impulse, follow these steps: Newton's second law stated:

$$F = ma \tag{1}$$

but *a* (acceleration) is equal to the change in velocity in a given time, namely,

$$a = \frac{v_1 - v_2}{t} \qquad (2)$$

So we may substitute for *a* in equation (1):

$$F = \frac{m(v_1 - v_2)}{t} \qquad (3)$$

and multiply both sides by *t*:

$$Ft = m(v_1 - v_2) \qquad (4)$$

$$Ft = mv_1 - mv_2 \qquad (5)$$

The right-hand side of equation (5) fits our definition for impulse; hence, the left-hand side must also represent impulse. This gives us a clue to the true character of impulse: It is a product of a force multiplied by the time that the force acts. The truck is capable of exerting such a large force upon the car that during the short interval of the crash we may expect almost total demolition of the car.

## ANGULAR MOMENTUM

The concept of momentum applies not only to objects moving in straight lines (linear momentum) but also to rotating objects (angular momentum). In the latter case, not only must mass and velocity be considered but also the manner in which the mass of the object is distributed with relation to its axis of turning. For example, a figure skater on ice starts a spinning motion rather slowly with her arms outstretched. By lowering her arms, thus redistributing a portion of her mass nearer the axis of turning, she rotates faster, conserving angular momentum.

## QUESTIONS

1. Why is a reproducible unit of length so important?
2. Which is a more reliable standard—a meter based on one ten-millionth of the distance from the earth's pole to the equator or a meter based on 1,650,763.73 times the wavelength of orange light of krypton-86?
3. What advantages and/or disadvantages do you see in the U.S. conversion to the SI system of units?
4. Which contains more liquid—a quart of Coca Cola or a liter of Coca Cola?
5. Define the basic unit of mass—the gram.
6. Which has the greater mass—two $cm^3$ of iron or one $cm^3$ of mercury?

motion

7. What is the difference between speed and velocity?
8. True or false? If the engine of a rocket were shut off, one might expect the rocket to slow down and come to rest even though it experienced no friction in the emptiness of space.
9. Newton discovered that as he exerted a certain force on an object, a definite acceleration resulted. What happened when he doubled the force? *acceleration would double*
10. The engine of a given car will produce a certain maximum acceleration. What could you expect if that same engine were placed in a car of twice the mass? *acceleration would be cut in half*
11. Explain, in terms of momentum and impulse, why a more massive car produces greater damage on a less massive car, i.e., a Mack truck hitting a VW.
12. Suppose you were traveling through empty space in a rocket at constant velocity; then you started throwing steel marbles out the rear of the rocket. What change in velocity could you expect? What source of energy would you be utilizing?
13. Explain how a steel marble, which possesses twice the mass of a glassy, falls side by side when dropped from a high tower.

gravity

Before the sixteenth century, celestial observers had been preoccupied by the question, "Who or what keeps the planets moving around in their paths?" Every experience had suggested that only as long as some motive force was actually in contact with a body could that body maintain its motion. Try pushing a block of wood on a smooth floor. As long as you maintain contact with the block, you may impart a motion to it; but as soon as you let go, it slows down and stops. Galileo had performed a similar experiment, and he found that as he reduced the friction between the block and the floor, the block when released would go farther before it came to rest. If he imparted the same velocity to a smooth, round ball, the ball would travel still farther. Galileo speculated that if friction could be reduced to zero, then the object would go forever. It was this kind of thinking that prepared the way for Newton's formal statement regarding inertia:

*An object in motion tends to remain in motion in a straight line and at the same speed unless acted upon by an outside force.*

The question, "What keeps the planets moving?" could now be rephrased as, "Why don't they move in a straight line?" Galileo prepared the way to answer this question also. He had experimented by releasing objects from high places and by allowing balls to roll down sloping ramps. By noting the distance traveled in equal intervals of time, Galileo observed that the object was accelerated; that is, it moved farther in each successive interval of time. His experiments showed the existence of a force that tends to move (to accelerate) a body without the "mover" being in direct contact with the body—i.e., producing an action at a distance. Galileo, however, did not conceive this force to be that which causes the planets to deviate from a straight path.

## THE CONCEPT OF GRAVITATION

It was Newton who one day while having tea in the garden noticed an apple fall to the ground. At that moment he asked the question, "Could the same force that caused the apple to fall toward the center of the earth also cause the planets to orbit the sun and the moon to orbit the earth?" (See Figure 4.1.)

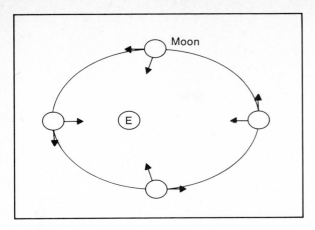

**Figure 4.1.** At any moment, the moon moves under the influence of inertia (its tendency to move in a straight line) and the force of gravity (its tendency to "fall" toward the earth).

In a very real sense the moon does fall toward the earth, for consider its location after a given amount of time (see Figure 4.2). Starting at *A*, if the moon experiences no external force, it would travel in a straight line to *B*; however, because it experiences a force toward the earth, it moves to *C*. Thus, the moon "fell" from *B* to *C* as a consequence of a force acting toward the earth. Was this the same force as that which made the apple fall?

Before Newton could conclude that it was the same force, he had to discover, for instance, what effect distance had on the force. Does it get stronger or weaker when the objects are moved apart? You can imagine the difficulty in trying to move an apple very much farther from the center of the earth than at sea level. Had Newton carried the apple to a very high mountain and measured its weight on a very sensitive spring scale (see Figure 4.3), he would have found the weight diminished. And had he compared the weight of the apple at sea level with its weight at different heights on the mountain, he might have discovered the idea that the force gets weaker as the square of the distance from the center of the earth (see

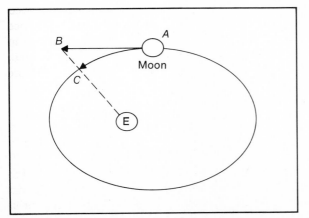

**Figure 4.2.** The moon continuously "falls" toward the earth, yet is maintained in its orbit around the earth.

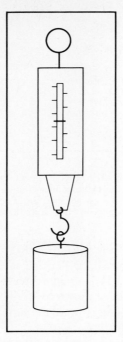

**Figure 4.3.** A spring scale, used to measure the force of gravity at any desired elevation above the earth.

Figure 4.4). This is to say—if the force of the apple is 1 newton (nt) when it is at sea level (approximately 6000 km from the center of the earth), it will weigh only 0.25 nt if it could be carried to a distance of 12,000 km from the center of the earth. Newton deduced that this would happen without being able to perform such an experiment; and in a sense, he "reached for the moon" to see what force it experiences. The moon is approximately 60 times as far from the center of the earth as is an apple on the surface (see Figure 4.5). Hence, if Newton's assumption

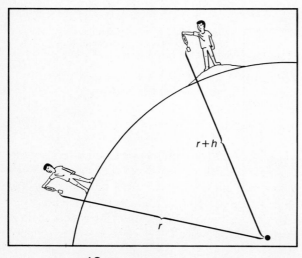

**Figure 4.4.** If a given mass is carried onto a high mountain, it will weigh less than at sea level.

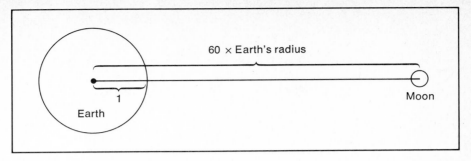

**Figure 4.5.** The moon is approximately 60 times as far away from the center of the earth as the surface of the earth is from the center of the earth.

is correct, then the force should be diminished by $(60)^2$ or 3600 times. This is to say that an apple, at the distance of the moon, would experience 1/3600th the force of attraction as compared to the same apple at the surface of the earth and would be accelerated proportionally. This Newton confirmed by calculating the distance the moon "fell" in a unit of time.

## THE UNIVERSAL NATURE OF GRAVITY

This calculation represents a major breakthrough in man's progress toward understanding that the same forces act in the solar system and in the universe at large as on the earth—the concept of universal gravitation was born. Gravity is the mutual attraction of any two objects for one another, and the strength of that attraction (the force due to gravity) is expressed by:

$$F = G \frac{m_1 \times m_2}{r^2}$$

where

$m_1$ = mass of object #1

$m_2$ = mass of object #2

$r$ = distance between centers

$G$ = a constant, called the *gravitational constant*

In 1798, Lord Cavendish, English chemist and physicist, devised a way to measure the tiny forces of attraction, such as between steel balls on earth. Two steel balls, fixed at either end of a rod, are suspended from a thin fiber, and the system is allowed to stabilize. Then large steel balls ($C$ and $D$) are moved nearby as shown in Figure 4.6. Because $A$ and $C$ exert a mutual force of attraction, as do $B$ and $D$, the suspended system twists a little. Since the force necessary to produce

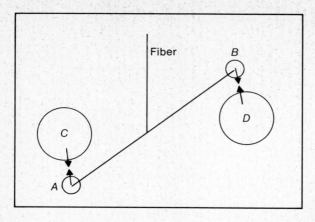

**Figure 4.6.** The Lord Cavendish experiment.

such a twist could be computed from the amount of twisting and because the masses ($m_1$ and $m_2$) and the radius ($r$) were also known, it was possible to solve for the gravitational constant ($G$) in the equation:

$$F = G \frac{m_1 \times m_2}{r^2}$$

$G$ is equal to $6.673 \times 10^{-11}$ $nt\text{-}m^2/kg^2$ (newton-meter squared per kilogram squared). Having evaluated $G$, it is now possible to predict the mutual gravitational force that any two masses ($m_1$ and $m_2$) will exert on each other at a distance ($r$) separating their centers.

In order to grasp the interrelationship between mass and distance in the gravitational formula, let us consider the following illustration, without regard to specific units.

$$F = G \frac{m_1 \times m_2}{r^2}$$

*Case 1:*   Let

$$m_1 = 4, \; m_2 = 9, \text{ and } r = 3$$

$$F = G \frac{4 \times 9}{(3)^2} = \frac{36}{9} G = 4G$$

Now suppose that the mass of $m_1$ is doubled:
*Case 2:*   Let

$$m_1 = 8, \; m_2 = 9, \text{ and } r = 3$$

$$F = G \frac{8 \times 9}{(3)^2} = \frac{72}{9} G = 8G$$

We see that when one of the masses is doubled, the force due to gravity is also doubled.

Now suppose that the mass of $m_1$ and $m_2$ are each doubled, as compared with Case 1.

*Case 3:* Let

$$m_1 = 8, \; m_2 = 18, \text{ and } r = 3$$

$$F = G \frac{8 \times 18}{(3)^2} = \frac{144}{9} G = 16G$$

We see that when both masses are doubled, the force due to gravity is quadrupled. Finally, going back to Case 1, let us move the two masses twice as far apart.

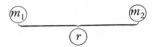

*Case 4:* Let

$$m_1 = 4, \; m_2 = 9, \text{ and } r = 6$$

$$F = G \frac{4 \times 9}{(6)^2} = \frac{36}{36} G = 1G$$

We see that when the distance between the two masses is doubled, the force due to gravity is only one-fourth.

In light of these findings, we may verbalize the gravitational relationship as follows:

*The force due to gravity is directly proportional to the product of the masses and inversely proportional to the square of the distance between centers.*

## ACCELERATION BECAUSE OF GRAVITY

Based on your present understanding, could you predict how two objects ($A$ and $B$) would behave if dropped simultaneously from a high place? (See Figure 4.7.) Let's suppose that object $B$ has twice the mass of object $A$. Aristotle had asserted that the more massive object would reach the ground first. First, let's see that the weight of any object is simply a measure of the force, due to gravity, between that object and the earth. Thus, it would be accurate to say that the earth exerts twice as much force on object $B$ as compared to object $A$. At first glance one might think Aristotle was correct in expecting object $B$ to fall quicker. However, Galileo was not willing to accept this idea without actually trying it. Galileo is reputed to have climbed the Leaning Tower of Pisa (in his hometown) and to have simultaneously dropped objects of different masses. To everyone's amazement, the objects fell side

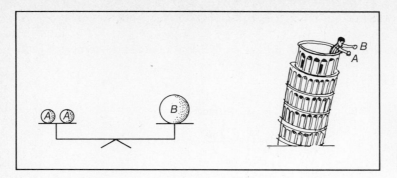

**Figure 4.7.** Galileo dropped objects of differing masses from the leaning tower of Pisa, yet they fell side by side, striking the ground at the same time.

by side, even though the earth exerted more force on one than on the other. In order to explain this result, we must recall Newton's second law. Suppose you placed one child in a wagon and pushed that wagon with a given force, a force that would accelerate the wagon to a velocity of 3 m/sec in 5 seconds. Now suppose you placed another child in the wagon, essentially doubling the mass of the system. You would have to push with twice the force in order to achieve the same velocity in 5 seconds. Likewise if the third child is added, then three times as much force is required to achieve the same acceleration. Newton expressed this relationship symbolically:

$$F = ma \text{ (force = mass} \times \text{acceleration)}$$

Now we have a logical explanation for the fact that the two bodies fell side by side; that is, although the earth exerts twice the force on object $B$, that is just exactly the force that is needed to produce the same acceleration as compared to that of object $A$. What would happen if you dropped two objects and the mass of one object was three times that of the other? They would still fall side by side because three times as much force is required to produce the same acceleration. Near the surface of the earth the acceleration of any object is approximately 9.8 m/sec². Remember, acceleration may be defined as the change in velocity per unit of time:

$$a = \frac{v_1 - v_2}{t}$$

Hence, if an object is dropped from rest and falls freely under the force of gravity, it will be traveling 9.8 m/sec² at the end of the first second, 19.6 m/sec² at the end of the second second, etc. It should be noted that the retarding force due to friction will alter the above results slightly.

Note further that the force necessary to support an object is equal to its weight. Hence, in the equation $F = ma$, we may set $F$ equal to the weight of the object and $a$ equal to the acceleration due to gravity ($g$), yielding:

$$W = mg$$

For example, the weight of an object that has mass equal to 2 kg is:

$$W = 2 \text{ kg} \times 9.8 \text{ m/sec}^2$$
$$= 19.6 \text{ kg m/sec}^2$$
$$= 19.6 \text{ nt}$$

This clearly illustrates that the weight of an object, although dependent on its mass, is not the same as its mass. The weight of an object is the force on it, due to gravity. The unit of force, called a *newton* (nt), is defined as the force just sufficient to accelerate a 1 kg mass 1 m/sec².

We have seen how a force applied in the direction of motion tends to accelerate the object in a positive sense, and a force applied in a direction opposite to the motion tends to accelerate in a negative sense. Now let's examine the effect of a force applied at right angles to the direction of motion. Since velocity is a vector, any change including direction represents an acceleration.

## CIRCULAR MOTION

Tie an object, say 100 g mass, at one end of a string, and holding the other end, swing the object in a circular path. If you let go of the string, the object will "fly off" on a tangent—traveling in a straight line. The only reason the object travels in a circular path is due to the force that you exert on it. You can feel that force as you swing the object. You might measure the force (at least roughly) by inserting a spring scale in the string as shown in Figure 4.8.

The numerical reading on the spring scale can be used to specify either of two forces: (1) that which is being exerted on the mass ($m_1$) directed toward the center of revolution (called the *centripetal force*), or (2) that which is being exerted on your finger away from the center of revolution (called the *centrifugal force*). It is the centripetal force on the whirling mass that actually produces its circular motion. This is a very good illustration of Newton's Third Law:

*For every action, there is an equal reaction* (on another body) *in the opposite direction.*

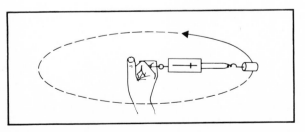

**Figure 4.8.** The tendency for an object to "fly off on a tangent" can be measured by this simple experiment.

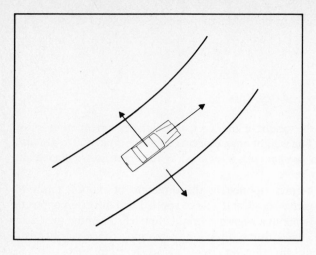

**Figure 4.9.** The car traveling around a curve exerts an outward force on the roadway.

When a car goes around a curve, it does so because it experiences a centripetal force directed toward the center of curvature. Can you tell what object experienced the centrifugal force in the opposite direction? It is the pavement of the road (and ultimately the earth to which the pavement is attached) (see Figure 4.9). Only as friction between the car's tires and the road prevents slippage can such a maneuver be performed. Therefore, it is easy to see the advantage of banking the road on a curve (see Figure 4.10).

The ultimate in banking a curve could be described as a *barrel effect*, sometimes utilized by motorcyclists in carnivals (see Figure 4.11). The barrellike track must be constructed so as to safely sustain a centrifugal force equal to the centripetal force needed to accelerate the motorcycle. Even if the rider is maintaining a constant speed, we still can specify his acceleration as the change in direction due to the centripetal force.

The force needed to produce circular motion in a given object depends on three basic factors: (1) the mass ($m$) of the object, (2) its velocity ($v$), and (3) its

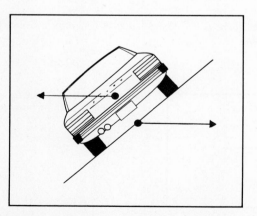

**Figure 4.10.** The "banking" of the curved roadway provides increased friction between the tires of the car and the roadway itself.

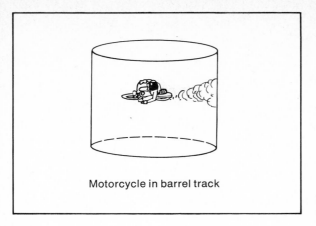

Motorcycle in barrel track

**Figure 4.11.** The ultimate in a "banked" curve.

radius of curvature ($r$). These factors are related by the statement:

$$F = \frac{mv^2}{r}$$

(See Figure 4.12.) Assuming this relationship to be true, can you predict the change in force necessary to accelerate twice the mass (other factors remaining constant)? (*Answer:* Twice the force.) What if the velocity is doubled? (*Answer:* Four times the force.) What if the radius of curvature were doubled? (*Answer:* One-half the force.)

Using this relationship, Newton concluded that the mutual force between the earth and the moon would be sufficient to maintain the moon in orbit. The most significant aspect of his discovery, however, was the universal nature of gravity and the fact that action may occur at a distance—that is, no direct contact need be maintained between the mover and the object moved.

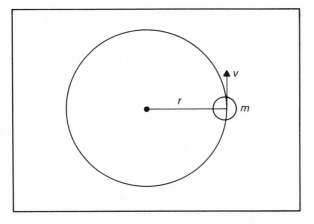

**Figure 4.12.** Centripetal force.

# ORBITING SATELLITES

When we consider the very practical problem of launching a satellite into an orbit that can be maintained without supplying additional energy, we will see an intimate relationship between centripetal force and gravitational force. Consider a device, located on top of a mountain, that is capable of projecting a missile at various speeds in a horizontal direction. If only a relatively slow speed is imparted to the missile, it will simply follow a parabolic path in its return to the earth at some distance ($d_1$). If a slightly greater speed is imparted, then a longer arc is achieved, and the missile falls to earth at a greater distance (see Figure 4.13). At what velocity will the missile achieve a circular orbit that can be maintained (neglecting the fact that friction tends to slow the flight)? When the centripetal force, caused by gravity, is just enough to cause the missile to curve in a path that is parallel to the curvature of the earth, such an orbit is possible. Because gravity provides the centripetal force that accelerates the rocket, the two quantities that follow may be equated:

$$\text{Centripetal force} \qquad \text{Gravitational force}$$
$$F = \frac{m_1 v^2}{r} \qquad F = G\frac{m_1 \times m_2}{r_2}$$

$$\frac{m_1 v^2}{r} = G\frac{m_1 \times m_2}{r^2}$$

$$v^2 = \frac{Gm_2}{r}$$

$$v = \sqrt{\frac{Gm_2}{r}}$$

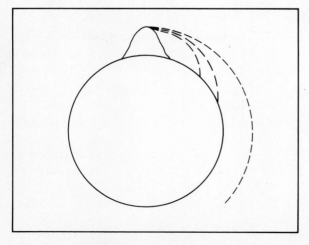

**Figure 4.13.** A missile, given sufficient horizontal velocity, will orbit the earth.

Where $m_1$ is the mass of the missile, $m_2$ is the mass of the earth, and $r$ is the distance between centers. Since $G$ and $m_2$ are fixed (known) quantities, you can see that the velocity necessary to maintain an orbit depends only on the distance between centers.

When rockets are fired from the earth's surface, their initial direction of flight is essentially vertical to gain height; but then they are diverted to horizontal flight, and sufficient velocity is imparted to sustain an orbit. Should a greater velocity be achieved, then the orbit will become an elongated ellipse; or if escape velocity is achieved, then the rocket may leave the vicinity of the earth altogether.

Most of the calculations used today regarding rocketry and orbits go back to the basic discoveries of Galileo and Newton.

## QUESTIONS

1. Why must a force be continually applied to maintain a given velocity in an automobile, whereas no force is necessary to maintain a given velocity in empty space?
2. Is it more natural for a rocket to move in a straight line or along a circular path if it is far from any other object in empty space?
3. Define three possible types of acceleration.
4. Explain the phrase "action (or force) at a distance."
5. In what sense does the earth fall toward the sun?
6. Why should a given steel cylinder (say of 1 kg mass) weigh less on top of a high mountain than at sea level?
7. A steel marble has twice the mass of a glass marble, even though the diameters of the two marbles are the same. Compare their weights.
8. Compute the force needed to accelerate a 2 kg steel marble 9.8 m/sec² and also the force needed to accelerate a 1 kg glass marble 9.8 m/sec². (Express force in newtons.)
9. Using the results from Problems 7 and 8, explain why a 2 kg steel marble would fall side by side with the 1 kg glass marble, if dropped from a tower.
10. If a mass is swung in a circular motion at the end of a string, and then released, it "flies off on a tangent." What basic principle of Newton does this illustrate?
11. Why doesn't a planet "fly off on a tangent"?
12. Why do the tires on a car wear out faster when driving around curves as compared to driving on straight roads?
13. A given velocity is required in order to keep a satellite in orbit at a certain distance from the earth. Does it make any difference as to the mass of the satellite?

# 5

energy

That which humans have termed *progress* is very closely allied with their use of energy. While still living a nomadic existence, humans relied on the natural processes of growth (plants and animals) around them—a result largely of the radiant energy of the sun being converted and stored in edible tissues. The major self-expenditure of energy came in their use of their muscles, such as in throwing a spear, etc. This represents a conversion of the sun's radiant energy, through eating, into chemical energy in the body, and then into mechanical energy defined by the flexing of the muscle. When humans discovered fire, they were able to release latent forms of solar energy that had been stored in fuels such as coal and wood. In the highly technological day in which we live, we have become dependent upon many other sources of energy. In fact, our standard of living is very closely correlated with our expenditure of energy. Why should energy be so important? The answer comes from the very definition of the word—energy is the ability to do work.

## WORK

Few words carry such a variety of meanings as the word *work*. Persons might say that they worked hard at the office, meaning they had many decisions to make. Others might express fatigue because of the work they did holding lumber in place while someone else drove the nails. Neither the office workers nor the individuals who held the lumber in place have done work according to the definition of work used by the physical scientist. The scientist would say that work only occurs when a force is exerted through a certain distance.

$$\text{Work} = \text{force} \times \text{distance}$$

$$W = fd$$

The persons who held the lumber in place exerted a force, but no change of position resulted; hence distance equals zero and no work resulted ($W = f \times 0 = 0$). On the other hand, the persons who drove the nails applied a force that moved the nails—i.e., they did work.

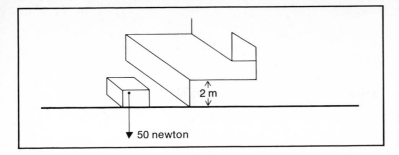

Figure 5.1. How much work is needed to lift the crate onto the dock?

## EFFICIENCY AND MECHANICAL ADVANTAGE

Suppose it is your task to lift a 50 newton crate onto a dock, a height of 2 meters. How much work is required? (See Figure 5.1.) You must exert a force of 50 newtons (nt) through 2 meters (m); therefore, $W = 50 \text{ nt} \times 2 \text{ m} = 100 \text{ nt-m}$, or 100 nt-m of work is required. Perhaps you would find it easier to use a ramp that is 4 m long. If the crate could be placed on a frictionless dolly, only 25 nt of force would be required to push the crate to the top of the ramp. Is less work required? (See Figure 5.2.) We see that the same work is required; hence the ramp did no work for you. However, it made an otherwise impossible task possible. Please note, however, that the work done in using the ramp (work in, 50 nt × 2 m) is equal to that which was accomplished (work out, 25 nt × 4 m), yielding an efficiency of 1 (100%).

$$\text{Efficiency} = \frac{\text{work out}}{\text{work in}}$$

It is rather naive to assume a frictionless cart, however, so let's rework the ramp problem assuming that 30 nt of force are required to move the crate up the ramp. Thus, $W = f \times d = 30 \text{ nt} \times 4 \text{ m} = 120 \text{ nt-m}$. Now we see an efficiency of:

$$\text{Efficiency} = \frac{\text{work out}}{\text{work in}} = \frac{100}{120} = \frac{5}{6} \ (83\tfrac{1}{3}\%)$$

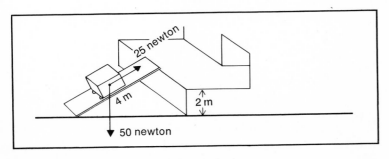

Figure 5.2. The use of a ramp makes the task easier, but no less work is required. Can you explain this statement?

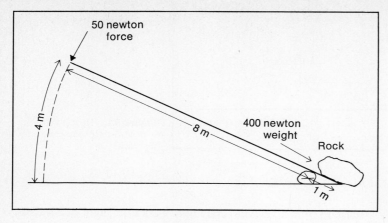

**Figure 5.3.** A lever makes possible an impossible task.

The lever is a high efficiency tool because of the virtual absence of friction. Let's apply our work definition to a task in which a lever is well suited—lifting a heavy rock partially embedded in soil. The lever may be a steel bar or large timber; the fulcrum (pivot point) is another rock (see Figure 5.3). By exerting a force ($F$) of 50 nt at the upper end of the lever, it will be possible to balance the weight of the rock (400 nt) at the other end. If, when the force ($F$) is applied, the lever end moves through 4 m, the rock will be lifted 0.5 m.

Work in = 50 nt × 4 m = 200 nt-m

Work out = 400 nt × 0.5 m = 200 nt-m

Efficiency = $\dfrac{\text{work out}}{\text{work in}} = \dfrac{200}{200} = 1$ (100%)

Did the lever do work for you? No! Did the lever allow you to accomplish a task otherwise impossible? Yes! How is that possible? By using the lever, you were only required to exert a force of 50 nt, which was easily possible; but it was necessary for you to exert that force through 4 m, whereas the rock only moved 0.5 m. What you gained in lowering the necessary force, you lost in distance. Before we leave this problem, let's see why only 50 nt of force were needed. Look at the distances on either side of the fulcrum; call them $d_1$ and $d_2$ (see Figure 5.4).

**Figure 5.4.** The forces are inversely proportional to the distances from the fulcrum.

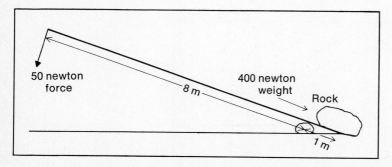

Think of the force exerted as $F_1$ and the weight of the rock as a force $F_2$. The following relationship is always true for levers:

$$F_1 \times d_1 = F_2 \times d_2$$

If $F_1$ were unknown, it could be found as follows:

$$F_1 \times 8 \text{ m} = 400 \text{ nt} \times 1 \text{ m}$$

$$\frac{F_1 \times 8 \text{ m}}{8 \text{ m}} = \frac{400 \text{ nt-m}}{8 \text{ m}}$$

$$F_1 = 50 \text{ nt}$$

The mechanical advantage of such a system is defined by comparing the force (weight) lifted ($F_2$) to the force exerted ($F_1$):

$$\text{Mechanical advantage} = \frac{F_2}{F_1} = \frac{400}{50} = \frac{8}{1}$$

The pulley is another simple machine (tool) that can help accomplish otherwise impossible (or difficult) tasks. However, pulleys do not do work for you, as you will see. First, consider only one pulley suspended from a point overhead (see Figure 5.5). This amounts to only a change of direction, which permits you to pull down to lift the weight—a matter of convenience with no mechanical advantage. On the other hand, the arrangements shown in Figure 5.6 do have mechanical advantages in addition to a change of direction. Can you see why the forces indicated are correct, assuming no friction in the pulleys (a fact that cannot be assumed in actual experience)? In (a) two lines support the weight; the third only represents a change of direction. In (b) three lines support the weight. In (c) four lines support the weight.

**Figure 5.5.** A change in direction is a convenience only.

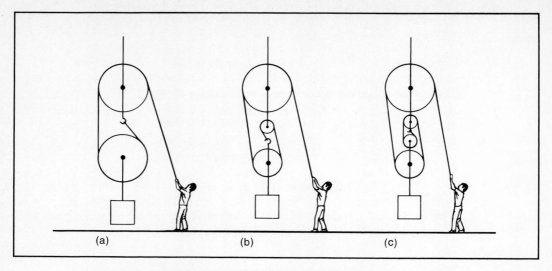

**Figure 5.6.** Three different mechanical advantages are shown. Can you give the mechanical advantage of each example, just by looking at the drawings?

In (a) when the force of 50 nt is exerted through 2 m, the weight will only rise 1 m. Hence, work in (50 × 2 = 100 nt-m) is equal to work out (100 × 1 = 100 nt-m). The mechanical advantage is 100/50 = 2 (assuming no friction). In practice, perhaps 60 nt are required to lift the weight, and the true mechanical advantage is

$$\frac{100}{60} = \frac{5}{3} = 1\frac{2}{3}$$

We can make the following general statement about pulleys: The (frictionless) mechanical advantage may be determined by counting only the number of lines directly supporting the weight.

## POWER

A concept that follows naturally from work is power—that is, the rate at which work is done:

$$P = \frac{W}{t} = \text{work per unit time}$$

The units we have used most often for work are the newton-meter (nt-m) or joule; hence a basic unit of power might be nt-m/sec or joule/sec. In addition, we will find it particularly advantageous to define the watt (and/or the kilowatt) as a unit of power:

$$1 \text{ watt} = 1 \text{ nt-m/sec}$$
$$1 \text{ watt} = 1 \text{ joule/sec}$$
$$(1 \text{ kilowatt} = 1000 \text{ watts})$$

The kilowatt is the unit of power most commonly used in terms of electrical power. Power may also be visualized further in mechanical terms by defining the unit called *horsepower*.

$$1 \text{ horsepower} = 746 \text{ joules/sec}$$

Today we are familiar with automobile engines rated at 300 horsepower. Can you imagine the work that such an engine could accomplish each second? Unfortunately almost 90% of this power is "lost" in overcoming the friction that is created by its multitude of moving parts. This leaves only 10% or 30 hp (horsepower) available for useful work in terms of moving the car forward.

$$30 \text{ hp} = 30 \times 746 \text{ nt-m/sec}$$
$$= 22{,}380 \text{ nt-m/sec}$$

## ENERGY—ITS MANY FORMS

We have defined energy as the ability to do work. Energy comes in many forms and can be converted from one form to another—a very useful consideration in light of the depletion of certain energy sources and the search for new ones. This search, in which we are presently engaged, is not a temporary one but will probably be with us for many generations. Perhaps by giving some thought to this matter, someone reading this book will be instrumental in finding solutions to current problems.

## POTENTIAL ENERGY

Potential energy results from the fact that some objects are higher than others, and the higher object can fall farther, giving up energy in the process. Potential energy (P.E.) is energy due to position and is equivalent to the weight of the object ($w$) multiplied by the height ($h$) through which it is capable of falling.

$$\text{P.E.} = wh$$

The classic example of such energy is that possessed by water behind a dam. Suppose the water level behind the dam is 200 m above the outlet on the downstream side (see Figure 5.7). Every cupful of water weighs approximately 2 nt; hence every cupful of water at that level possesses potential energy in the amount of

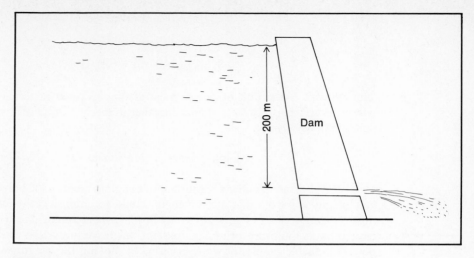

**Figure 5.7.** Water behind the dam possesses potential energy.

2 nt × 200 m = 400 nt-m. If, on the other hand, the water level had been only 100 m above the discharge point, then each cupful would have a potential energy of only 2 nt × 100 m = 200 nt-m. This illustrates that potential energy is always related to the heights through which it can potentially fall and should not be thought of as absolute.

We can express potential energy in a more general way if we recognize that the weight ($w$) of an object is equivalent to its mass ($m$) times the acceleration ($a$) due to gravity ($F = ma$ or $w = mg$) where $g$ is the acceleration due to gravity and $h$ is height, yielding

$$\text{P.E.} = mgh$$

This form for potential energy emphasizes the fact that such energy is actually due to gravity; hence could be called *gravitational energy*.

However, when we think of gravitational energy, we are usually thinking on a larger scale—that of the universe. In the spaces between the stars, we find gas clouds called *nebulae*. In such nebulae each atom is located at a certain distance from a point where a star may form in the future; let's say at point $P$ in Figure 5.8. Therefore, each atom has the potential of being drawn by mutual gravitation to that point. The gravitational energy can then be specified for each atom in relation to that point. Upon collapse of such a cloud, the potential energy is converted to other forms of energy, i.e., heat, kinetic energy, and so forth. It would be safe to conclude that a great deal of potential energy still remains in the universe, since so much material exists in a noncondensed state (nebulae). Even a planet going around the sun has potential energy in relation to the sun, for it has a specific distance it would "fall" if halted in its orbit. A planet also has a form of energy called *kinetic energy* due to its motion.

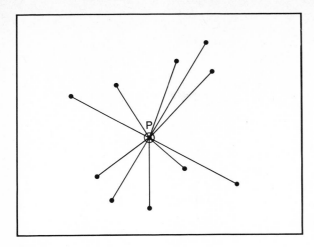

**Figure 5.8.** The atoms of a nebula possess potential energy.

## KINETIC ENERGY

Kinetic energy is the ability to do work by reason of an object's motion. One would not hesitate to stand in front of the pins of a bowling alley if the bowler only intended to rest the ball in front of him, but certainly it is doubtful that anyone would stand there under normal bowling conditions. What is it that the same bowling ball possesses when traveling down the alley at 10 m/sec that it did not possess when at rest? The answer is kinetic energy—energy due to motion. That energy can be computed using the formula:

$$\text{Kinetic energy} = \tfrac{1}{2} mv^2$$

Where

$$m = \text{mass}$$
$$v = \text{velocity}$$

Thus, if the bowling ball has a mass of 20 kg and a velocity of 10 m/sec, its kinetic energy is:

$$\text{Kinetic energy} = \tfrac{1}{2} \times 20 \times (10)^2 = 1000 \text{ kg-m}^2/\text{sec}^2$$

Let's see how kinetic energy can be derived from one basic assumption. Suppose a mass ($m$) were carried to height ($h$) above the surface of the earth. Potential energy = $mgh$. If dropped, the potential energy of the object is converted entirely to kinetic energy as it reaches its lowest point. Hence kinetic energy = $mgh$, but the height ($h$) of a falling body is related to time ($t$) of fall by:

$$h = \tfrac{1}{2} gt^2$$

So:

$$\text{Kinetic energy} = mg(\tfrac{1}{2}gt^2) = \tfrac{1}{2}m(gt)^2$$

Velocity ($v$) of a falling object is related to time ($t$) by

$$v = gt$$

Therefore, let ($gt$) be replaced by ($v$) and:

$$\text{Kinetic energy} = \tfrac{1}{2}mv^2$$

Furthermore, kg-m$^2$/sec$^2$ is equivalent to newton-meters when we realize that a newton is the force necessary to accelerate a mass of 1 kg 1 m/sec$^2$.

## POTENTIAL ENERGY VERSUS KINETIC ENERGY

A good example of the conversion of one form of energy to another is shown by the pendulum (see Figure 5.9). When at position $A$ or $C$, the pendulum is momentarily stopped; hence kinetic energy $= \tfrac{1}{2}mv^2 = 0$ because $v = 0$. However, potential energy is at a maximum; that is, potential energy $= mgh_1 = wh_1$. When at position $B$, the pendulum has zero potential energy because it can drop no lower, but maximum kinetic energy $= \tfrac{1}{2}mv^2$ ($v$ is at a maximum). Thus, there is a continual conversion back and forth between these forms of energy. At point $D$, the mass would possess some kinetic energy due to its velocity and some potential energy due to its height ($h_2$). The sum of potential and kinetic energy at $D$ just equals the kinetic energy at $B$ or the potential energy at $A$. This suggests that energy is not created nor destroyed as it is converted—a principle called *conservation of energy*. On the other hand, a pendulum will not swing forever; so some portion of the total energy must be dissipated on each swing. Can you think of a possible source of

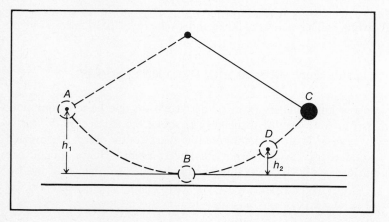

**Figure 5.9.** A swinging pendulum continuously converts potential energy to kinetic energy and visa versa.

such dissipation? The mass (of the pendulum) experiences a little friction in moving through the air molecules, and additional friction in the supporting string or cable. Energy losses because of friction usually show themselves as heat losses. If we could account for potential energy, plus kinetic energy, plus energy converted to heat, on each swing, we would have a truer picture of what is happening. Eventually when the pendulum stops swinging, we can assume that the potential and kinetic energy have been converted to heat.

## TEMPERATURE VERSUS HEAT

Often temperature is treated as though it were the same thing as heat. We must differentiate between heat and temperature. Heat is a measure of the *total* random molecular kinetic energy, whereas temperature is a measure of the *average* random molecular kinetic energy. The faster molecules move, the higher the temperature reads. This subject will be treated more fully in Chapter 12, so only the following illustration will be presented here.

The molecules of air in a room at 20°C (70°F) have an average velocity of about 1600 km/hr (1000 mph). An increase of temperature, say to 25°C (79°F), would result in an increase in average velocity of the molecules to about 1627 km/sec (1017 mph). The use of a thermometer allows us to measure the temperature of an object rather than relying only on estimates by our sense of touch. Since mercury expands with the increase of temperature, a thermometer may be constructed by introducing mercury (or other substances) into a narrow passage of glass tube, then placing marking on the glass tube according to some standard agreed upon. Since water is so abundant, it certainly makes sense to let the temperature at which water freezes be called zero, and the temperature at which it boils be called 100°, with 100 equal divisions between these two temperatures (see Figure 5.10), thus producing the Celsius (formerly called *centigrade*) scale, which is used in most scientific work.

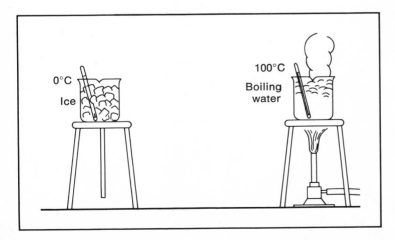

**Figure 5.10.** The freezing and boiling temperatures of water provide a convenient scale of temperatures.

Earlier another scale was defined in England; a scale that we have used for everyday specification of temperature—the Fahrenheit scale. This scale reads 32° when water freezes and 212° when water boils—with 180 steps separating the two values. The two scales may be compared by the formulae below. With 100 steps on the C scale and 180 on the F scale, perhaps you can see why the factor of 5/9 or 9/5 is present in these conversion formulae (180/100 = 9/5; 100/180 = 5/9):

$$C = \tfrac{5}{9}(F - 32)$$
$$F = \tfrac{9}{5}C + 32$$

However, simply the visual comparison shown in Figure 5.11 will serve to give us approximate values. Note that Figure 5.11 in addition shows the Kelvin scale. It was discovered that as the temperature of −273°C was approached, the average velocity of molecules approached zero; hence, the term *absolute zero*. This too seemed a very logical place to begin a temperature scale since no lower

**Figure 5.11.** Comparison of temperature scales: Fahrenheit, Celsius, and Absolute (Kelvin).

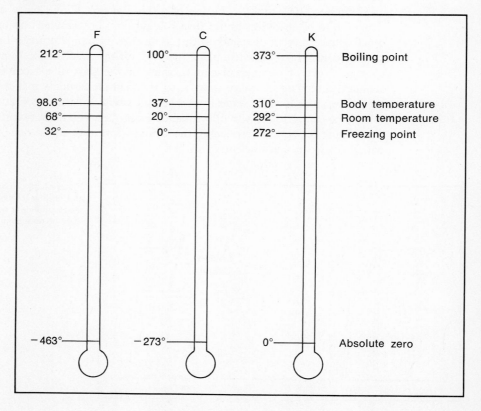

temperature may be obtained. The Kelvin scale uses the same divisions as the Celsius scale; therefore, water freezes at 273°K and boils at 373° K. Your body temperature is very nearly 310° K. The Kelvin scale ultimately will serve us better than either of the other two.

## HEAT, A FORM OF ENERGY

If heat is really a form of energy, we should be able to express it in foot-pounds (ft-lb) or joules (nt-m), and in the next section we will show that this is possible. It will be more convenient, however, to define a unit of energy particularly associated with heat, namely, the *calorie*. The calorie (cal) is the amount of heat needed to raise the temperature of one gram of water one degree Celsius (or Kelvin). This definition embodies three factors that are essential in computing the heat of a body: its mass, its material, and its change in temperature.

The quantity of heat $Q$, in calories, can be found by the equation:

$$Q = ms\,(t_2 - t_1)$$

where

$$m = \text{mass in grams}$$
$$s = \text{specific heat}$$
$$t_2 - t_1 = \text{change in temperature}$$

This equation contains one undefined term, $s$, the specific heat of a given material. The specific heat of water is 1 cal/g°C, based on our definition of calorie. The specific heat of any substance represents the amount of heat needed to raise 1 gram of that substance 1°C. Note in Table 5.1 that only 0.12 cal is needed to raise the temperature of iron 1°C.

**Table 5.1**  Specific Heats

| SUBSTANCE | SPECIFIC HEAT (cal/g °C) |
|---|---|
| Water | 1.00 |
| Ice | 0.5 |
| Iron | 0.12 |
| Aluminum | 0.22 |
| Brass | 0.094 |
| Copper | 0.093 |

How many cal of heat are required to raise 500 g of iron 20°C?

$$Q = ms(t_2 - t_1) = (500 \text{ g}) (0.12 \text{ cal/g°C}) (20°C)$$
$$q = 1200 \text{ cal}$$

How much heat could 2 kg (2000 g) of water give up if its temperature is lowered from 94°C to 70°C?

$$Q = (2000 \text{ g}) (1 \text{ cal/g°C}) (94° - 70°C)$$
$$= 48{,}000 \text{ cal } (48 \text{ kcal})$$

where one kilocalorie (kcal) = 1000 cal.

This last illustration shows how energy may be given up by the cooling of water or any other substance. Energy is the ability to do work; hence, if we first have water at a high temperature with the possibility of it cooling to a lower temperature, we should be able to make that energy do work for us. But how much work?

## HEAT, MECHANICAL EQUIVALENT

One of the most convincing demonstrations that heat is a form of energy and can be equated to the mechanical forms we have studied is shown in Figure 5.12. A weight ($w$), which possesses a certain potential energy (P.E. = $wh$), is allowed to "fall" through a distance, $h$, and through the system of pulleys shown in Figure 5.12 turns the paddle wheel inside an insulated chamber filled with water. The molecules of water experience friction as the paddle wheel is turned, and the friction produces heat and a consequent rise in temperature. When James P. Joule (English physicist) first performed this experiment, over one hundred years

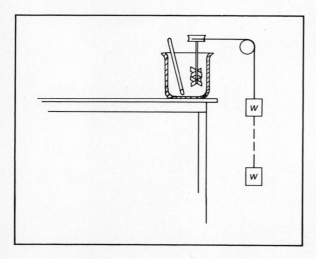

**Figure 5.12.** The equivalency of heat and energy is shown by this experiment. As the weight ($w$) falls, energy is converted to heat by the turning paddle wheel.

ago, he discovered that 1 calorie of heat is equivalent to $4.18 \times 10^7$ ergs of work, or 4.18 joules. Thus, the 4800 cal mentioned earlier would be capable of producing 20,064 joules of work ($4800 \times 4.18 = 20{,}064$).

## STEAM ENGINE

One way in which heat energy may be converted to kinetic energy is by means of the steam engine. Water is heated to the boiling point, providing a source of

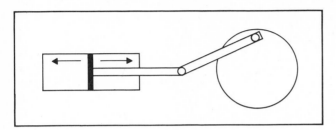

**Figure 5.13.** Heat energy (in the form of steam) is converted to kinetic energy by means of the cylinder and linkage of a steam locomotive.

**Figure 5.14.** The steam turbine. (Brown, Boveri, and Comp. Ltd., Badew, Schweiz)

expanding steam that is introduced first on one side of a movable piston and then on the other side, moving it back and forth in the cylinder. A mechanical linkage transfers the force to the wheels, as in the case of the steam locomotive (see Figure 5.13). Typically, steam engines operate at low efficiency (15% or less) because so much of the heat is dissipated into the air surrounding the engine.

Somewhat more efficient is the steam turbine shown in Figure 5.14. A jet of steam is directed toward the fanlike structure, which reacts by rotating. Steam leaving the first set of rotors produces additional reaction on the second set, etc., until the mechanical (rotating) system possesses a significant part of the energy originally in the form of heat.

## LAWS OF THERMODYNAMICS

We have already been working with the first law of thermodynamics:

*Energy is neither created nor destroyed, but may only be changed from one form to another.*

This fact is known as the conservation of energy.

The second law may be stated in several ways:

*Heat naturally flows from a warmer body to a cooler body, and work may be extracted only if this lowering of temperature occurs.*

Stated alternately:

*Work is required to cause heat to flow from the cooler to the hotter body.*

## ENTROPY—A MEASURE OF DISORDER

Consider a system composed of two tanks of water, one at considerably higher temperature than the other. We call this an ordered system because it is possible to extract work from the system as energy flows from the hot tank to the cold tank. We define the entropy of a system to correlate directly with its disorderliness, hence the entropy of the system just mentioned is low. On the other hand, as energy flows from the hot water to the cold, the orderliness of the system continually decreases (it is less able to do work), hence we say that entropy is increasing. Eventually the temperature of both tanks of water is the same and no further work can be extracted. Entropy is now at a maximum for this system.

Using the concept of entropy, we may define the second law of thermodynamics as follows: In a closed system (with no outside influence—no energy flowing in), entropy tends to increase. We can think of the increase in entropy as a time arrow pointing in only one direction. Consider the event of a baseball breaking a window. The glass crashes to the ground in small pieces as the ball passes through the window. Can you imagine turning time backward, the small pieces of glass reassembling themselves into a window, and the ball returning to its original point? No; we have never seen such a thing and cannot conceive of it as possible. We believe that time runs in only one direction. In the same way, the increase of entropy correlates with the passage of time. At any given point in time the universe possesses a certain order or arrangement of matter. Because of this order, work can be done, but along with that work comes a less orderly arrangement—an increase in entropy. Ultimately, if no event restores an orderliness to the universe, work will no longer be possible. This possibility is sometimes thought of as the "heat death of the universe."

## QUESTIONS

1. List as many different forms of energy as you know and describe each.
2. What kind of energy is possessed by a car standing still at the top of a hill?
3. What kind of energy is possessed by a car traveling 85 km/hr? (55 mi/hr?)
4. Why is the efficiency of an automobile so low?
5. Can you think of any way to improve the efficiency of the automobile.
6. In what way is the physicist's definition of "work" better than an ordinary dictionary definition?
7. Find the efficiency of a ramp 6 m long if a force of 40 nt is required to roll a 100-nt crate to a height of 2 m.
8. Why would anyone use a lever or pulley system if in reality such a device can do no work?
9. Draw the picture of a frictionless pulley system such that a force of 40 nt will lift a weight of 120 nt.
10. Which do you think is more efficient—the lever or a system of pulleys? Why?
11. What other factor must be taken into consideration when we relate work to power?
12. Using the formula for conversion of Fahrenheit temperature to Celsius, compute your body temperature in Celsius, assuming that $F = 99°$.
13. Which substance requires less heat to raise the temperature $20°C$, 50 g iron or 50 g of brass?
14. Describe several ways by which heat energy may be transferred to mechanical energy.
15. Tell in your own words why time seems to run in only one direction. Can you think of any event that might illustrate time running backward?

# 6

waves and sound

The phenomenon of waves pervades our entire sensory experience. Most easily visualized are water waves—distortions of the smooth surface of a pond caused by a pebble thrown into it, or the swells of an ocean driven by the wind. But waves are also a part of sight and sound, of radio and TV, of earthquakes, and of the very atoms of which everything is composed. A wave is a mechanism whereby energy may be carried from one point to another without the actual transfer of material over that same distance. Consider the fisherman who sits in his boat for long periods at a time. His boat rises and falls as waves pass him by, yet he does not move with those waves. The waves impart energy to each boat in their path, yet not even the molecules of water travel with the waves. If one could keep track of a single molecule, its motion would be primarily one of rising and falling, even as is true of the boat itself.

When the motion of the particles, through which a wave is passing, is oriented perpendicular to the wave motion, we speak of this as a *transverse wave*. To further illustrate this kind of wave, consider a rope attached to a post, as in Figure 6.1. If the free end is moved vertically (up and down), a wave pattern will be seen; yet each particle of the rope simply moves up and down. Again we see that

**Figure 6.1.** The traverse wave.

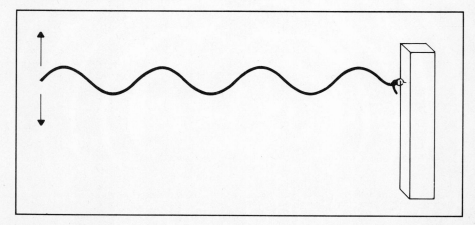

energy may be delivered to the post end from the free end without the actual transmittal of mass from one end to the other.

Waves are typically generated by a vibrating body—one that oscillates back and forth (up and down). In the case of visible light or radio waves, which are examples of electromagnetic waves, no visible wave pattern exists, as in the case of the rope, for such waves need no medium through which to travel. We still find it productive to visualize light and radio as waves, and in the following section you will see a general description for all transverse waves.

## THE GRAPH OF A VIBRATING OBJECT

Consider a weight hung at the end of a spring. Allow it to seek its own level, and mark that position as zero. Now pull downward on the spring a distance of 10 cm and let go. You will see (Figure 6.2) that the weight moves upward through the zero point to a position approximately 10 cm above, then back downward again to the −10 cm position, etc. How can we describe the location of the weight at any given instant? We could make a table in which we record the location of the weight every tenth of a second (see Table 6.1).

**Table 6.1** A record of weight location with respect to time

| TIME | 0 | 0.1 | 0.2 | 0.3 | 0.4 | 0.5 | 0.6 | 0.7 | 0.8 | 0.9 | 1.0 | 1.1 | 1.2 | 1.3 | 1.4 | 1.5 | 1.6 |
|---|---|---|---|---|---|---|---|---|---|---|---|---|---|---|---|---|---|
| LOCATION | −10 | −9 | −7 | −4 | 0 | +4 | +7 | +9 | +10 | +9 | +7 | +4 | 0 | −4 | −7 | −9 | −10 |

This looks like just so many numbers and has little meaning, but note how much more descriptive it would be if we show the same information on a graph. Let "time" run to the right, and "location" (displacement) of the weight run up and down on our graph. Plot the respective points; then join the points with a smooth curve (see Figure 6.3). We recognize that the shape of the graph is much like waves in the rope; however, the graph is simply a plot of the position of a vibrating weight

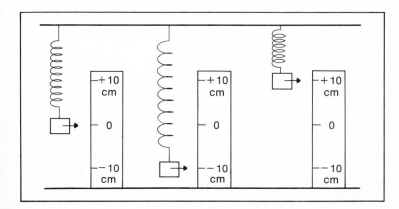

**Figure 6.2.** A vibrating weight.

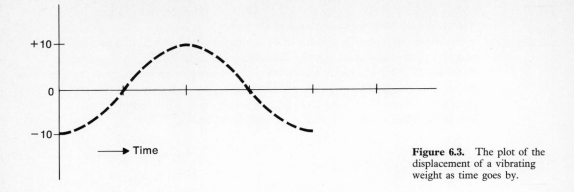

**Figure 6.3.** The plot of the displacement of a vibrating weight as time goes by.

on a spring as time goes by. This idea has much more general application, for now we can think of light as being generated by an oscillating charged particle and plot the changes in the electrical field as having a similar format (see Figure 6.4). This idea will be developed more fully in Chapter 8.

Consider the simple linkage of the drive wheel of a locomotive to the piston in the steam chamber (see Figure 6.5). As the piston is forced back and forth in the cylinder, the wheels are driven, with the point $A$ on the wheel describing a circular path. When point $A$ is at position 1, the piston is at a central location in the cylinder. Call this the zero (0) position. When point $A$ moves to position 2, the piston is as far forward as possible. Call this position $+1$. When point $A$ moves to position 3, the piston is again at 0, and when point $A$ moves to position 4, the piston is in the extreme back position. Call that a $-1$ position. The wavelike graph of Figure 6.6 accurately describes the motion of the piston, and because it is associated with the circular motion of the wheel, it is called *simple harmonic motion*.

Anyone who has studied trigonometry will recognize this graph as a sine curve, and we will see its many applications in the study of vibrating objects. The curve generated by the weight on the spring was also a sine wave, and its motion

**Figure 6.4.** The plot of the electrical field of a vibrating charged particle.

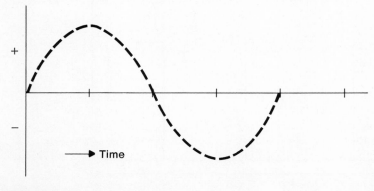

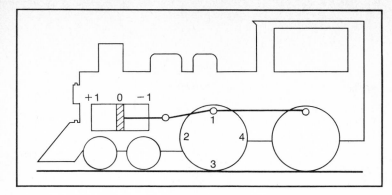

**Figure 6.5.** A locomotive.

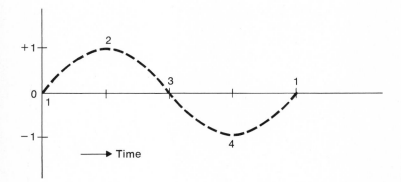

**Figure 6.6.** The plot of the piston's displacement as time goes by.

may also be described as simple harmonic motion. The pendulum of a clock is another example, as is the vibration of a violin string.

## PROPERTIES OF A WAVE

Once again, let's visualize water waves in order to define a number of concepts that are common to all wave phenomena. Suppose a mechanical bobbler (wave maker) is designed to create waves in a shallow layer of water. This water if illuminated from above will cast a pattern on the screen below (see Figure 6.7). An enlarged cross section of these waves would look like Figure 6.8. The distance between two successive wave tops (crests) is called the *wavelength*, indicated by the Greek letter $\lambda$ (lambda). In fact, this length may be measured from any point on a wave to the corresponding point on the next wave. The maximum vertical displacement of the wave above or below the normal level of the water is called the *amplitude* ($a$) of the wave. The time required for the generation of one complete wavelength or cycle is called the *period* ($T$), and the number of cycles generated in one unit of time is called the frequency ($f$). This is summarized in Table 6.2.

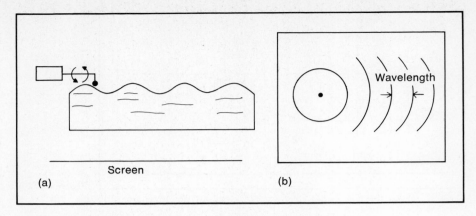

**Figure 6.7.** (a) A ripple tank in which water waves are being generated; (b) top view of the waves created within the ripple tank.

**Figure 6.8.** Wavelength and amplitude of a wave.

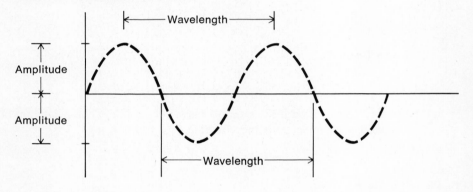

**Table 6.2**

| TERM | SYMBOL | UNITS |
|---|---|---|
| Wavelength | $\lambda$ | Meters |
| Amplitude | $a$ | Meters or cm |
| Period | $T$ | Seconds |
| Frequency | $f$ | Cycles per second |

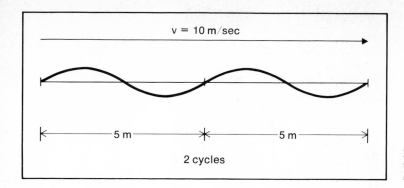

**Figure 6.9.** Relationship between wavelength, velocity, and frequency.

There is a simple relationship between period ($T$) and frequency ($f$), namely:

$$f = \frac{1}{T} \quad \text{or} \quad T = \frac{1}{f}$$

For example, if the period of a wave is one-half second, then two (2) such waves can be generated in one second; hence the frequency is 2 cycles per second.

Another very useful relationship among all kinds of waves involves the velocity ($v$) with which they travel. Suppose a given water wave travels at the rate of 10 m/sec and two waves are generated every second, then each wave must have a wavelength of 5 meters (see Figure 6.9). This illustrates the relationship:

$$v = \lambda f$$

(10 m/sec = 5 m/cycle · 2 cycles/sec)

Thus, if any two of these quantities is known, the third may be found.

## REFLECTION

Water waves may also be used to illustrate the way most waves react when they hit a barrier; i.e., they are reflected in a very definite and predictable way. The angle of reflection is always equal to the angle of incidence, as measured from a normal line that is drawn perpendicular to the barrier (see Figure 6.10).

## REFRACTION

In Figure 6.11 you will see that waves are flowing into shallower water as they travel from $A$ to $B$, and they are slowing down as a result. Accompanying their slowing is also a change in direction, and this is called *refraction*. Henceforth, regardless of the specific kind of wave we discuss, you may expect a change in direction whenever the wave experiences a change in velocity. The concepts of reflection and refraction, in relation to light, are more fully developed in Chapter 8.

**Figure 6.10.** Principle of reflection. (Educational Development Center, Inc., Newton, Mass.)

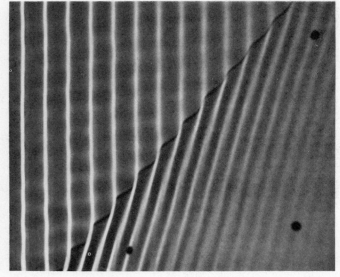

**Figure 6.11.** Principle of refraction. (Educational Development Center, Inc., Newton, Mass.)

## INTERFERENCE

If two sources of waves should be introduced into a ripple tank, and let's suppose that they have the same frequency, there will be locations where the waves reinforce each other, called *constructive interference,* and other places where the waves tend to destroy each other, called *destructive interference* (see Figure 6.12). Constructive interference occurs when the two waves arrive at a given point in step (in phase) with each other, as shown in Figure 6.13 (a); and destructive interference occurs when the waves arrive at a point out of step (out of phase) with each other, as shown in Figure 6.13 (b).

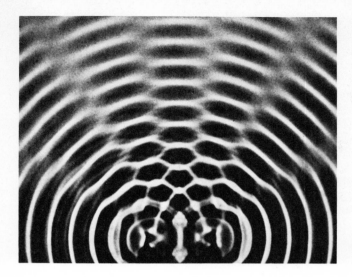

**Figure 6.12.** Principle of interference. (Educational Development Center, Inc., Newton, Mass.)

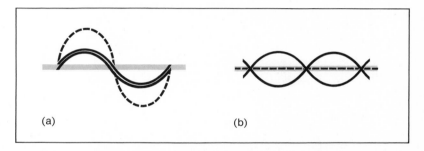

**Figure 6.13.** (a) Constructive interference; (b) destructive interference.

## DIFFRACTION

Whenever a wavelike disturbance passes an obstacle, that obstacle absorbs some of the energy of the wave and then reemits that energy in all directions, making it appear that the obstacle has energy of its own (see Figure 6.14). This phenomenon is called *diffraction*. It also occurs when a wave passes through an opening, each edge of that opening acting like an obstacle, and in this sense the wave is able to bend around that edge (see Figure 6.15).

In our discussion thus far, we have been thinking only of transverse waves; however, we will see that many of the attributes of transverse waves carry over into an entirely separate class of waves (*longitudinal* waves), and sound is an example of such.

**Figure 6.14.** Principle of diffraction. (Educational Development Center, Inc., Newton, Mass.)

**Figure 6.15.** Diffraction in a small slit. (Educational Development Center, Inc., Newton, Mass.)

## SOUND

There exists an age-old question about sound. If a tree fell in the forest and no one was there to hear it, would there still be the sound of its falling? The physicist would say yes, for he defines sound in terms of the disturbance of a medium, whether anyone is present or not to sense that disturbance. Sound is created by a vibrating source, such as the bowed string of a violin or the clashing cymbal of an orchestra or the tuning fork illustrated in Figure 6.16. As the fork is struck and set to vibrating, the air that surrounds it experiences periodic increases and decreases in pressure. Imagine a horizontal column of air off to the right of the fork. The air

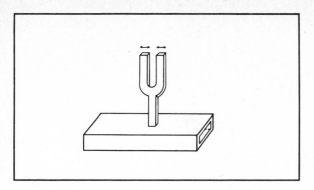

**Figure 6.16.** The tuning fork, on a hollow (open-ended) sounding box, producing sound waves.

molecules that were fairly evenly spaced before the fork was struck are now pressed together in certain regions (called *condensations*) and pulled apart in other regions (called *rarefactions*) (see Figure 6.17). If we could keep track of a given molecule, it would seem to have a very limited back-and-forth motion parallel to the direction of travel. When this is true, we speak of such waves as being *longitudinal*. If a long spring is stretched and then a quick back-and-forth motion is produced at one end, parallel to the length of the spring, one can see a condensation travel as a wave down the length of the spring (see Figure 6.18).

If sound is produced by a vibrating object, but without a transmitting medium in which a sequence of alternate condensations and rarefactions may be carried from the source to the observer, the sound will not be heard. A simple

**Figure 6.17.** Consensations (C) and rarefactions (R) in sound waves.

**Figure 6.18.** Condensations (C) and rarefactions (R) in a coil spring.

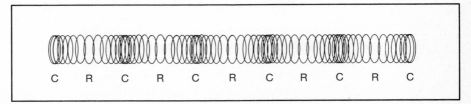

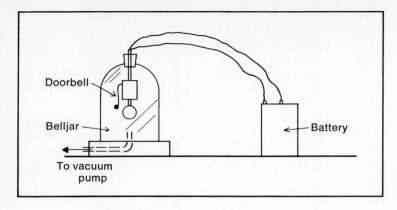

**Figure 6.19.** Sound cannot be transmitted through a vacuum.

experiment will illustrate this fact. Place a doorbell inside a glass dome, as shown in Figure 6.19. With the battery connected, the bell will be heard because the vibrations pass through the air in the dome to the glass wall, through the glass wall to the air on the outside, and through the air to the observer's ear. Now pump out the air from inside the dome, using a vacuum pump, and while the bell can be seen to vibrate, no sound is heard. This demonstrates that the air in the dome was needed to transmit the sound.

## SPEED OF SOUND

Not only does sound require a medium in which to travel, but it travels at differing speed in different media. The fact that sound travels quite slowly in air is made very clear when we see lightning and then hear the thunder several seconds later. We see the lightning almost without delay because light travels at the very rapid pace of 300,000 km/sec, yet it is necessary to wait approximately 5 sec for each mile traveled by the sound. This reveals the speed of sound to be only 331 m/sec or 1192 km/hr (740 mi/hr) at 0°C. The velocity of sound is slightly increased with increasing temperature. For instance, at 20°C, the speed of sound is 343 m/sec. Can you estimate the distance to a lightning flash if it requires 10 sec additional time to hear the thunder, assuming a temperature of 20°C?

Another phenomenon that is directly related to the speed of sound is the sonic boom that is experienced whenever aircraft fly at supersonic speeds—faster than the speed of sound. Figure 6.20 shows, using water waves, the effect of a source traveling at a slightly greater speed than its waves travel. Note the V-shaped buildup of waves that results. The cone of energy causes a shock wave that not only makes a loud report but may break windows because of the sharp change in pressure that this wave represents. Remember that in sound we are talking about condensations and rarefactions rather than transverse waves, and these are alternate regions of high and low pressure. The sonic boom can be greatly diminished if an aircraft will remain subsonic in its flight until it has reached higher altitudes. The

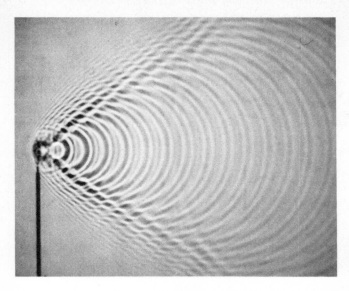

**Figure 6.20.** The sonic boom, illustrated by water waves. (Educational Development Center, Inc., Newton, Mass.)

sonic boom has been a great deterrent to the flight of the SST (supersonic transport) because it flies at supersonic speeds near the ground.

Typically sound travels much faster in a liquid or solid; i.e., in water at 15°C, the speed of sound is 1450 m/sec. One of the very significant uses of sound in water is the *sonar* (*so*und *na*vigation *a*nd *r*anging). By sending out a sound signal and then listening for its echo, ships at sea may determine the contour of the ocean bottom (to limited depths) and thereby escape running aground. Depths are judged by the round-trip time of the signal, knowing the velocity of that signal. For instance, a round-trip time of one second would indicate a depth of water just over 700 m. Even fishermen may obtain a device that reflects sound impulses from fish to tell their depth (see Figure 6.21). Perhaps the most natural use of this principle is by the bat. A bat emits a sound of such high frequency that it cannot be heard by the human ear. By timing the echo of a given pulse, the bat is able to detect objects as small as a flying insect, then intercept and devour it, without the aid of sight.

**Figure 6.21.** Using sound waves to measure the depth of water. (Wesmar Marine Systems).

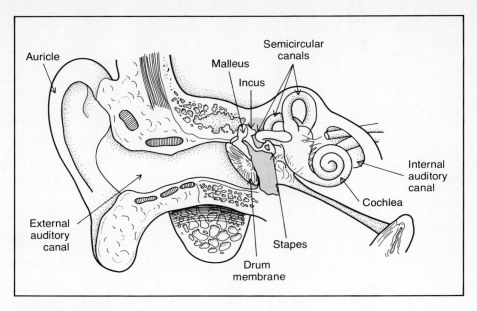

**Figure 6.22.** The human ear.

## THE HUMAN EAR

The ear is a device that can respond in a sensitive and accurate manner to the condensations and rarefactions that constitute sound. In Figure 6.22 we visualize sound entering the ear canal, reacting with the eardrum, and causing it to vibrate. This in turn sets each of a series of three linked bones into exactly the same vibration, thus transmitting the vibrations to the inner ear, where they are converted to electrical impulses that can be interpreted by the brain. The ear of the average young adult is sensitive to a range of frequencies from 20 to 20,000 cycles per second. With aging, most adults loose their ability to hear the highest notes in this range. The pitch of a musical note is dependent on the frequency of the vibration that produces the note. For instance, middle C is produced by a vibrating string or air column that produces 256 cycles per second.

## SOURCES OF MUSICAL SOUNDS

Fundamental to our understanding of the way a musical instrument produces a desired note is an understanding of the *standing wave*. Imagine a rope of infinite length. Moving one end up and down will generate a wave disturbance that will move along the rope without limit. However, if the rope is of limited length and one end is fixed to a post, then waves reaching that end will be reflected, causing waves to travel backward along the rope. If a backward moving wave should meet an oncoming wave in such a way that at certain points the two waves reinforce each other constructively (see A in Figure 6.23), and at other points they react to destroy

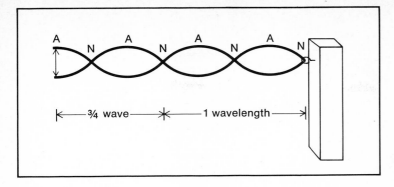

**Figure 6.23.** Standing waves ($1\frac{3}{4}$ wavelengths).

one another (see N in Figure 6.23), then a pattern of standing waves is created. The points of destructive interference are called *nodes* (N)—points of no activity, and those points of maximum constructive interference are called antinodes (A)—points of maximum activity. Of course the free end of the rope must be an antinode because this is where the wave is generated; however, the fixed end can have no activity, and hence must be a node.

Figure 6.23 shows $1\frac{3}{4}$ standing waves fitting into the length of 945 cm; hence the wavelength must equal 540 cm (945 divided by $1\frac{3}{4}$). But note that $1\frac{1}{4}$ waves could also fit so as to have an antinode at the free end and a node at the fixed end (see Figure 6.24). In this case the wavelength is 756 cm (945 divided by $1\frac{1}{4}$). Likewise $\frac{3}{4}$ waves could fit, producing a wavelength of 1260 cm. Sketch this yourself. Also find the wavelength if you fit $2\frac{1}{4}$ waves into this length of 945 cm. (*Answer:* 420 cm) Could you fit a wavelength of 500 cm into this length so as to have an antinode at the free end and a node at the fixed end? Figure 6.25 shows that this is not possible, for an antinode does not appear at the free end. This should convince you that only certain wavelengths can produce standing waves in a given length of rope. The wavelengths that can produce standing waves are called *resonant* wavelengths.

**Figure 6.24.** Standing waves ($1\frac{1}{4}$ wavelengths).

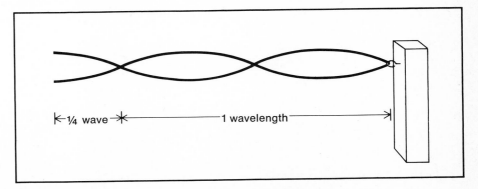

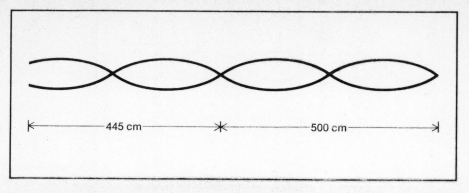

**Figure 6.25.** An antinode does not coincide with the open end, hence this wavelength will not fit the given length.

A closed-end organ pipe, 945 cm long, reacts in a manner very similar to that of the fixed-end rope. A node must appear at the closed end and an antinode at the open end; therefore only the indicated wavelengths can produce standing waves (see Figure 6.26). The violin string also produces certain notes by creating standing waves; however, the violin string is fixed at both ends and therefore a node must reside at each end—points of no activity. If a violin string is plucked or bowed near one end, the indicated standing wave patterns will naturally develop (see Figure 6.27).

If the string is 24 cm long, it is easy to see that standing waves of (a) 48 cm

**Figure 6.26.** The closed-end organ pipe.

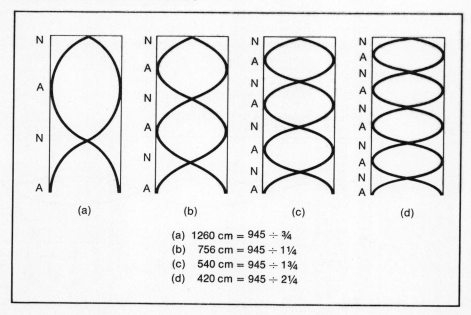

(a) 1260 cm = 945 ÷ ¾
(b)  756 cm = 945 ÷ 1¼
(c)  540 cm = 945 ÷ 1¾
(d)  420 cm = 945 ÷ 2¼

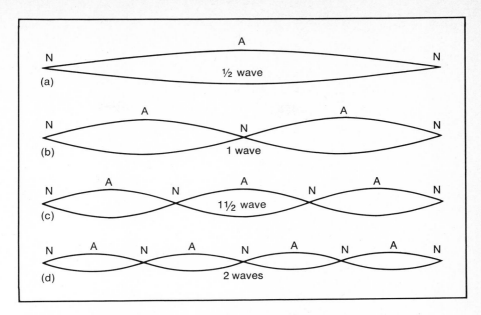

**Figure 6.27.** The violin string. A node must occur at each end.

$(24 \div \frac{1}{2})$, (b) 24 cm $(24 \div 1)$, (c) 16 cm $(24 \div 1\frac{1}{2})$, and (d) 12 cm $(24 \div 2)$, etc., will be created. A 24 cm organ pipe, with both ends open, will react in much the same way, except that antinodes must occur at open ends of the pipe (see Figure 6.28).

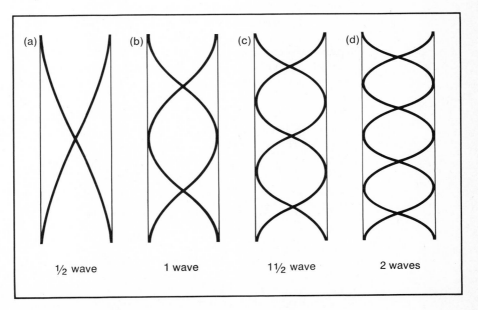

**Figure 6.28.** The open-end organ pipe.

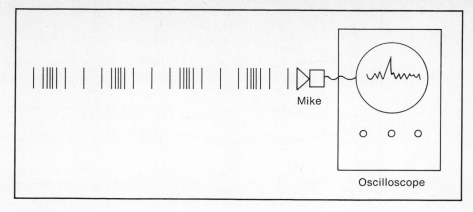

**Figure 6.29.** "Seeing" sound.

## QUALITY OF SOUND

In a violin string the frequency of the fundamental note that is produced is dependent not only on the length of the string but also on its tension. Let us suppose that the length and tension of a given string are such as to produce a vibration of 256 cycles per second. We define this musical note as middle C; however, the same string will simultaneously vibrate at twice the frequency (half the wavelength) or 512 cycles per second, and three times the frequency or 768 cycles per second, etc. These additional frequencies are called the *overtones* or *harmonics* of the fundamental note. It is the presence of these additional related notes with each fundamental that produces a pleasing sound to the ear. In fact, this is the essential difference between musical sounds and noise. Noise is a collection of unrelated sounds that are not pleasing to the human ear but rather may constitute a serious form of pollution. Persons who are regularly subjected to high levels of noise not only experience loss of hearing but may suffer adverse psychological reactions.

Let's compare noise and musical sound by graphically displaying these disturbances on an oscilloscope. Here sound waves are detected by a microphone, where they are converted to electrical waves that can be used to deflect the electron beam in the display tube (see Figure 6.29). The repeating (periodic) nature of the sound wave is not seen in the noise pattern; furthermore, the musical sound wave can be analyzed as simply the sum of a fundamental note and its overtones—those that are produced by a vibrating string or air column (see Figure 6.30).

## HIGH FIDELITY REPRODUCTION OF SOUND

Now suppose we "strike" a high note on a musical instrument, for example, one having a frequency of 2000 cycles per second. Its overtones will also be produced, i.e., at 4000, 6000, and 8000 cycles per second, etc. You can see how it is easily

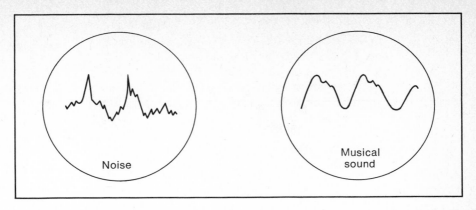

**Figure 6.30.** Noise versus musical sound.

possible to produce notes that extend to the upper limits of the audible range (20,000 cycles/sec). When such notes are reproduced on a record or tape, and then played through an amplifier and speaker, much of the character of that sound may be lost due to distortion by the system itself. If the higher overtones are lost or severly distorted, then the concert-hall quality of sound is lost. A good quality high-fidelity system is designed to faithfully reproduce sounds up to 20,000 cycles/sec and thereby make the listener feel that he is in a concert hall rather than just listening at home. This emphasizes once again that the real beauty in musical sounds is a function of the overtones that are produced whenever a fundamental note is generated.

## SOURCE OF VIBRATION IN WIND INSTRUMENTS

As unlikely as it seems, the same reaction that makes a flag flap in the wind makes a clarinet reed or a column of air in an organ pipe vibrate (see Fig. 6.31). As air

**Figure 6.31.** Vibrations set up by the flow of air over: (a) a flag and (b) an organ pipe.

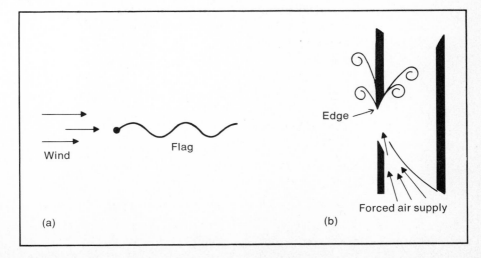

passes an obstacle (the flagpole, the edge of a reed, or the edge that is a part of an organ pipe), the air flow is diverted first on one side and then on the other. Figure 6.31(a) shows the ripple effect on a flag, and Figure 6.31(b) shows the alternate eddy currents that develop in the organ pipe. The eddies inside the pipe establish standing waves that correspond to the fundamental note or its harmonic, as determined by the length of the pipe.

Now as we move into electricity, magnetism, and electromagnetic disturbances, you will see how our study of waves will aid us in the description of these phenomena.

## QUESTIONS

1. List as many phenomena as you can think of that can be explained as wave phenomena.
2. In the case of the steam locomotive, at what point(s) is the piston moving most rapidly in the cylinder, and at what point(s) is it standing still?
3. Two waves can come together to reinforce each other or to destroy each other. Describe these two conditions.
4. Find the frequency of a wave that has a period of 0.1 sec.
5. Find the wavelength of a wave if it has a frequency of 50 cycles per second and moves with a velocity of 100 m/sec.
6. What can you expect ocean waves to do as they approach shallower water near the beach?
7. When water waves pass a small obstacle, that obstacle becomes a new source of waves. What is this called?
8. What is the difference between transverse and longitudinal waves?
9. Suppose a car could be placed in a large vacuum chamber and all the air removed. Could its horn be heard outside the chamber?
10. When we experience a thunderstorm, we see the lightning before we hear the accompaning thunder. Why is this true?
11. What condition produces a sonic boom by a fast-flying aircraft?
12. Characterize the difference between musical sounds and noise.
13. How does the length of an organ pipe influence the note it is capable of producing?
14. What is it that one can hear in a concert-hall performance of a good violinist as compared with the same concert reproduced on poor recording equipment?
15. Most musical sounds originate from a very common phenomenon. What is that phenomenon?

# electricity and magnetism

Electricity represents a form of energy that is utilized in hundreds of ways in our daily lives—such as lighting the night, cooling the home, and cooking our meals. What is at the heart of its ability to do work for us? This question was answered partially as early as 600 B.C., when Thales of Miletus demonstrated the fact that a piece of amber, after being rubbed, was capable of attracting bits of straw. Today we would demonstrate the same feat by rubbing a hard rubber rod with a piece of fur. Such a rod would easily attract bits of paper; however, this experiment may or may not prove anything about electricity. A more definitive experiment may be performed by suspending a rubbed rod by a silk thread, as illustrated in Figure 7.1. Bringing another rubbed rod near one end of the suspended rod causes that end to be repelled. Two facts are clear thus far: (1) We are not seeing simply a gravitational force, for gravitation is an attractive force. (2) The force we are seeing is much stronger than a gravitational force. (The gravitational force between two rod ends would be so weak that it would not be detected by such a crude experiment.)

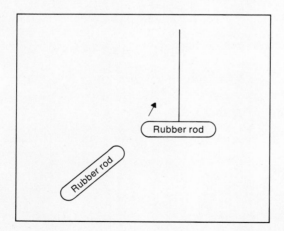

**Figure 7.1.** Like charges repel.

## ELECTRICAL FORCES

In our search for understanding the new force, let us try two glass rods (one suspended), both having been rubbed by silk. Again we see a repulsion force at

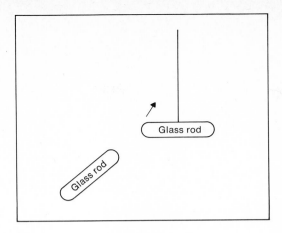

**Figure 7.2.** Like charges repel.

work (see Figure 7.2). As a third experiment, bring the rubber rod (having been rubbed by fur) near the suspended glass rod. The two rods are now attracted to one another. What can we conclude? (See Figure 7.3).

An electrical charge can be created by rubbing different materials. When these charges are alike, they repel; when different, they attract. We might arbitrarily speak of the glass rod as possessing a positive charge and the rubber rod a negative charge. Only after some further explanation of atomic structure can we come to a better understanding of electrical charges. It is known today that all nature is composed of positively and negatively charged particles; and when neutral particles are found, they represent a balance of positive and negative charges. The atom is modeled as possessing a positive nucleus with negative electons moving about it, held together by the attractive forces of unlike particles. Chapter 10 will be devoted to a more complete description of the atom.

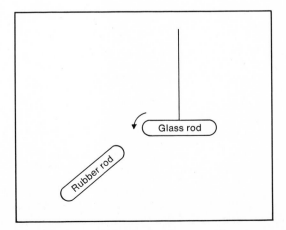

**Figure 7.3.** Unlike charges attract.

# THE ATOMIC MODEL OF A CHARGED BODY

Using the atomic model, we would explain the negative charge on the hard rubber rod as being due to an excess of negative electrons as compared to positive protons. In the rubbing process, electrons were removed from the fur and deposited on the rod. Of necessity then, the fur was left with a deficiency of electrons and hence is positive in charge. In the case of the glass rod, the electrons had a greater tendency to remain with the silk cloth, leaving a deficiency on the glass rod. You will see that it is the organization of an excess of negative electrons in one place and a deficiency in another that makes work possible. Relate this statement to the second law of thermodynamics (and to the concept of entropy). After all negative particles have found their way to neutralize the positive particles, entropy is at a maximum and no work is possible. However, if we can reestablish an excess of negative particles at one point and a deficiency at another point, then a flow of electrons is possible and work can be done by that flow. We can compare such an organization of negative charges to that of water at a high point, capable of flowing to a lower point and doing work on the way. Thus, we may speak of electrical energy as we would the potential energy of water at a high point. We will see later that a cell (battery) provides such a potential through chemical action.

## THE ELECTROSCOPE

One of the tools that is helpful in detecting electrical charges is the electroscope. This instrument is composed of a vertical metal rod suspended from the top of a boxline insulated case, with two strips of gold leaf attached to the base of the rod. The rod is topped by a metal sphere or knob. When a negatively charged hard rubber rod is brought near the spherical knob (but not touched to it), electrons are repelled down the rod to distribute themselves on the gold leaves. Since like charges repel, the gold leaves swing as shown in Figure 7.4 (a). As soon as the

**Figure 7.4.** The electroscope and a charged rubber rod.

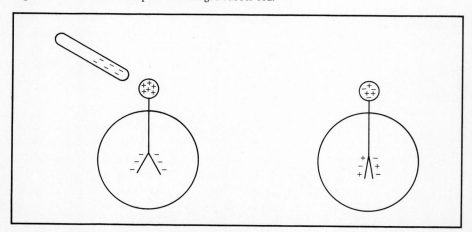

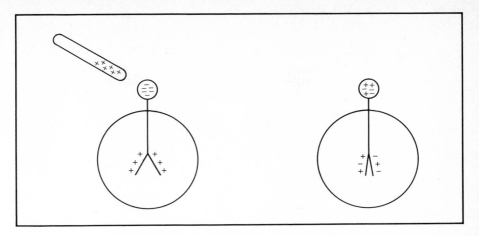

**Figure 7.5.** The electroscope and a charged glass rod.

rubber rod is removed, the positive and negative charges redistribute to neutralize each other [see Figure 7.4 (b)].

Trying the same experiment with the glass rod, one will notice a similar reaction but for a slightly different reason. This time the positive glass rod will attract the electrons to the spherical knob, leaving the gold leaves positively charged. They will spread apart again because like charges repel [see Figure 7.5 (a)]. When the rod is removed, the charges neutralize again [see Figure 7.5 (b)].

## CHARGING AN ELECTROSCOPE

If the negatively charged rubber rod is brought into contact with the knob of the electroscope, some of its excess electrons will be deposited on the knob, rod, and leaves. When the rod is removed, the electroscope remains negatively charged, as shown by the spread leaves (see Figure 7.6). Likewise, the positively charged glass rod can draw off electrons from the knob and from the leaves when touching the knob. Removing the rod will leave knob and leaves deficient in electrons, and the leaves will stand apart (see Figure 7.7).

## CHARGING BY INDUCTION

Suppose you had a rubber rod (negatively charged), but you wanted to leave a positive charge on the electroscope. This may be done by a three-step process. First, bring the rubber rod near the knob, repelling negative electrons to one side [see Figure 7.8 (a)]. Then touch the far side of the knob, allowing your body to drain off excess electrons. Remove your finger; the rod and electroscope will remain deficient in electrons, hence positively charged. This is called *charging by induction*.

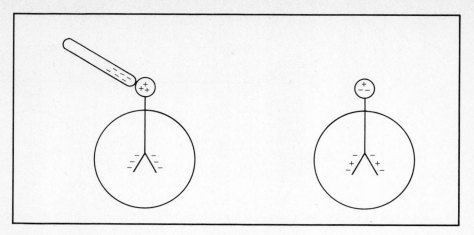

**Figure 7.6.** Charging an electroscope.

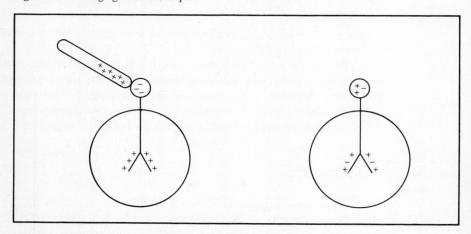

**Figure 7.7.** Charging an electroscope.

**Figure 7.8.** Charging an electroscope by induction.

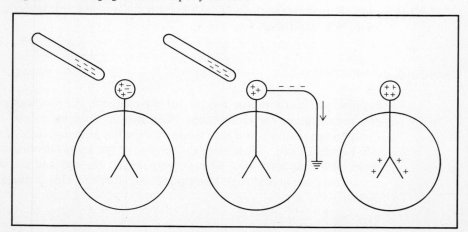

# COULOMB'S LAW OF CHARGES

In the late eighteenth century Charles Coulomb (1736–1806) succeeded in quantifying the force due to electrical charge. This means that he was able to associate numbers with the amount of charge and the distance and to discover a relationship between such quantities. His device resembled our suspended rod, but with important refinements. He suspended two interconnected spheres on a fiber that if left alone (protected from disturbance by an enclosure) would come to rest in a certain position (see Figure 7.9). A charge was placed on sphere $A$, and then another fixed sphere $B$ with a similar charge was placed close to $A$. The suspended system rotated as $A$ experienced a repulsion from $B$. The fiber support was twisted at the top to restore $A$ to the original nearness to $B$. From the amount of twisting, the force between the two spheres could be computed. Coulomb found that when he doubled the charge on one sphere (holding the other constant and holding the distance constant), the force was doubled. When the charge was tripled, the force was likewise tripled. Thus, he said that the force of electrical repulsion is directly proportional to the charge, or to the product of the two charges if both are varied. Then he held the charges constant and allowed distances to vary. He found that at twice the distance, the force is reduced to one-fourth of the original value; and if the distance is tripled, the force is reduced to one-ninth. Therefore; the force varies

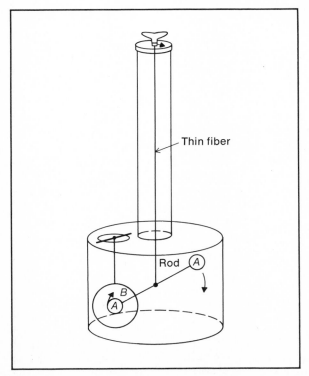

**Figure 7.9.** Coulomb's experiment to quantify the force due to an electrical charge.

inversely as the square of the distance. These relationships sound identical to those of gravity and are expressed in the same form:

$$F = K \frac{q_1 \times q_2}{r^2}$$

where

$q_1$ = charge on one sphere

$q_2$ = charge on the other sphere

$r$ = distance between centers

$K$ represents a constant, even as $G$ in the gravitation expression:

$$F = G \frac{m_1 \times m_2}{r^2}$$

In honor of Charles Coulomb, his name is given to the unit that measures the quantity of charge—the coulomb. We will define this quantity in a more practical way when we study the flow of electricity a few sections hence. Suffice it to say that when $q_1$ and $q_2$ are measured in coulombs and $r$ is measured in meters, then $K = 9 \times 10^9$ nt m$^2$/coul$^2$ and the force ($F$) will come out in newtons.

## ELECTRICAL AND GRAVITATIONAL FORCES

It should be interesting if we compare the electrical force (of attraction) between the electron and proton of a hydrogen atom with the gravitational attraction of the same two bodies, using the appropriate formulae given above. The gravitational force, which depends on the masses of the two particles and their separation, yields an answer in the order of $4 \times 10^{-47}$ nt; whereas the electrical force, which depends on the charges and separation, yields an answer in the order of $8 \times 10^{-8}$ nt. This electrical force is $2 \times 10^{39}$ (2,000,000,000,000,000,000,000,000,000,000,000,000,000) times as strong. It seems clearly evident that the electrical force is responsible for holding the electron in the proximity of the proton. Thus, it is electrical attraction that holds atoms together in molecules.

This computation may beg the question, If gravity is so weak, how is it so effective in the larger scale of man and in the macroscale of the universe? Whereas electrical charges may produce either attractive or repulsive forces, gravity between ordinary matter is always attractive. This single state of attraction seems to more than make up for the weakness of the force.

## ELECTRICAL FORCE FIELD

Surrounding any charged particle one may visualize a field (region) within which a positive test particle would experience a force. Lines shown in Figure 7.10 indicate the direction of force on such a positive test particle. Thus, if both a positive and

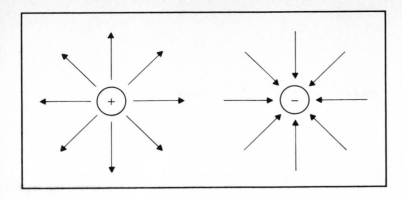

**Figure 7.10.** The electric field.

a negative particle occupy the same general region, their combined field can be visualized as shown in Figure 7.11.

We may think of the field extending to an infinite distance; however, from a practical point of view, we treat it only out to the point where it can be measured and assigned a value.

The intensity ($E$) of the field of any given point within the field is defined as the force that would be exerted on a positive test particle of one coulomb charge. Thus, the force ($F$) on a given charge ($q$) at a point where the field has intensity ($E$) is:

$$F = qE$$

If a charged body, containing an excess of electrons, is connected to a body that has a deficiency of electrons by means of a copper wire, the electrons will flow from the negative body to the positive body until the charges are equalized. We might consider this an electric current, but it is only a momentary current, for once equalized, no further current will flow. Is it possible to create a supply of electrons that will provide a continuous flow? Yes; the ordinary flashlight battery represents such a device. Let's see its origin.

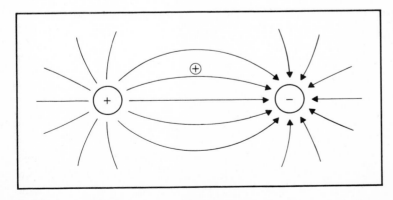

**Figure 7.11.** The lines of an electric field indicate the direction of force that would be experienced by a positive test particle.

## THE VOLTAIC CELL

In the late eighteenth century Italian physicist Alexander Volta (1745-1827) discovered that a continuous supply of electrons can be provided by immersing two different materials in a solution called an *electrolyte*. When certain substances are dissolved in water, their component atoms separate in such a manner that one atom carries off an extra electron (or two), giving it a negative electrical property. Such a charged atom is called a *negative ion*. The other atom of the original substance separates with a deficiency of electrons and hence becomes a *positive ion*. This ionic nature of an electrolyte makes it possible for an electric current to be conducted, through the electrolyte, from one electrode (e.g., zinc) to the other electrode (e.g., carbon).

These two electrodes are the ones often used in a flashlight battery. The zinc electrode, which is usually in the form of the "can" that contains the electrolyte, gradually dissolves (ionizes) in the electrolyte, with each zinc ion leaving two electrons behind. In order to maintain an overall neutrality in the electrolyte, its positive ion attracts electrons (positive). Thus, by this double chemical reaction, an excess of electrons accumulate at the zinc electrode and a deficiency of electrons at the carbon electrode. If these two electrodes are connected by a copper wire, electrons will flow and work can be done; i.e., a flashlight bulb may be lighted. As long as the chemical reaction is possible, the supply of electrons is replenished and the current can continue to flow.


## ELECTROMOTIVE FORCE

As a charge passes through a cell, it is given a certain amount of energy. For instance, a $1\frac{1}{2}$ volt cell gives $1\frac{1}{2}$ joules of potential energy to each coulomb charge that passes through the cell. Suppose that four such cells are connected in a manner that allows the charge to flow through each one in succession. Then one coulomb should pick up 6 volts ($4 \times 1\frac{1}{2} = 6$). We speak of such a combination of cells as a *battery*. With such a source of electrical energy at hand we can conceive of electrons flowing in a conductor at some steady rate, that is, an electric current. Some materials are good *conductors*, like copper and silver, for they possess electrons that are loosely bound; therefore, the electrons can move easily among the atomic nucleii. Other elements, like sulfur, have such tightly bound electrons that an electric current will not flow through them. Such materials are called *insulators*. Still other materials fall somewhere between these extremes.

Even a good conductor presents some resistance to the flow of electricity. In the case of ordinary wire, the resistance depends on three factors: the length of the wire, its cross-sectional area and its composition. For instance, copper wire will present less resistance to the flow of electricity if it is short and of long cross section. Extension cords used to provide electrical power should always be as short as possible and have large enough conductors to carry the current needed.

# ELECTRIC CIRCUITS

Let's suppose that a 6-volt battery is connected to a given resistor as shown in Figure 7.12. The direction of current as indicated by the arrows appears to be opposite to the flow of electrons discussed earlier. Before the discovery of the electron, currents were simply assumed to flow from (+) to (−) even as water flows from higher to lower points. This convention carries over into today's work so that current may still be specified from (+) to (−); however, electron flow must be thought of as (−) to (+). How may we specify the amount of current that will flow in such a system? First, we may define a unit of current, called an *ampere*, as equaling one coulomb per second. We have already defined the volt, our unit of electromotive force, as representing one joule per coulomb; therefore, we will now specify the unit of resistance—the ohm.

A resistance of one ohm will allow a flow of one ampere under a potential difference of one volt. This relationship is expressed in Ohm's Law:

$$V = IR \quad \text{or} \quad R = \frac{V}{I} \quad \text{or} \quad I = \frac{V}{R}$$

where

$$V = \text{potential difference in volts}$$
$$R = \text{resistance to ohms}$$
$$I = \text{current in amperes (amp)}$$

Find the current that will flow through a resistance of 3 ohms if $V = 12$ volts.

$$I = \frac{V}{R} = \frac{12 \text{ volts}}{3 \text{ ohms}} = 4 \text{ amp}$$

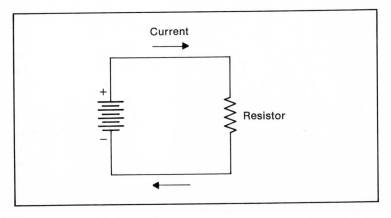

**Figure 7.12.** The flow of current through a resistor.

You can see that the current in such a circuit is directly proportional to the potential difference and inversely proportional to the resistance.

## SERIES CIRCUITS

When several resistors are connected in such a way that the entire flow of current must pass through each in turn, the arrangement is called a *series circuit*. Under such circumstances the individual resistances add to yield a total resistance (see Figure 7.13).

$$R_{total} = R_1 + R_2 + R_3$$
$$R_{total} = 2 + 3 + 4 = 9 \text{ ohms}$$

The current that will flow in such a circuit is:

$$I = \frac{V}{R} = \frac{18 \text{ volts}}{9 \text{ ohms}} = 2 \text{ amp}$$

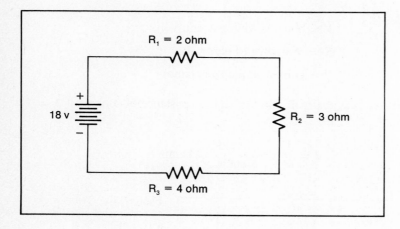

**Figure 7.13.** Resistances in series.

## PARALLEL CIRCUITS

When resistances are arranged in such a way that only part of the total current flows through each, we have a parallel circuit (see Figure 7.14). The total resistance is computed by:

$$\frac{1}{R_{total}} = \frac{1}{R_1} + \frac{1}{R_2} + \frac{1}{R_3} \cdots$$
$$\frac{1}{R_{total}} = \frac{1}{10} + \frac{1}{15} = \frac{5}{30} = \frac{1}{6}$$

$$R_{total} = \frac{6}{1} = 6 \text{ ohms}$$

$$I_{total} = \frac{V}{R} = \frac{24}{6} = 4 \text{ amps}$$

Current through 10-ohm resistor:

$$I = \frac{V}{R} = \frac{24}{10} = 2.4 \text{ amp}$$

Current through 15-ohm resistor:

$$I = \frac{V}{R} = \frac{24}{15} = 1.6 \text{ amp}$$

Checking, we see the total current:

$$2.4 + 1.6 = 4 \text{ amp}$$

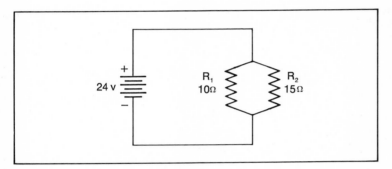

**Figure 7.14.** Resistances in parallel.

## A PRACTICAL CONSIDERATION

The electrical circuits of your home are wired so that each electrical device that is plugged into an outlet (say in one part of the house) is in parallel with each other. Suppose a single kitchen circuit is capable of carrying 20 amp of current, as determined by the size of wire used. If you plug in a toaster that draws 2 amp, a coffee pot that draws 5 amp, a portable oven that draws 10 amp, and a mixer that draws 5 amp, what may you expect to happen? (*Answer:* A blown circuit breaker or fuse.) In a parallel circuit the currents in each branch add to produce a total greater than 20 amp. If the circuit breaker failed to interrupt service, then a dangerous heating of the wire might result in a fire. Never replace a fuse or circuit breaker with a larger value than specified for a given wire size.

## MAGNETISM

We have seen that gravity acts, at a distance, to produce mutually attractive forces. On the other hand, charged particles may produce either attractive or repulsive forces depending on the nature of the charges. Magnetism is another basic force that occurs in nature. Ancient observers discovered that a certain kind of rock, if suspended by a thread so that it could turn freely, would consistently come to rest in a north-south alignment. Most rocks do not show such a tendency. What conclusion may be drawn from the alignment of what is now called *magnetite* (lodestone)? We must conclude that an invisible force field must exist on the earth, a field much stronger than gravity and one that is not dependent on a net electric charge, for the rocks tested were all neutral electrically. We speak of this field as being of magnetic nature, and we can discover some of the properties of magnets by simple experiments. For example, choose several pieces of magnetite, suspending each in turn by a thread. Mark the north-seeking pole of each by an *N* and the south-seeking pole of each by an *S*. Now bring the two suspended stones near each other so that the two like poles are closest. Such poles will be seen to repel each other. Alternately, bring the unlike poles close and they will show an attractive force. In this sense they behave in a way similar to charged particles.

Until the early nineteenth century, a number of theories were put forth regarding the origin of magnetism, but none recognized its close association with electricity. In 1820, The Danish physicist, H. C. Oersted (1777-1851) discovered that as he caused a current to flow in a wire, a compass nearby was disturbed (see Figure 7.15).

If electrons are directed to flow in a wire [as indicated in Figure 7.15 (a)] across the top of a compass that normally points toward the north, the needle would point west. If the electron flow is reversed, the compass needle would point east. The magnetic field is created by the flow of electrons, and its direction can be predicted by using your left hand as follows: pointing your thumb in the direction of electron flow, wrap your hand around the wire, and the direction of your fingers show the direction of the magnetic field. Thus, the direction of a magnetic field is defined by the direction a compass would point at a given location. The term *field* is used in the same sense as an electrical field—the region surrounding a magnetic pole in which a test pole (a compass) would experience a force.

A bar magnet may be made by striking a steel bar repeatedly while keeping it

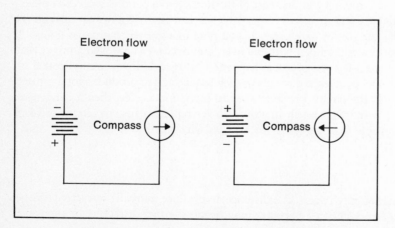

**Figure 7.15.** The magnetic property of a flowing electric current.

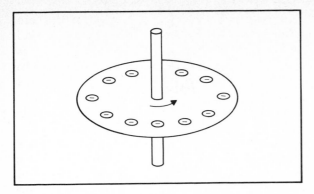

**Figure 7.16.** The magnetic property of a rotating charged plate.

oriented north and south. This fact would suggest that something within the bar can be oriented by such a process. This experiment together with Oersted's discovery suggested the possibility that something within the bar had an electrical quality that produced the magnetic field.

One additional experiment, which you can perform with simple equipment as shown in Figure 7.16, would give us a further clue. Rub a hard rubber rod with a piece of fur and transfer the negative charge to the metal discs, as shown in Figure 7.16. Then spin the cardboard disc by rolling the dowel between your hands. A nearby compass will react because you are thereby creating a magnetic field.

We believe that within the atom, electrons spin on an axis somewhat like our experimental disc above. If the spins of a significant number of electrons in the steel bar can be aligned, then their integrated effect will produce a magnetic field.

## THE ATOMIC MODEL OF MAGNETISM

In an unmagnetized steel bar these "atomic" magnets are randomly oriented, as shown in Figure 7.17 (a); their individual fields tending to cancel one another. Once oriented in the same general direction, one end of the bar takes on the aspect of a north pole and the other end a south pole. In the central portion of the bar, the north pole of one atom is adjacent to the south pole of its neighboring atom; hence these adjacent fields tend to cancel. Only at the ends do the north or south poles exist separately, and thus these are concentrations of magnetic influence. There-

**Figure 7.17.** The bar magnet: (a) demagnetized; (b) magnetized.

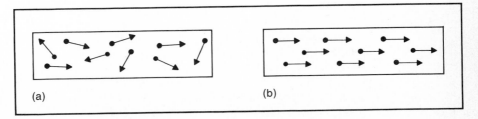

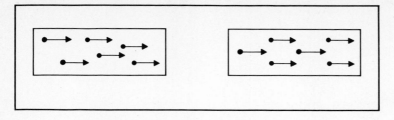

**Figure 7.18.** The severed bar magnet.

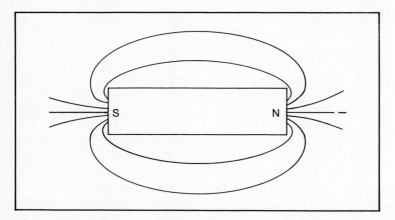

**Figure 7.19.** The magnetic field.

fore, if a bar magnet is cut in half, two separate sets of north-south poles will be formed (see Figure 7.18).

If a bar magnet is placed beneath a glass plate (or cardboard) and soft iron filings are sprinkled over the surface, the field of that magnet may be visualized (see Figure 7.19). The small iron filings behave as small magnets, each one pointing in the direction of the field at any point. Thus, we visualize the field as sets of magnetic lines of force. However, these are only a visualizing technique, not something that exists in the sense of seeing or feeling the lines.

## THE SOLENOID

We have seen that a wire carrying an electrical current creates a magnetic field. This field may be concentrated by forming the wire into a coil. Think of the field at each point in the coil, and you will see how these individual fields complement each other to produce a stronger field through the center of the coil (see Figure 7.20). The coil (solenoid) acts as a bar magnet with a definite north and south pole.

There is an interesting note regarding the naming of a pole. When we make a compass needle by magnetizing a slender piece of metal, we usually put an arrowhead on the end that seeks the north. Do you think that both the arrowhead of a compass and the magnetic pole of the earth toward which it points can be called a north pole? No, for we know from experiments with two magnets that like poles repel each other. Therefore, we should define these two poles more clearly so

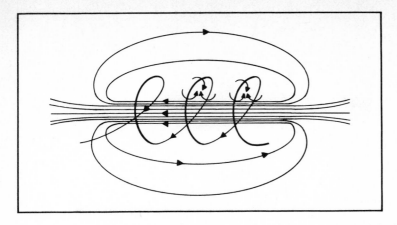

**Figure 7.20.** The solenoid.

as to prevent this ambiguity. Let the magnetic pole of the earth that is closest to the geographic north pole of the earth be called the *magnetic north pole*. Then we may think of the arrow head of the compass as the *north-seeking* pole (actually a south pole in our scheme of definitions).

The earth's magnetic field has some interesting properties that may give us a clue as to the origin of that field. First, the magnetic north pole of the earth is not located at its geographic pole (as defined by its axis of rotation) but is displaced over 1000 miles. Furthermore, the magnetic pole of the earth moves about rather drastically, even in a period of several hundred years. Figure 7.21 shows the present location of the magnetic pole, and the zigzag pattern suggests how it has changed over the years. Does this shifting about suggest that great quantities of lodestone are buried in the northeastern part of Canada, creating the earth's magnetic field? No, for we could not explain the motion of the poles by the drifting about of lodestone. We look for another explanation, something akin to the source of magnetism in the atom itself. The earth is naturally rotating; hence if it possesses a predominance of one kind of charge in a given region, either positive or negative, a magnetic field may be generated by its rotation. A refinement of this

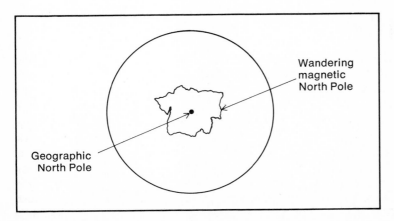

**Figure 7.21.** The earth's wandering magnetic pole.

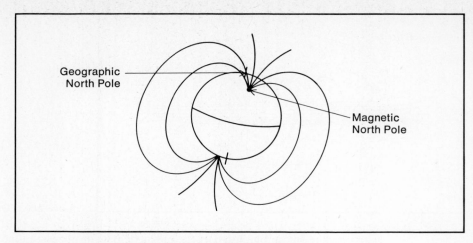

**Figure 7.22.** Geographic and magnetic North Poles of the earth.

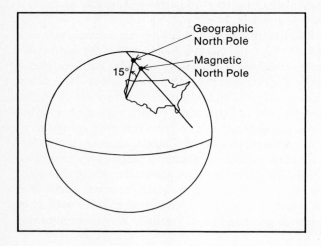

**Figure 7.23.** Magnetic declination.

idea suggests that on a smaller scale charged particles within the liquid outer core of the earth may be set into rotation by the rotation of the earth, and the motions of smaller groups of charged particles in that liquid core create individual fields. The integration of these lesser fields creates the total field of the earth (see Figure 7.22).

Because the magnetic pole is displaced from the geographic pole, we must correct our compass reading to acquire the true north. This is referred to as the *magnetic declination* of a given location on the earth's surface (see Figure 7.23).

## VAN ALLEN BELTS

The sun radiates all forms of electromagnetic radiations (radio, infrared, visible light, ultraviolet, X-rays, gamma rays, etc.), and it also pours out charged particles

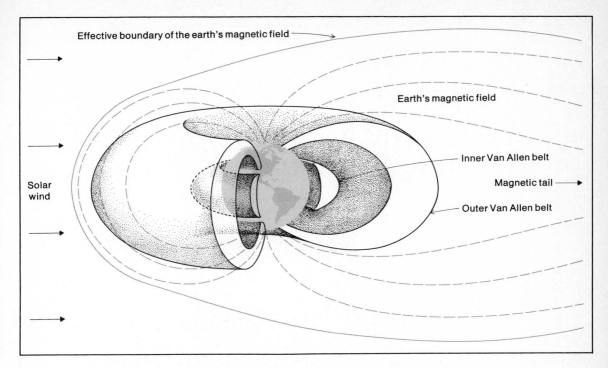

**Figure 7.24.** The Van Allen belts of the earth.

to the tune of about 1 million metric tons per second. These charged particles consist of protons, electrons, and combinations of protons and neutrons, for instance, helium nuclei consisting of two protons and two neutrons (alpha particles). As these charged particles enter the magnetic field of the earth, they experience a sideways force (perpendicular to their direction of travel and perpendicular to the magnetic lines of force). Thus, they are entrapped by the magnetic field of the earth. These areas of fast-moving charged particles are known as the *Van Allen belts*. You can see how these belts dimple in toward the poles (see Figure 7.24). Here some particles find their way into the atmosphere, producing bands or streamers of light, known as the northern (and southern) lights. This process will be detailed later.

## TORQUE—IN A MAGNETIC FIELD

Because the sideways force on a charged particle is so useful to us, let's define it a little more specifically. Visualize a straight wire carrying a conventional current toward you (out of the page in Figure 7.25). The magnetic lines of force are shown—the resulting force is up.

If a current flows in a loop of wire in a magnetic field, then one side of the loop experiences an upward force, while simultaneously the other side experiences

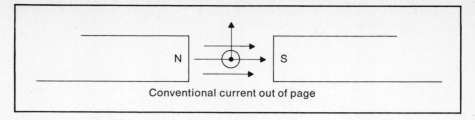

**Figure 7.25.** A current flowing in a magnetic field.

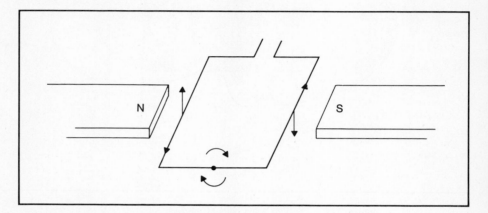

**Figure 7.26.** The principle of the electric motor.

a downward force. This provides a twisting force called a *torque* (see Figure 7.26). This is the basis for the electric motor—an electric current in a magnetic field providing a torque (twisting force), which can drive many devices. Can you see that if the current would continue to flow in the same direction, no useful torque would exist after 90° of rotation, so a mechanism must be provided to reverse the current on each half turn. Other refinements are also employed to provide a uniform torque throughout a full rotation, but we will not discuss that refinement here.

## ALTERNATING CURRENT GENERATOR

Let's suppose that instead of passing a current through the loop in Figure 7.26, we simply attach a crank in place of the pulley and turn the loop through the magnetic field. When a conductor moves across magnetic lines of force, charged particles (electrons) in the conductor experience a force perpendicular to the lines of force and perpendicular to the direction of motion of the conductor, causing the electrons to flow along the conductor (the loop). Thus, we see that an electrical current can be generated with such a device. If the current is conducted away from the loop with a slip ring, as shown in Figure 7.27, a very interesting thing happens—the

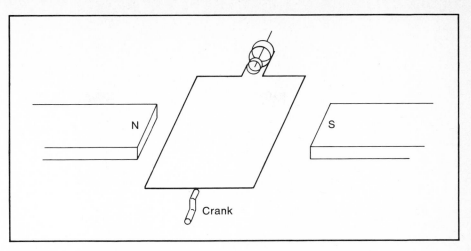

**Figure 7.27.** The principle of the generator.

current flows one direction during one-half of a rotation and the opposite direction during the other half.

Let's trace one revolution, following a given conductor as it turns through $A$, $B$, $C$, and $D$. At position $A$, no magnetic lines are being cut; hence no current. At $B$, the maximum numbers of lines are being cut per unit time; hence maximum current is generated here. Again at $C$ no lines are being cut; hence current is zero. But at $D$ the maximum numbers of lines per unit time are cut, and maximum current is generated but in the opposite direction from that generated at $B$. When the current is plotted on a graph, as time goes by, we see a sine wave form as defined in trigonometry (see Figure 7.28). One may accurately think of it as a plot of the number of magnetic lines that are being cut per unit time as the wire loop turns through one rotation. Since the current generated is proportional to the number of lines being cut per unit time, the plot of current is the same as that of the numbers of lines being cut.

**Figure 7.28.** The alternating current generator produces a current of the form shown in (b). The shape of the curve is that of a sine wave.

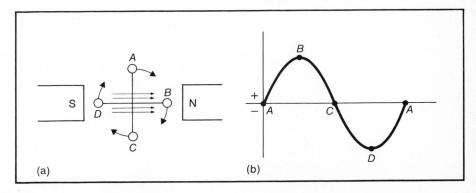

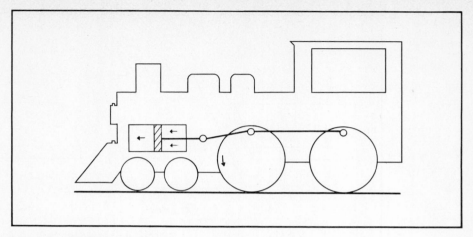

**Figure 7.29.** The locomotive.

We may generalize this concept to say that whenever circular motion is translated to energy production or to longitudinal motion (as in the locomotive illustrated in Figure 7.29), a sine wave describes its output. As the piston is driven forward and backward by the injection of steam on the appropriate side, the wheel is driven in a circular motion. If you were to analyze the motion of the piston in the cylinder and plot it on a graph as time goes by, then it will appear as a sine wave (see Figure 7.30).

As either the wheel of the locomotive or the loop of a generator turns through one complete rotation, one cycle is generated. In the case of alternating current used in the households of the United States, the timing of the generator is carefully controlled to provide 60 cycles per second. Your electric clock experiences 60 bursts of energy (say at the beginning of each cycle) every second. It is these regular bursts that regulate its time-keeping rate. If a power company should

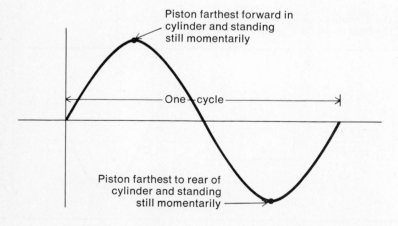

**Figure 7.30.** The sine wave describes the motion of the piston.

become slack in regulating their generators' rate of turning, then your clock would run fast or slow as a result.

To turn the ac generators that provide our electrical power, a source of potential energy is needed. Water that is stored at a high elevation may be released to fall a given distance, thus converting its potential energy to kinetic energy; the moving water molecules will then exert a force against the blades of a turbine, which are coupled to a generator (see Figure 7.31). Thus we see potential energy converted to kinetic energy, then to mechanical energy, and finally to electrical energy.

Another common source of energy for generating electricity is found in the chemical form of coal or oil or natural gas, which may be burned to produce heat energy, which turns water into steam. The steam is then introduced into the chambers of the turbine, where its expansion turns the turbine (energy now mechanical), which is coupled to turn the generator and produce electrical energy.

Two of the great advantages of electrical energy are the ease with which it can be transmitted to distant locations and the ease with which it can be controlled.

**Figure 7.31.** The modern generator.

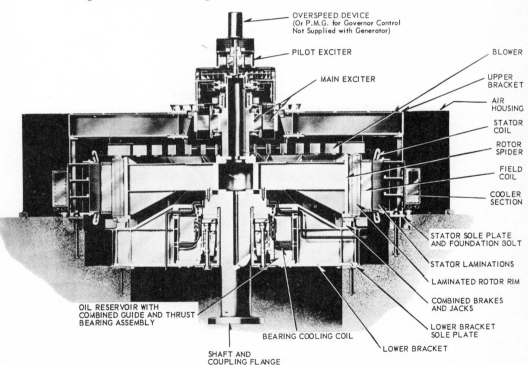

# TRANSMISSION OF ELECTRICAL ENERGY

A given alternating current generator may be capable of delivering a flow of 2500 amps at an average voltage of 2000 volts. Since power is equal to the current multiplied by the voltage ($P = IV$), the power output of this generator is 5,000,000 watts or 5000 kilowatts. Now suppose the transmission lines over which this current must flow have a resistance of 10 ohms; heating will result, and in turn a power loss will be experienced due to that heating, equivalent to $I^2R$.

$$\text{Power loss} = I^2R$$
$$= (2500 \text{ amp})^2 \times (10 \text{ ohms})$$
$$= 6,250,000 \times 10$$
$$= 62,500,000 \text{ watts}$$

This represents a greater loss of energy due to heat than the energy available at the generator; hence this would present an impossible situation. However, if the voltage could be stepped up, say to 50,000 volts, the current would be reduced as follows:

$$P = IV, \; I = \frac{P}{V} = \frac{5,000,000 \text{ watts}}{50,000}$$
$$I = 100 \text{ amp}$$

The power loss will be reduced drastically as follows:

$$\text{Power loss} = I^2R$$
$$= (100)^2(10) = 100,000 \text{ watts}$$
$$= 100 \text{ kilowatts}$$

This loss amounts to only two percent ($100/5000 = 0.02 = 2\%$) of the total energy involved. This saving in heat loss certainly motivates the engineer to perfect a device to accomplish the change in voltage. The device is called a *transformer*. It works on the principle that when the current varies in one conductor, a changing magnetic field is generated, and this changing field induces a current in any conductor in the vicinity, much as a moving magnet induces a current in a coil.

The changing magnetic field is equivalent to the moving magnet (see Figure 7.32), for as the field builds up, magnetic lines of force cut across the conductor. Once the current is established at a given level, the magnetic field becomes static and no longer induces a current in the nearby conductor; however, if the current falls off, the magnetic field collapses, and once again magnetic lines cut the nearby conductor, causing a current in the opposite direction (see Figure 7.33).

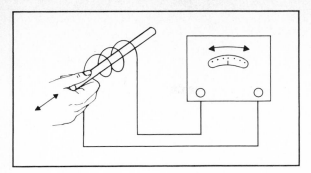

**Figure 7.32.** A changing magnetic field induces a current in the wire loop.

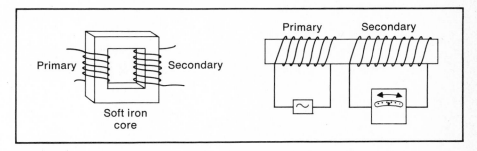

**Figure 7.33.** The transformer.

An alternating current is continually changing, building up to a maximum in one direction, falling back to zero, building to a maximum in the opposite direction, returning to zero, etc. Therefore, if this current is attached to a coil, the magnetic field is continuously changing and creates a similar alternating current in the secondary coil. The soft iron core helps to concentrate the magnetic field in the vicinity of the secondary coil, thus improving the efficiency of the transformer. Notice that the secondary coil has twice as many turns as the primary coil; hence each magnetic line generated by the primary will cross twice as many turns in the secondary, resulting in a doubling of the voltage. This concept is generalized by the expression:

$$\frac{V_s}{V_p} = \frac{N_s}{N_p}$$

where

$V_s$ = voltage in secondary

$V_p$ = voltage in primary

$N_s$ = number of turns in secondary

$N_p$ = number of turns in primary

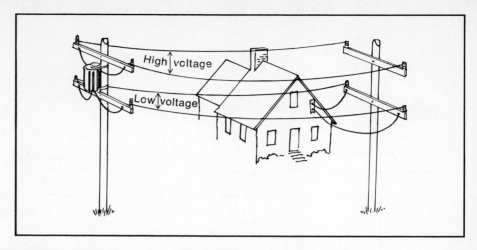

**Figure 7.34.** The losses of power are greatly reduced by high voltage transmission.

If the efficiency is high, then we may assume the power out (in the secondary) to be equal to the power in (in the primary).

$$P_s = P_p$$
$$P = IV$$
$$I_s V_s = I_p V_p$$
$$\frac{I_s}{I_p} = \frac{V_p}{V_s}$$

This relationship indicates that when voltage goes up, current goes down (assuming the same power). Hence, the relationship between the number of turns and the current is an inverse relationship.

$$\frac{I_s}{I_p} = \frac{N_p}{N_s}$$

Thus, the transformer provides the means for reducing the current in transmission lines, by raising the voltage. A typical arrangement for the transmission of electrical energy to a household is illustrated in Figure 7.34.

## DIRECT-CURRENT GENERATORS

Although most of the lightening, heating, cooling, and appliances we use are designed to utilize alternating current, a few applications may demand current that always flows in one direction, that is, direct current (dc). We have already discussed

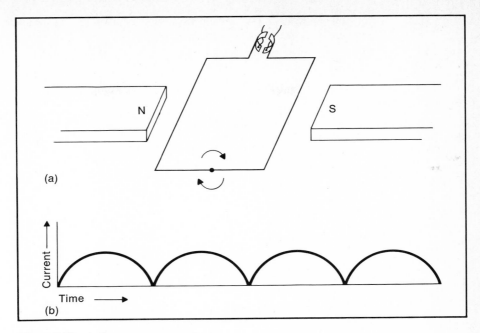

**Figure 7.35.** A direct current generator.

the cell (battery), which supplies an excess of electrons at one electrode and a deficiency of electrons at the other. Hence, the cell is capable of supplying a direct current. A pulsating direct current can be provided by a generator like the one that produces alternating current, with only a slight modification of the slip rings. Note the split ring in Figure 7.35 (a), which reverses the contacts with the loop each half turn. The output graphs like that shown in Figure 7.35 (b).

The intimate relationship between electricity and magnetism will be dramatically expanded in the next chapter in the subject of electromagnetic disturbances.

## QUESTIONS

1. When two charged rubber rods are suspended by thread, they repel each other. Why would one suspect an electric force to be responsible for this repulsion, rather than a gravitational force?
2. True or false: Like charges attract and unlike charges repel.
3. How does your answer to Problem 2 explain the fact that whether the electroscope is charged positively or negatively, the gold leaves stand apart.
4. True or false: When an electroscope is charged by induction, the resulting charge on the scope is always opposite to that of the charging rod.
5. Electric force and gravitational force react in a similar way when objects are

moved apart. True or false: When the distance between the objects is doubled, the force between them is one-fourth.

6. What property of an electrolyte allows an electric current to be conducted through it?
7. How does a flashlight battery provide a continuing flow of electrons to light the bulb?
8. How is it possible for the very weak force of gravity to be so effective in the universe as a whole?
9. What voltage is necessary to cause a current of 6 amp to flow through a resistance of 2 ohms?
10. Find the total resistance of three resistors connected in series; they are 2 ohms, 3 ohms, and 4 ohms, respectively.
11. Find the total resistance of the three resistors in Problem 10, if connected in parallel.
12. Describe a magnetic "field" in your own words. What do the lines used to show a magnetic field represent?
13. Describe ways in which electricity and magnetism are interrelated.
14. What is the suspected source of the earth's magnetic field?
15. What kinds of particles are entrapped in the Van Allen belts?
16. Describe the output of an alternating-current generator by use of a graph (or verbally).
17. State the advantage of transmitting alternating current at very high voltage.
18. Characterize the differences between alternating and direct current.

# electromagnetic spectrum

In Chapter 7 we saw that an electrically charged particle creates an electrical field, in which a test particle would experience a constant force. Likewise, we saw that a magnetic pole creates a magnetic field in which a test pole would experience a constant force. Now suppose we allow a charged particle to oscillate. It should be obvious that a test charge would no longer experience a constant force but a changing force, and the test particle would respond by oscillating. Hence, we say the oscillating charge creates a changing field (see Figure 8.1).

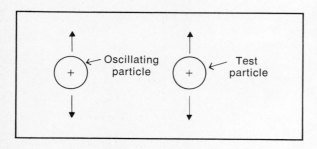

**Figure 8.1.** An oscillating charged particle creates a changing electric field.

## AN OSCILLATING CHARGED PARTICLE

Furthermore, as we saw in Chapter 7, whenever an electric field varies, a changing magnetic field is also created. Thus, the oscillating charged particle generates both a changing electrical and a changing magnetic field simultaneously. If we were to show the changing electrical field at a given point as time goes by, it would graph as shown in Figure 8.2 (solid curve). Varying in a similar way but oriented at right angles is the plot of variations in magnetic strength and direction. Since a changing electrical field is always accompanied by a changing magnetic field, it seems proper to refer to this phenomenon as an *electromagnetic disturbance* (wave).

It is true that a plot (graph) of either the electric or magnetic disturbance has a wave form; however, one should not conclude that electromagnetic disturbances go waving through space. An electromagnetic disturbance differs from both sound and water waves in that it requires no medium in which to travel; hence in no sense could it have a literal wavelike form. Only the graph of the changing electrical and/or magnetic field, as time goes by, yields this wavelike form (see Figure 8.3).

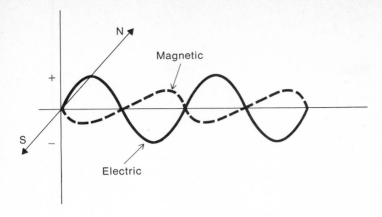

**Figure 8.2.** A graphical representation of changing electric and magnetic fields.

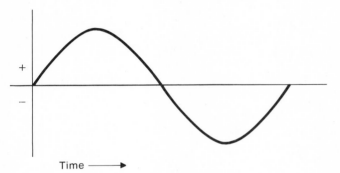

**Figure 8.3.** The wave form of the plot of a changing electric or magnetic field.

## THE SPEED OF LIGHT

Even as you read these words there are several forms of electromagnetic disturbances that you can sense, such as infrared (radiant heat) from the sun or a lamp, or visible light, or ultraviolet light that produces a suntan. And with the proper detector (receiver) one might receive any of a number of communications via radio (simply by turning the dial). Less apparent are X-rays and gamma rays. All of these forms combined constitute the electromagnetic spectrum—none requires a medium in which to travel, and all travel at the same speed in a vacuum.

The speed of an electromagnetic disturbance, namely that of light, was approximated (in air) in 1924 by American physicist Albert A. Michelson (1852-1931). He caused an intense light to reflect from one of the mirrored faces of an eight-sided wheel, travel 35 km to a flat mirror, and then return to the wheel (see face #7, Figure 8.4) and into a viewing tube. By rotating the eight-sided wheel at high speed, it was possible for Michelson to "catch" the flash of light that left face #1 with face #8; the wheel having turned one-eighth of a revolution while the light traveled to the flat mirror and back. Subsequently, the flash from #2 was caught by #1, etc. He found that it was necessary to turn the wheel 529.37 times per second in order to see light in the viewing tube. This meant that 8 × 529.37 flashes were transmitted per second, each traveling the round trip distance of 70 km

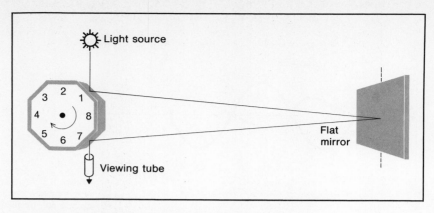

**Figure 8.4.** Michelson's rotating mirror experiment.

before the next flash left. Thus, two factors were known: the distance traveled and the time required. Hence, it was possible to compute the velocity ($v$):

$$v = 529.37 \text{ turns/sec} \times 8 \text{ flashes/turn} \times 70 \text{ km/flash}$$
$$= 299{,}729 \text{ km/sec}$$
$$= \text{approximately } 300{,}000 \text{ km/sec } (186{,}000 \text{ mi/sec})$$

## WAVELENGTHS

One aspect that all electromagnetic disturbances have in common is that they travel through empty space at the speed of light. Then what is it that makes radio waves different from light waves? It is a property called *wavelength*. For instance, if a given radio wave travels one meter while going through one cycle, as shown in Figure 8.5, then we say its wavelength is one meter. Radio waves range in wavelength from 1000 m to 1 mm (0.001 m) approximately. By contrast, blue

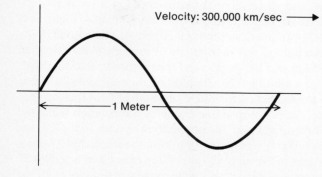

**Figure 8.5.** An electromagnetic disturbance, showing one cycle.

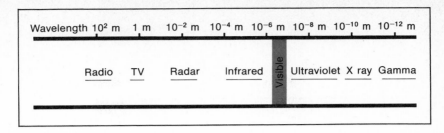

**Figure 8.6.** The electromagnetic spectrum.

light, for example, has a wavelength of only 0.00000045 m. Figure 8.6 shows the entire electromagnetic spectrum with wavelengths indicated for each portion of that spectrum.

When dealing with wavelengths between $10^{-3}$ m (0.001 m) and $10^{-6}$ m (0.000,001 m), it is more convenient to specify the short length in microns, where $10^{-6}$ m = 1 micron. Thus, a particular wavelength in the infrared range such as $4.5 \times 10^{-6}$ m could be more simply stated as 4.5 microns.

For wavelengths still shorter, as in visible light, we may use a unit called an *angstrom* (Å) unit, where 1 Å = $10^{-10}$ m. Thus, the wavelength of blue light written 0.00000045 m could be written simply as 4500 Å. The range of wavelengths in visible light extends from about 8000 Å (red light) to 4000 Å (violet light). Wavelengths for ultraviolet, X-rays, and gamma rays are successively shorter.

We know that the sun and stars in general emit electromagnetic energy in the full range of wavelengths as depicted in Figure 8.6. In the section that immediately follows, we will be discovering additional properties of light, but remember that these properties apply equally to the entire electromagnetic spectrum.

## LIGHT TRAVELS IN A STRAIGHT LINE

On a bright sunny day almost any object casts a shadow, and the edges of that shadow are sharp (well-defined) (see Figure 8.7). Does such an observation seem to suggest that light travels in a straight line? Would you be surprised if the light from a flashlight traveled around the corner of a building to illuminate a door as in Figure 8.8? The fact that we would not expect this result from our experience with an uninterrupted beam (ray) of light points toward the fact that light travels in a straight line. This is true whenever light travels in empty space and far from any objects. However, every day we utilize the fact that light may be bent in a number of ways. Looking at oneself in a mirror is an obvious example. Rays of light leave your face, travel to the mirror surface, are bent by reflection, and return to be detected in your eye. Furthermore, reflection occurs in a predictable manner, depending on the angle at which the incident (incoming) ray strikes the mirror surface.

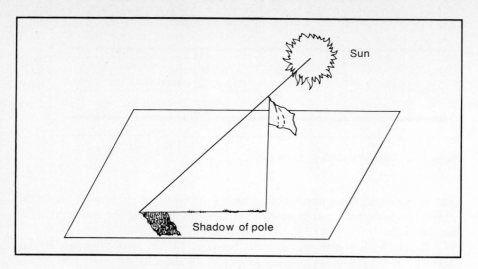

**Figure 8.7.** Light travels in a straight line (in empty space).

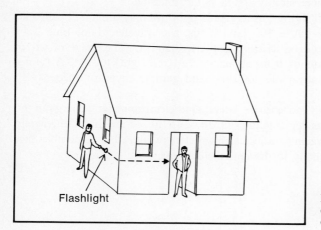

**Figure 8.8.** Light will not travel around the corner of a building nor can we see around corners.

**Figure 8.9.** The law of reflection.

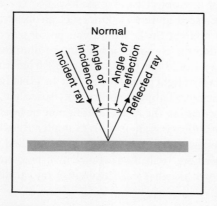

## THE LAW OF REFLECTION

In order to specify this angle of incidence, let's first define the normal to the surface. The *normal* is a line that stands perpendicular to the surface at a given point (see Figure 8.9). The reflected ray will make an angle with the normal equal to that of the incident ray, and it will lie in the plane defined by the incident ray and the normal. Thus we may state a law of reflection:

*The angle of reflection always equals the angle of incidence.*

## REFRACTION

Light may also be bent by refraction as it passes from one medium into another medium in which its velocity is different. For instance, suppose that light travels from air in which its velocity is 300,000 km/sec (186,000 mi/sec) into water in which its velocity is 225,000 km/sec (140,000 mi/sec); one may expect the ray to bend toward the normal as it passes into the water (see Figure 8.10). This is clearly illustrated by picturing a marching band practicing on dry ground but then marching into muddy ground and experiencing the retarding influence of the mud (see Figure 8.11). You can see how the direction of the band was changed because the band members on the right-hand side of the page enter the mud before those on the left. A wave front striking a retarding medium at an angle will experience a similar result.

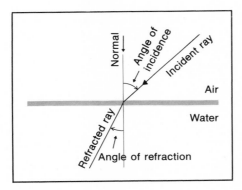

**Figure 8.10.** The law of refraction.

## THE REFRACTOR TELESCOPE

In 1610 Galileo recognized the potential for using refraction to focus light from a distant object into a point. By grinding a lens to the proper shape he was able to refract the light to a single focus, or very nearly so (see Figure 8.12). Galileo's telescope was plagued by the fact that not all colors came to the same focal point; that is, blue light came to a focus before the red light, with the other colors spread out between (see Figure 8.13).

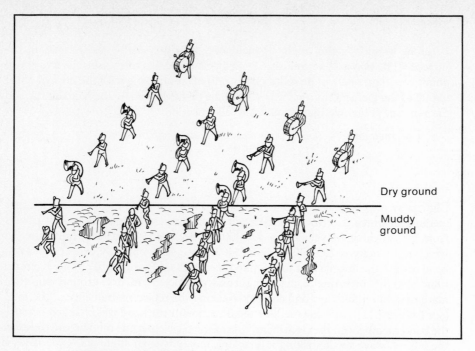

**Figure 8.11.** Refraction effect of slowing in muddy ground.

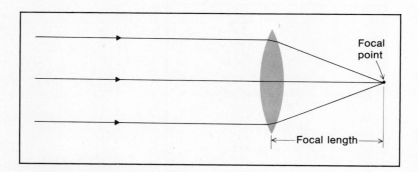

**Figure 8.12.** The simple convex lens.

**Figure 8.13.** Chromatic aberration in the simple lens—a color defect.

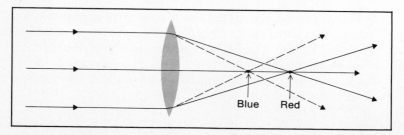

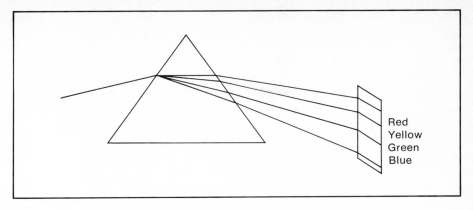

**Figure 8.14.** Dispersion of light as it passes through a triangular prism.

## DISPERSION OF LIGHT

Sir Isaac Newton discovered the explanation for this phenomenon. When he passed light through a triangular piece of glass (a prism), he noticed a full rainbow of colors cast onto the screen (see Figure 8.14). He was able to distinguish the colors tending toward the violet bend more than those tending toward the red—due to a phenomenon called *dispersion*. The differing colors have different wavelengths, and the fact that the shorter waves experience more slowing in the denser medium of the glass than those of longer wavelength causes a separation of color. This dispersion causes a defect in simple refracting telescopes and also in simple cameras. However, the same phenomenon makes possible one of the most significant tools of science—the spectroscope. (See Figure 8.15)

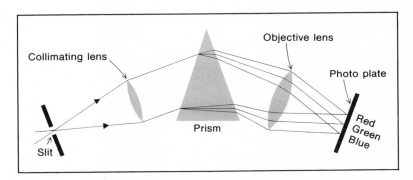

**Figure 8.15.** The principle of the spectroscope.

## THE SPECTRUM

When light, for example, from an ordinary incandescent bulb, is passed through a prism, a full rainbow results. On the other hand, if a tube containing only

electro-
magnetic
spectrum

hydrogen gas (at low pressure) is excited (by a high voltage supply) and the light of the gas is passed through a spectroscope, only a few bright lines are produced. As you can see in Figure 8.15, light from a given source passes through a slit, is refracted into parallel beams by a lens, is dispersed by the prism and then refocused onto a photographic plate. If the light of helium were used, a different set of lines would appear, and so each element has its own set of spectral lines. You can see the great potential of such a tool. If the light of a star is collected in a telescope and passed through a prism, by noting what lines its spectrum contained the astronomer can identify what elements are present in the star.

## THE BOHR MODEL OF THE ATOM

Why should a given element produce only certain lines in its spectrum and why should different elements produce differing spectra? In 1913, the Danish nuclear physicist Niels Bohr produced a model of the hydrogen atom that seemed to be consistent with its observed spectrum. He pictured the atom as shown in Figure 8.16. The central mass of the atom is a positively charged particle called the *proton*, about which moves a negatively charged particle called the *electron*. Normally one would be most likely to find the electron somewhere on or near the first energy level. However, if that same atom had been excited, e.g., by an electrical current or by heating, the electron might be given enough energy to take up a new motion on or near another energy level, for example, number six. Bohr assumed that only certain energy levels exist in a given atom and that one would not expect to find the electron in between these levels. This idea is similar to that of an empty china cupboard with several shelves (see Figure 8.17). A dish may be placed on shelf #1, #2, #3, #4, or #5, but one could not place a dish halfway between #3 and #4 and expect it to stay there. When we place a dish on a higher shelf (e.g., #5), it has more potential energy (it took more energy to put it there and likewise it can give up more energy if it should fall to shelf #1). Similarly, the electron when at energy level #5 has more energy, and that energy

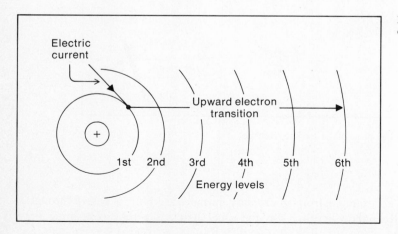

**Figure 8.16.** The Bohr model of the hydrogen atom.

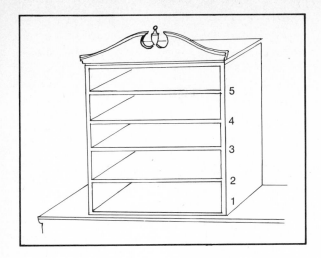

**Figure 8.17.** A china cupboard.

may be given up as the electron makes transitions (jumps) downward. In fact, Bohr explained the production of visible light by the downward transitions of electrons, which stop at least temporarily at energy level #2. Figure 8.18 shows specific downward electron transitions, marked $\alpha$, $\beta$, $\gamma$, $\delta$, which are believed to produce the corresponding spectral lines, labeled $H_\alpha$, $H_\beta$, $H_\gamma$, $H_\delta$ in the spectrum of hydrogen (see Figure 8.18).

With this model we are prepared to answer the question: Why are only certain lines produced in the spectrum of an excited gas? Only certain energy levels are available in a given atom; hence only certain transitions are possible. Thus,

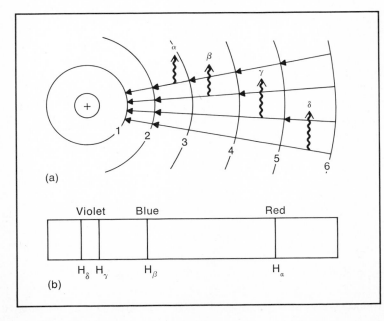

**Figure 8.18.** Downward electron transitions, which stop temporarily at energy level #2, produce visible lines in the spectrum (b).

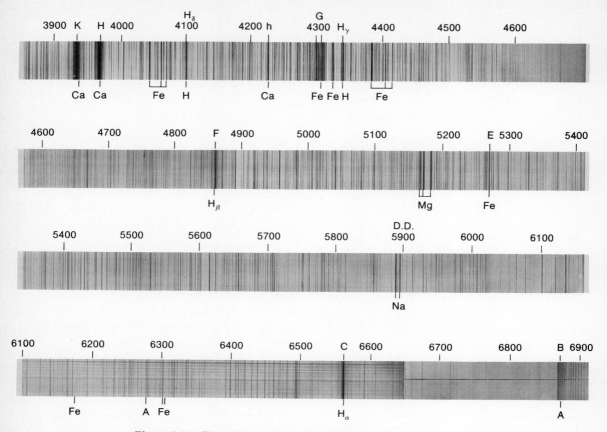

**Figure 8.19.** The solar spectrum. (Hale Observatories)

only if all transitions were possible would we obtain a continuous spectrum (full rainbow). Of course, we do obtain a full spectrum in the case of a heated solid (like the filament of an ordinary light bulb) or of a heated liquid. Why should the spectrum of helium gas, when excited, be different from that of hydrogen? Because the energy levels available in the helium atom are different from those in the hydrogen; hence different transitions are possible and different lines (colors) are produced. In fact, each element has its own characteristic set of lines. The spectrum of the sun has thousands of lines, some of which have been identified in Figure 8.19.

## THE SUN'S SPECTRUM

Of the 92 elements that naturally occur on earth, over 70 have been identified in the solar spectrum. But notice that in the case of the sun, the lines are dark on a colorful background. We speak of this kind of spectrum as an *absorption spectrum*,

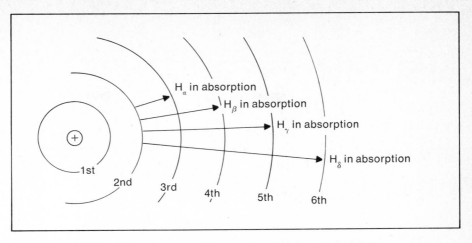

**Figure 8.20.** Upward electron transitions in the hydrogen atom produce dark lines against the continuous spectrum.

and it is produced when light of a continuous nature (as from a heated liquid or solid) passes through a low-pressure gas. Upward transitions occur in the atoms involved, producing dark lines in the same location as the bright lines of an emitting atom (see Figure 8.20).

Thus, the scientists may utilize either emission (bright line) or absorption (dark line) spectra in identifying elements present in a given gas cloud. Doctors may pass white light through a blood sample and by the absorption bands observed can tell the oxygen content of the blood. Almost any substance can be vaporized and excited by heat, or by other means, to produce a spectrum. For instance, if a spark is made to jump between two iron electrodes (as used by a welder), the spectrum of iron is clearly visible in a spectroscope (see Figure 8.21).

It should be noted that atoms and molecules also emit characteristic wavelengths in the radio, infrared, and ultraviolet ranges, equivalent to the visible spectral lines we have seen. In addition to recognizing the composition of a given object by its spectrum, one may also detect its motion.

## DOPPLER EFFECT

Consider a bobbler (wave maker) placed in the center of a small pond. The waves that it generates spread in all directions, and regardless of where you might measure the wavelength, it would be the same (see Figure 8.22).

Now suppose the bobbler moved to the right as it made waves (see Figure 8.23). Observer $A$ would see the wavelengths shortened because each time the bobbler puts out a new wave, it is closer to the observer. Observer $B$ would see the wavelengths lengthened because each time the bobbler puts out a new wave, it is farther from the observer. This is called the *Doppler Effect* and may be defined as the change in wavelength due to the motion of the source or of the observer or

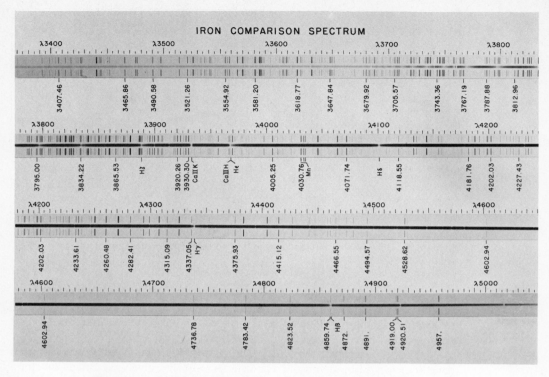

**Figure 8.21.** The spectrum of iron, used as a standard against which the spectral lines of a star may be calibrated. (Hale Observatories)

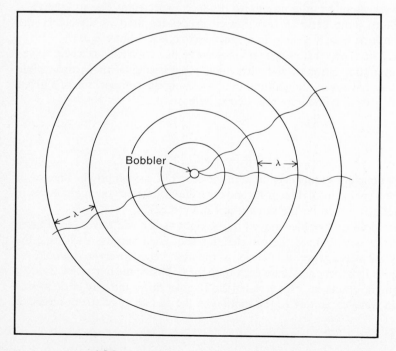

**Figure 8.22.** Circular water waves, created by a bobbler in a ripple tank.

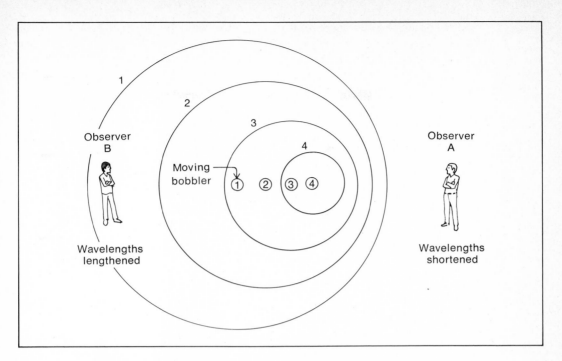

**Figure 8.23.** The Doppler effect, created by a moving source.

both. Thus, if one knows what wavelength to expect, he may judge the motion of the source relative to himself. A shortening of wavelengths indicates motion toward the observer, and a lengthening indicates motion away from the observer. The same effect occurs in sound and in electromagnetic disturbances as in water waves.

We have seen that the spectral lines of hydrogen occur normally (in the laboratory) as shown in Figure 8.24 (a). What would you surmise if these same

**Figure 8.24.** The Doppler effect, as seen in the spectrum of a star that is moving away from the observer.

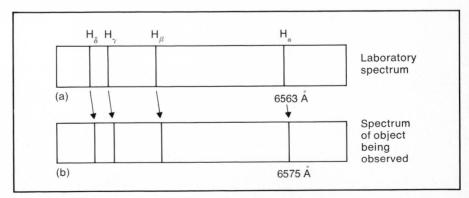

lines occurred as in Figure 8.24 (b), in the spectrum of a distant object? As you can see, the wavelengths have been lengthened (shifted toward the red); therefore, you would conclude that the object is moving away. Furthermore, there is a direct relationship between the amount of shift (change in wavelength) and the velocity at which the object is receding from the observer, expressed by:

$$v = \frac{\Delta\lambda}{\lambda} \times c$$

$$v = \frac{\text{change}}{\text{standard}} \times \text{speed of light}$$

$$v = \frac{6575 - 6563}{6563} \times 300{,}000 \text{ km/sec}$$

$$v \approx 600 \text{ km/sec}$$

where

$v = $ velocity

$\Delta\lambda = $ change in wavelength

$\lambda = $ standard wavelength

$c = $ speed of light

Had all the spectral lines of the object been shifted toward the blue (wavelengths shortened), then we would have concluded that the object was moving toward us.

## TOOLS FOR OBSERVATION—THE HUMAN EYE

Virtually all science depends on visual observation—ultimately an interaction between light (or other electromagnetic disturbances) and matter. We "see" because of such interaction. The essential parts of the human eye are shown in Figure 8.25. Enlarging or contracting the iris regulates the amount of light entering the eye—usually less than 10 mm (0.4 in.) in the daytime but enlarging to 20 mm (0.8 in.) or more for nighttime vision. The lens is similar in shape to that of Galileo's first telescope; however, the shape of the human lens may be changed by muscles so as to focus on near or far objects. In any simple lens, the image formed is inverted (as shown), yet the brain is capable of interpreting the image as right side up. Sometimes the lens of the eye forms an image ahead of the retina, resulting in a blurred image. Can you see in Figure 8.26 that it is possible to design a lens (eye glasses) that diverges (spreads out) the light rays to correct for nearsightedness? Similarly, if the natural eye tends to focus the image beyond the retina, this condition of farsightedness may be corrected by a lens that converges (brings together) the light rays.

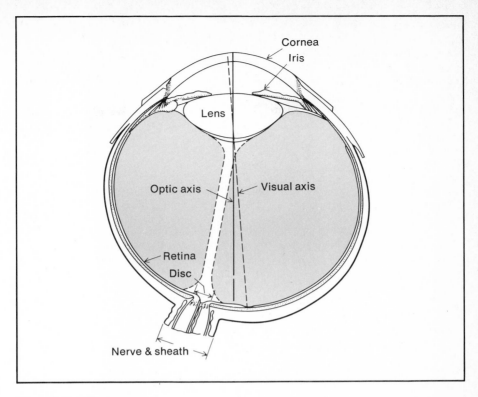

**Figure 8.25.** The human eye.

## THE CAMERA

Note that the simple camera is essentially of the same design as the human eye: the lens focusing the rays on the film, the iris adjusting the amount of light that enters the lens (see Figure 8.27). The shutter of the camera further regulates how long the light falls on the film.

## THE TELESCOPE

As an aid to viewing objects that are too dim to be seen with the naked eye, telescopes are still being perfected. Whereas Galileo's simple refractor suffered from dispersion (explained earlier in this chapter), Sir Isaac Newton was able to design a telescope that skirted this problem. Newton used a curved mirror to reflect rays of light to a common focus. His success lay in the fact that light, when reflected, is not dispersed into its various colors. His basic design produced an instrument still in use today. The parabolic shape of the mirror brings all rays from a given point, say on the surface of the moon or from a given star, to the same focal point (see Figure 8.28). Because the prime focus point is not a convenient

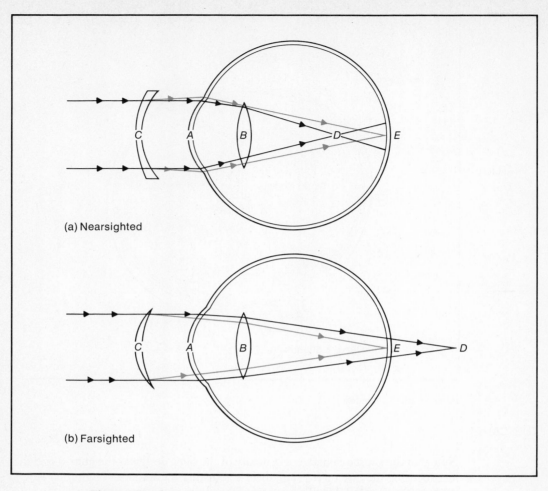

**Figure 8.26.** Corrective eyeglasses for the (a) nearsighted; (b) farsighted. (Key: A, cornea; B, crystalline lens of the eye; C, corrective eyeglass; D, image of distant point formed by uncorrected eye; E, image of distant point formed by corrected eye.)

place to put one's eye, Newton imposed a small diagonal mirror that reflected the light rays through a tube in the side of the telescope to the eye. On larger instruments like the 200 in. Hale telescope on Mt. Palomar the astronomer may actually ride with the telescope at the prime focus (see Figure 8.29).

## THE BINOCULAR

Mass production makes the binocular available as a valuable observational tool to almost everyone today. Although it represents basically a pair of simple refracting telescopes, it has the added advantage of presenting images erect (right side up).

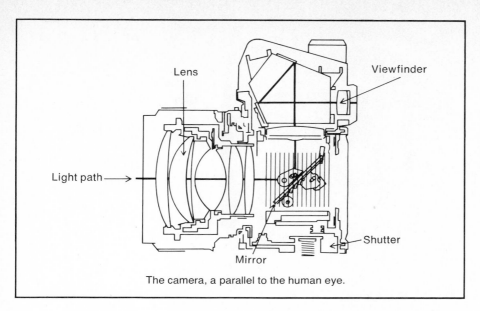

**Figure 8.27.** The camera, a parallel to the human eye.

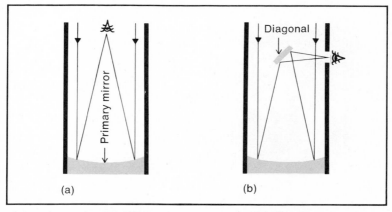

**Figure 8.28.** The reflecting telescope: (a) prime focus design; (b) Newtonian focus design.

Figure 8.30 shows how this is accomplished through multiple reflection in the corner prisms.

Binoculars are available in various sizes and powers:

7 × 35, meaning 7 power with 35 mm objective lens
7 × 50, meaning 7 power with 50 mm objective lens

The "7 power" means that objects appear seven times larger than to the unaided

**Figure 8.29.** The Palomar 200-inch (6-meter) telescope. (Hale Observatories)

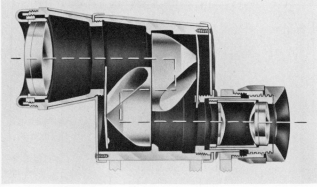

**Figure 8.30.** The binocular, showing the light path. (Bushnell Optical Co.)

eye. The 50 mm objective will gather approximately twice as much light as the 35 mm lens; however, the human eye can take advantage of this additional light only when dark-adapted—that is, during nighttime viewing. It should be noted that a large number of celestial objects are easily viewed in binoculars, i.e., the major features of the moon, the disc like appearance of Jupiter and Saturn, the beauty of open clusters like the Pleiades, and certain nebulae (gas clouds).

THE MICROSCOPE

To the scientist who studies the microworld (the world of the very small), the microscope is an indispensable tool. Figure 8.31 shows the refraction effects of the

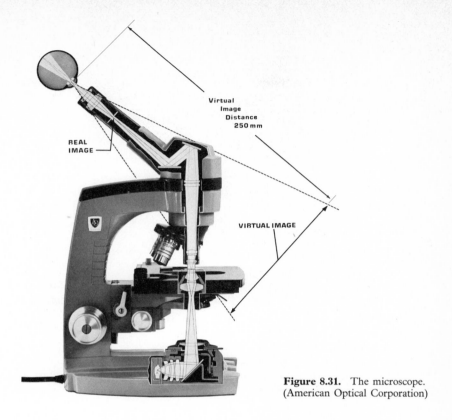

**Figure 8.31.** The microscope. (American Optical Corporation)

two simple lenses of a microscope. The objective lens focuses rays from the object to form the real image, labeled *Image 1*. The eyepiece lens forms a virtual image, labeled *Image 2*. You can see that each "imaging" produces an enlarged view of the original object. It is accurate to think of the eyepiece, whether it is used on a microscope or on a telescope, as simply a magnifying glass.

ELECTRON MICROSCOPE

By the very nature of light itself, that is, due to its wavelength, optical microscopes are limited in their potential for "seeing" the very small. A general rule of thumb says that one cannot expect to see details smaller than the wavelength used to react with the material observed. For example, the wavelength of light is greater than the diameter of even the largest atom; hence it would be futile to try to see an atom.

On the other hand, even the larger atoms may succumb to a technique that employs the electron itself. What we visualize as a particle is thought to have a wave phenomenon associated with it, and the wavelength of an electron is considerably shorter than that of light. In place of optical lenses, magnets are used to focus electron beams with a resulting magnification factor as high as 100,000

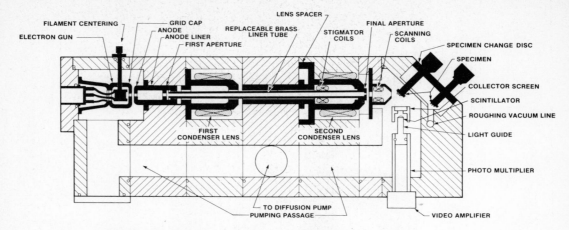

**Figure 8.32.** The electron microscope. (Bausch & Lomb)

times. Generally we must be content with magnification factors of 50, 100, or 500 times in optical telescopes (see Figure 8.32 photo and schematic).

## QUESTIONS

1. When a positive test particle is brought near a stationary charge, the test particle experiences a constant force. However, when the stationary charge is set into oscillation, how does the test particle react?
2. List as many forms of electromagnetic radiation as you know. Why should sound not be included in this list?
3. Michelson had to measure two factors in order to determine the speed of light—time and distance. Which factor was determined by the rate at which the mirror turned?
4. If a rocket could travel around the earth (at the equator) at the speed of light, how many trips could it make in one second? The circumference of the earth is approximately 40,000 km.
5. True or false: The law of reflection applies only to flat mirrors.
6. What basic principle did Galileo use in making his telescope in 1610? (reflection or refraction)
7. When white light is passed through a prism, the colors are separated. Which bends most—blue light or red light?
8. When the light of a star is collected by a telescope and passed through a prism, certain spectral lines appear. What can be determined from these lines?
9. When a low-pressure gas is excited, e.g., by an electric current, bright lines appear in the spectrum of that gas. Describe what is thought to transpire in the atom to produce these bright lines.

electro-
magnetic
spectrum

10. When all the lines of the spectrum of a star are shifted toward the red end of the spectrum (toward longer wavelengths), we know that the star is either receding or approaching. Which?

11. True or false: The shape of the lens or mirror of a telescope is a critical factor in its functioning.

12. In what ways may a $30 pair of binoculars be a better investment than a $60 (refracting) telescope.

13. True or false: We depend on the magnifying ability of a microscope in order to see objects too small to see otherwise; however, in the case of the telescope, we depend on its ability to reveal objects too dim to see otherwise.

# 9

relativity

Humans are continually refining their description of the universe, if not turning it upside down. The picture we have presented in Chapters 3 through 8 is essentially that of classical physics, based on the work of Sir Isaac Newton. The physical laws that he set forth seemed to ring true for over two hundred years and still serve to describe our everyday experiences. Newton had asserted, "I hold time and space to be absolute." By this he meant that two (or more) observers would always get the same answer when measuring the length of a rod or the time interval between two events, even if the observers were moving with respect to each other. He thought that time and distance between two events were not dependent on the relative motion of the observers. Let's see why these assumptions are incorrect.

RELATIVE MOTION

Today we live with many aspects of relativity in our ordinary experiences. Consider a bullet fired from a hunter's gun. The bullet leaves the muzzle of the gun with a velocity of 200 m/sec. Now suppose that the same hunter was traveling in a car at 25 m/sec and fired his gun in the same direction as the car's travel. What velocity would an observer riding in the car assign to the bullet? He would get an answer of 200 m/sec because he is moving with the gun. What velocity would an observer on the ground assign to the bullet? Within the limitations of his ability to measure, it would appear that the answer to this question is simply the algebraic sum of the two velocities, 200 m/sec plus 25 m/sec, yielding 225 m/sec. Indeed this illustrates the Newtonian relationship mentioned above, and only as we approach very high velocities (those that represent a significant fraction of the speed of light) will it become obvious that this simple algebraic approach is incorrect.

  Why did the two observers describe the velocity of the bullet differently? Because they were in motion relative to each other; that is, they had different frames of reference. The observer in the car thinks of the moving car as his frame of reference. In fact, he could just as well think of the car as "at rest," with the ground moving toward the rear of the car at 25 m/sec. The observer on the ground choses to think of the earth's crust as his frame of reference, and he describes the car as moving 25 m/sec in a forward direction. Neither observer should be reprimanded for his view, for either one is acceptable. Indeed these two frames of reference do not exhaust the possibilities.

Consider the fact that a point near the equator of the earth has a velocity of approximately 400 m/sec due to the earth's rotation and that the earth moves in its orbit around the sun at approximately 30 km/sec. An observer stationed in a rocket ship at a point in space "above" the solar system could assume an entirely different frame of reference from those mentioned previously. The velocity of the bullet, as measured by this observer, within the limits of his ability to measure such, would still *seem* to be a simple algebraic sum of the four velocities mentioned thus far. What experiment would show that this answer is incorrect? Let the gun become a powerful searchlight in the nose of a rocket and let the car become that rocket ship (see Figure 9.1). Light travels 300,000 km/sec, and let's suppose the rocket ship is traveling toward earth at the rate of 100,000 km/sec (although this is not yet possible under modern technology). According to our ordinary experience, we would think that an observer on earth would measure the velocity of the light from the rocket as 400,000 km/sec, the algebraic sum of the two velocities. However, when the velocity of the light is measured by the observer on the ground, it is found to be 300,000 km/sec, the same as that measured by an observer in the rocket. The velocity of the vehicle that carried the source of light added nothing to the measured velocity of that light. To see how this was discovered, consider the late nineteenth-century experiment of American physicist Albert A. Michelson (1852-1931) and American chemist Edward W. Morley (1838-1923).

Observers of the eighteenth and nineteenth centuries, upon realizing that light is a wave phenomenon, were convinced that electromagnetic disturbances needed a medium through which their wavelike nature could be propagated. They

**Figure 9.1.** Both the observer in the rocket and the observer on the ground measure the same speed of light.

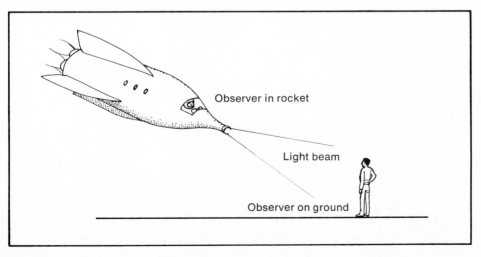

called this medium the *ether* and thought of it as a frame of reference through which the earth (and other celestial bodies) moved.

## MICHELSON-MORLEY EXPERIMENT

In 1881, Michelson tried to measure the velocity of the earth through the ether. He repeated his experiment with Morley in 1887.

To illustrate the idea behind the experiment, consider a boat that makes different speeds, relative to the land, depending on its direction of travel on a river (see Figure 9.2). Suppose the river flows steadily at 3 km/hr and that the boat can make 5 km/hr in still water. When going upstream, the boat will make only 2 km/hr (5 − 3), relative to the land. Going downstream, it will make 8 km/hr (5 + 3), relative to the land. Going across the stream, the boat has to head slightly upstream to compensate for the drift, and its velocity is computed by the Pythagorean relationship, $v^2 = (5)^2 - (3)^2 = 25 - 9 = 16$; $v = 4$ km/hr. These velocities are born out by actual experiments, and the relationships that are expressed thereby seem to hold true for velocities encountered in such a situation.

Note this interesting fact: If the boat made a trip of 8 km upstream and then returned, 5 hr would be required:

$$\text{Time upstream} = \frac{8 \text{ km}}{2 \text{ km/hr}} = 4 \text{ hr}$$

$$\text{Time downstream} = \frac{8 \text{ km}}{8 \text{ km/hr}} = 1 \text{ hr}$$

$$\text{Total time of round trip} = 5 \text{ hr}$$

**Figure 9.2.** When a boat moves upstream, its speed is slowed by the current; when it moves downstream, its speed is increased by the current, relative to the ground. In crossing the river, the boat must head slightly upstream in order to travel directly across the river.

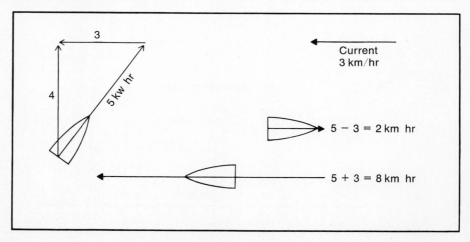

However, if the same distance were covered across the river and back, it would require only 4 hr:

$$\text{Time across} = \frac{8 \text{ km}}{4 \text{ km/hr}} = 2 \text{ hr}$$

$$\text{Time back} = \frac{8 \text{ km}}{4 \text{ km/hr}} = 2 \text{ hr}$$

Total time of round trip = 4 hr

Michelson and Morley, taking this same line of thought, devised an instrument whereby they treated the rate at which the earth "flows" through the ether as the rate of the river and the speed of light as the speed of the boat in still water. In Figure 9.3 you can see that they caused the light to reflect back and forth and in certain cases to be partially transmitted through half-silvered mirrors. If the ether really does exist and the earth moves through it, then the total time required for the upstream and downstream round trip should be longer than the across-stream round trip. As the two beams are brought back to the eye, they interfere with one another, producing alternate light and dark bands. If the device is rotated 90°, interchanging the roles of the two beams, then one would expect a change in the interference pattern because the time required for each beam to travel the set distance has been changed. To everyone's amazement, the interference pattern remained unchanged, in spite of the fact that the device was capable of showing the expected change. The failure to see any change established the fact that the velocity of light is constant—that it is not affected by the relative motion of the source or of any observer. The results of this experiment produced a revolution of thought, eventually leading to the full-blown theory of relativity.

**Figure 9.3.** The Michelson-Morley experiment.

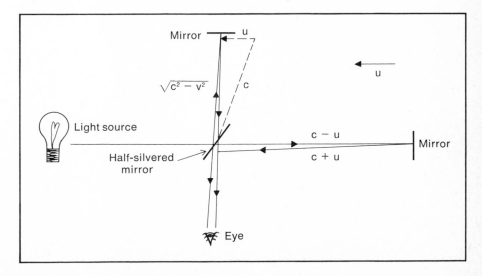

## THE LORENTZ CONTRACTION

On the other hand, there were those who were bent on preserving the idea of the ether. Typical of this school were George F. Fitzgerald (1851-1901) of Ireland and Hendrik A. Lorentz (1853-1928) of Holland, who, working independently, suggested that any device moving through the ether was shortened in the direction of its motion by just the amount necessary to allow the light to travel up and back in the same time as across and back. This is spoken of as the *Lorentz contraction* and is expressed mathematically as follows:

$$l = l_0 \sqrt{1 - \frac{v^2}{c^2}}$$

where $l_0$ is the length "at rest," relative to the earth, $v$ is the velocity of the earth through the ether, and $c$ is the speed of light. Lorentz and Fitzgerald held such changes in length to be real, whereas we will see that Albert Einstein took quite a different view.

## ALBERT EINSTEIN

Einstein's name is almost synonymous with the term *relativity*, and rightly so, for he was truly interested in separating aspects of observation that are relative from those that are absolute. Einstein felt no compunction about preserving the idea of the ether; rather, he first postulated that:

> *The speed of light in a vacuum is constant, regardless of the motion of the source or of the observer.*

This, in effect, says that it is impossible to detect the motion of the earth in its orbit by means of an experiment using light; and it also would have explained the negative results of the Michelson-Morley experiment.

However, Einstein's thinking extended this same concept in a more fundamental assumption:

> *There is no experiment that can be performed to demonstrate the motion of an object if that object is moving uniformly (at a constant speed) in a straight line.*

To illustrate this concept, suppose you were riding in a train car that has no windows, and further, that the car rides so smoothly that no bumping or lurching can be felt. How would you know that you are moving with respect to the ground? One might try bowling a ball down the aisle to strike a set of pins at the end of the car; however, nothing about the car's motion would be revealed by such an experiment, for the ball would seem to move in precisely the same way as if the car were standing still, and the pins would be struck with the same force. One might conclude that the laws of nature, inside a moving car, are just the same as those experienced in a stationary car. This was Einstein's second postulate:

*The laws of nature are the same for observer* A *and for observer* B, *even though they are moving uniformly in relation to one another.*

Now if this assumption is to be true and the speed of light is to remain constant, the measures of certain other quantities must be different for observers who are moving relative to one another. One such quantity is time.

## A THOUGHT EXPERIMENT

Consider the following thought experiment, even as Einstein must have done. Because light is the carrier of information, if you could travel away from a clock tower at the speed of light, no new information could reach you about that clock. The hands of the clock would seem frozen in place. Time would have stopped for you, as measured by that clock. If this is true, then an observer traveling at a speed of a little less than the speed of light, relative to the clock tower, will see the hands of that clock move more slowly than does the man who stands at its base.

On the other hand, if the man who is moving rapidly away from the clock tower has a watch on his wrist, he will see that watch running at its usual pace; whereas the man at the base of the tower will see that watch running slowly. Which one is correct? Both men are correct, for there is no absolute time by which these observed times can be compared. Rather, the rate at which time is measured, in any given frame of reference, is relative to the motion of the observer with respect to that frame. If an observer moves with a clock, the time it keeps can be labeled its "proper time." However, if the observer moves in uniform motion, with velocity $v$, with respect to the clock, its rate will appear to be slowed, its minutes and hours lengthened (dilated) as follows:

$$t = \frac{t_0}{\sqrt{1 - \frac{v^2}{c^2}}}$$

where $t_0$ is proper time, and $c$ is the speed of light.

Based on his fundamental postulates, and working independently of Lorentz, Einstein also derived a relationship between the "proper" length of a rod and the length that would be measured by an observer who was moving uniformly at velocity $v$ with respect to that rod. His results were in agreement with those of Lorentz:

$$l = l_0 \sqrt{1 - \frac{v^2}{c^2}}$$

where $l$ is the measured length, $l_0$ is the "proper" length, and $c$ is the speed of light. In contrast to Lorentz, Einstein said that the "true" length of a rod cannot be known; hence his emphasis was on measured lengths and measured times. Einstein said simply that two observers, moving uniformly relative to each other, will *measure* the passage of time and the distance between two points differently; hence time and distance are relative, whereas the speed of light and the laws of nature are not relative.

For example, how much shorter is the measured length of a given rod, for an observer who is moving past it at one-half the speed of light, as compared to its proper length, as measured by an observer moving with the rod? Let $v = \frac{1}{2}c$.

$$l = l_0 \sqrt{1 - \frac{v^2}{c^2}}$$

$$l = l_0 \sqrt{1 - \frac{(\frac{1}{2}c)^2}{c^2}}$$

$$l = l_0 \sqrt{1 - \frac{1}{4}}$$

$$l = l_0 \sqrt{\frac{3}{4}} = 0.866 \, l_0$$

(*Answer:* It measures approximately 13% shorter.)

If you could move down a street at half the speed of light, all the buildings would appear skinny and tall, for in the direction of your travel, the widths of the buildings would appear diminished by 13%; however, the heights of the buildings would appear normal to you. Thus the apparent contraction effect applies only to the direction of relative motion.

## THE EQUIVALENCY OF MASS AND ENERGY

One further quantity must be considered—that of mass. It has been effectively demonstrated that the measured mass of an object increases with higher and higher velocities relative to the observer. Electrons have been accelerated to almost the speed of light, and moving at such speeds they become more and more difficult to accelerate, indicating that their measured mass is significantly greater than when they were "at rest" relative to the observer. At velocity $v$, with respect to the observer, measured mass $m$ is given by:

$$m = \frac{m_0}{\sqrt{1 - \frac{v^2}{c^2}}}$$

where $m_0$ is the "rest mass" of the object and $c$ is the speed of light; hence mass is also a relative quantity.

What is the source of this increase in measured mass? To continuously apply a force $F$, energy must be continuously expended. As the object gains velocity, relative to the observer, it also gains kinetic energy. Einstein recognized an equivalency between energy and mass when he stated: $E = mc^2$. We see that equivalency here in that an object moving at velocity $v$, relative to the observer, not only possesses its rest mass but gains mass due to its energy of motion. As its velocity approaches the speed of light, the term

$$\sqrt{1 - \frac{v^2}{c^2}}$$

tends toward zero; that is,

$$1 - \frac{c^2}{c^2} = 1 - 1 = 0$$

and hence $m$ tends toward infinity (a very large value). As the mass approaches infinity, more and more energy is required to produce any additional velocity. The speed of light, thus, seems to be a limit beyond which no material object can travel.

The equivalency of mass and energy is also demonstrated in the process whereby the sun converts a portion of its mass to energy every second (see Chapter 19). In a much less obvious way, we add to the mass of our watch every time we wind it, and it looses mass as it runs down.

## QUESTIONS

1. If a train travels eastward at 80 km/hr and a passenger runs down the aisle toward the rear of the train (westward) at 6 km/hr, how will someone on the ground describe the passenger's motion?
2. How will someone seated in the train describe the runner's motion in Problem 1?
3. Why are the answers to Problems 1 and 2 different?
4. If modern rockets (and/or space probes) can only approach speeds such as 10 km/sec, what fraction of the speed of light does this represent?
5. When the Michelson-Morley apparatus was rotated 90°, the roles of the beams were interchanged. Explain what this sentence means.
6. What changes did Lorentz predict in moving objects?
7. Einstein said that it is impossible to determine the motion of an object in a smooth-riding railroad car with no windows by performing an experiment in that car. What would happen if one threw a ball straight upward in a moving car? Wouldn't it fall toward the rear of the car? If not, why not?
8. Einstein's correction for the observed length of a rod, moving with respect to the observer, appears to be the same (mathematically) as that of Lorentz.

What is different, however, about their interpretation of the meaning of that correction?

9. Determine the "measured" length of a rod that has a "proper" length of one meter if the observer is moving at three-fourths the speed of light ($\frac{3}{4} c$) with respect to the rod. (*Answer:* $\sqrt{\frac{7}{4}}$.)

10. How does the equation $E = mc^2$ relate to the sun's source of energy?

# 10

the atom

Early Greek history reveals that man sensed that all matter had some common base from which it was constituted. Thales (ca. 600 B.C.) thought it was water. Anaximenes (ca. 550 B.C.) thought it was air. Heralitus (ca. 500 B.C.) particularly noticed change and concluded that all things were manifestations of fire. It was Democritus (ca. 400 B.C.) who took the more general approach: that if matter were repeatedly cut into smaller subdivisions, eventually a particle would be reached that was uncutable. To this particle he assigned the word *atomos*. Thus in a purely speculative way the concept of the atom was born—the atom being the fundamental building unit from which all matter was constructed, simply by different arrangements.

As we develop the modern picture of the atom and see that it can be subdivided further, let's realize that we are no closer to actually seeing the parts of an atom than was Democritus. In spite of the tremendous advances in the technology of microscopes, nature seems to have provided its own limitation on the direct observation of such subatomic particles. On the other hand, modern scientists have succeeded in photographing some of the larger atoms, although these photographs do not reveal structure. Thus, we are suggesting that the models of the atom that you will encounter represent just that—models and no more. The models help us to visualize the atom and to predict its behavior under certain conditions. Therefore, models are very useful, but they will require change from time to time.

## MODERN ATOMIC THEORY

In coming forward in time from Democritus approximately 2000 years, we should not forget the seventeenth-century work of British physicist and chemist Robert Boyle (on gases), the eighteenth-century work of Swiss physicist Daniel Bernoulli (on fluids), and others whose work influenced John Dalton (1766–1844), English chemist and physicist. It was Dalton who made a significant breakthrough regarding the atomic structure of matter. Dalton's work included experiments in which various substances were given opportunity to combine. The fact he discovered is that certain specific quantities of one substance would combine completely with certain specific quantities of others. If too much of one substance is provided, it will simply be left in an uncombined state. Dalton reasoned that if only certain ratios combine, it must be due to fundamental units of matter—the atoms—which combine in that same ratio. Thus he assumed that any sample of a substance in its pure state is an assemblage of discrete units called *atoms*. He viewed the atom as indestructible, only changing in its combination with other atoms according to definite ratios (proportions). He knew that water was composed of hydrogen and oxygen atoms in a weight ratio of 1:7, that is, approximately. Today we would measure the ratio as approximately 1:8. On the other hand, Dalton did not realize that two hydrogen atoms were involved in each molecule of water ($H_2O$). We will see refinements of this sort in the next chapter.

## RADIOACTIVE ATOMS

One of the first indications of a substructure within the atom itself came in 1896 when the French physicist A. H. Becquerel (1852–1908) was experimenting with phosphorescence—the production of light by certain substances after being excited by radiation such as sunlight or by invisible X-rays. He wrapped a photographic plate with a protective paper cover and placed a pile of uranium salt on the plate. Upon development, the plate showed an outline of the salt. He repeated the experiment in total darkness and found that the uranium salts required no excitement from an outside radiant source but emitted radiation spontaneously. He had discovered the fact that certain atoms are radioactive.

Madame Marie S. Curie (1867–1934), whose life and work exude an almost unbelievable determination, made a systematic search for other radioactive elements, discovering thorium, polonium (named for her native Poland), and radium. She measured the rate of emission from various sources by placing samples between two metal plates that were maintained at differing charge levels. As some samples were placed between the condenser plates, a small current was seen to flow (see Figure 10.1). Perhaps this experiment hinted at the nature of the radiation.

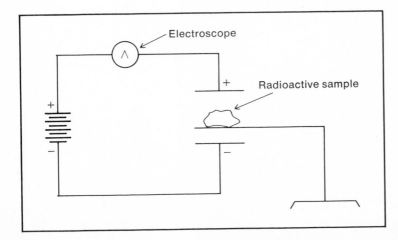

**Figure 10.1.** Madame Curie's radioactivity detector.

It was the British physicist Baron Ernest Rutherford (1871–1937) who made the next significant discovery. He placed thin sheets of aluminum foil between the source of radioactivity and the condenser plates and discovered a gradual diminishing of the current with each additional sheet up to a certain point. Beyond that point, additional sheets made very little difference in the flow. What would you conclude from such an experiment? Obviously there must be at least two kinds of emanation, one that is interrupted more easily than the other. Rutherford named these $\alpha$ (alpha) and $\beta$ (beta) rays; the $\alpha$ rays being those stopped more easily. Rutherford also reasoned that these emanations were able to ionize the air between the condenser plates, permitting the flow of current.

## GAMMA RAYS

It was not long afterward that the nature of the "rays" was better understood by means of a rather simple experiment, as described below.

A radioactive source is encased in lead with only a small hole that allows the "rays" to be emitted in one direction. This beam is directed between the plates of a condenser (see Figure 10.2) and is detected by dots on a fluorescent screen. Three distinct dots are present: one is deviated toward the negative plate, another toward the positive plate, and a third is not deviated at all. The conclusion drawn is that the beam attracted toward the negative plate must be composed of positive particles, the beam attracted toward the positive plate must be composed of negative particles (for unlike charges attract), and the undeviated beam must consist of uncharged particles. This undeviated beam is defined as $\gamma$ (gamma) rays. This experiment reveals that the atom has a structure composed of positive and negative parts that are spontaneously ejected in the case of radioactive elements.

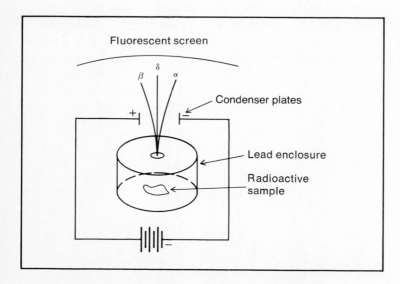

**Figure 10.2.** Detection of alpha, beta, and gamma emissions.

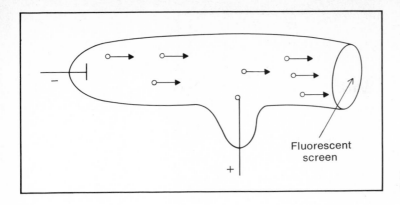

**Figure 10.3.** The cathode-ray tube.

## THE ELECTRON

It has been only a few years earlier (1897) that the British physicist Sir Joseph J. Thomson (1856–1940) had demonstrated that cathode rays were actually the negatively charged particles that we call *electrons*. The negatively charged particles that are emitted by radioactive elements behave as though they are also electrons; hence today we view the $\beta$ particles as electrons. Let's look at Thomson's experiments, which reveal some of the properties of the electron.

A simple cathode ray tube can be constructed by sealing off two electrodes as indicated in Figure 10.3 and removing the air from the tube. A source of high voltage is required. When the negative source is connected to the cathode and a positive source to the anode, the end of the tube begins to glow as a result of the flow of electrons from the cathode. If a small cross is erected in the path of the electrons, a shadow of the cross is seen on the screen indicating the straight-line path of the electrons (see Figure 10.4).

The fact that cathode rays (electrons) are really negatively charged particles is revealed by their deflection toward a positive plate in or near the tube and/or the deflection at right angles to a magnetic field, even as other moving negative charges are deflected (see Figure 10.5). Can you see that the work of Thomson, almost a

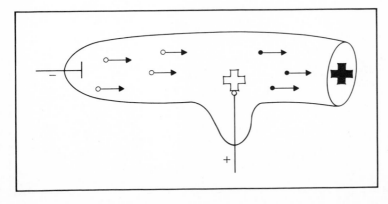

**Figure 10.4.** The nature of an electron flow.

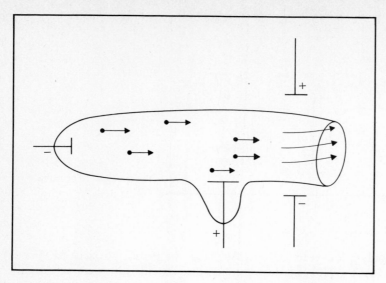

**Figure 10.5.** Controlling the flow of electrons.

hundred years ago, is only a little different from the TV tube of today? A beam of electrons is directed toward a fluorescent screen, but that beam is bent back and forth as it scans up and down to build up the picture (see Figure 10.6).

Thomson studied the amounts by which the electron beam is deflected in an electric field as compared to the hydrogen ion, for instance, and found the mass of the electron to be about 1/2000th that of the hydrogen ion (the proton). He concluded that an electron is not just another kind of atom but a part of an atom. (Today we measure this mass ratio at approximately 1 : 1800, very nearly that found by Thomson.) Thomson constructed a model of the atom sometimes called the

**Figure 10.6.** The principle of the television tube.

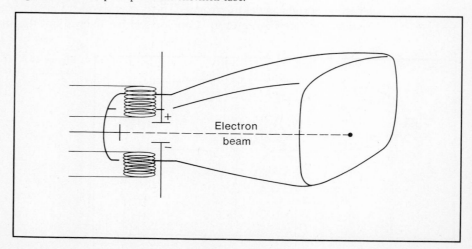

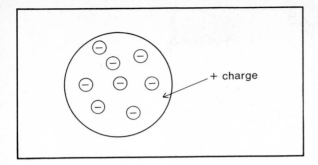

**Figure 10.7.** The raisin pudding model of the atom.

*raisin pudding model*, the raisins being the electrons embedded throughout the positively charged pudding (see Figure 10.7).

## THE NUCLEUS

In 1911 Rutherford set out to test Thomson's raisin pudding model of the atom. Rutherford knew that $\alpha$ particles carried a positive charge because of the way they were deflected in an electric field. He used a radioactive source of these particles and directed them through a thin foil (gold). He reasoned that if Thomson's raisin pudding model was correct, only slight deviations would result as $\alpha$ particles "plowed" through the positive pudding. Much to his surprise, large deviations resulted, and he concluded that the positive charge of the atom must be highly concentrated in a small volume, perhaps as small as $10^{-12}$ cm (see Figure 10.8).

This experiment drastically altered the model of the atom to resemble that of a planetary system with the nucleus being similar to the sun and the electrons to the planets. Atoms are often represented this way today, but let me hasten to warn that gross misconceptions may result if one is not careful to distinguish the similarities and the lack thereof. Perhaps the most serious pitfall would result if we thought

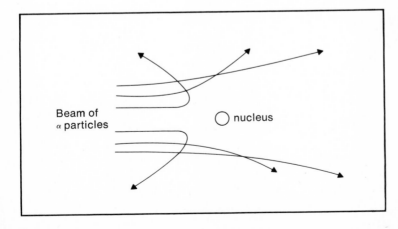

**Figure 10.8.** The Rutherford experiment showed a very small, compact nucleus.

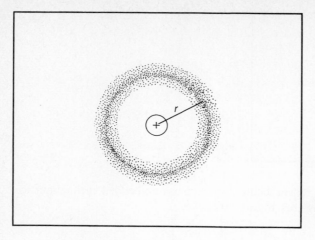

**Figure 10.9.** A model of the atom based upon a compact nucleus and probable locations of the electrons.

that the position and momentum of the various electrons could be predicted (or even known at a given instant) as can be done with the planets. The similarity does imply that most of the mass of the atom is concentrated in the nucleus and that a great amount of empty space is defined by the possible positions of the electrons.

If you could plot the locations of the electron in a hydrogen atom at the rate of a million locations per second, the plot might look like Figure 10.9. The darkest appearing sphere represents the most probable location of the electron at a distance $r$ from the nucleus. Only in this sense of high probability do we think of the sphere of radius $r$ being the electron's orbital (sometimes called its *ground state of energy*).

## THE QUANTUM THEORY OF RADIATION

You will recall that in Chapter 8 we discussed the spectrum of the hydrogen atom, which was composed of specific color bands (lines), each relating to a specific wavelength. We presented a rather superficial explanation at that occasion in terms of available energy levels. However, let's now ask a more fundamental question: Why are only certain energy levels available in the hydrogen atom? Early in this century, the brilliant Danish physicist Neils Bohr (1885–1962) asked this question, and in answering it he sought direction from the work of Albert Einstein and German physicist Max Planck (1858–1947).

Einstein had already introduced the idea that light behaves as though it were packets (quanta) of energy when he demonstrated that a certain frequency of light, namely, ultraviolet, carried sufficient energy to knock electrons from a polished zinc plate, whereas ordinary visible light did not (see Figure 10.10).

Max Planck had also found that a similar assumption seemed to explain the radiation of a heated body. In fact, he assumed a direct relationship between the frequency ($f$) of radiation and the energy ($E$) of a single quantum at that frequency ($f$), stated:

$$E = hf$$

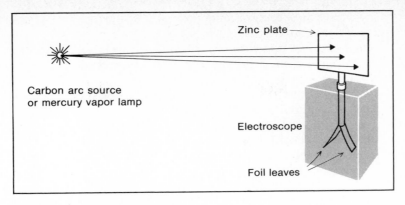

**Figure 10.10.** The photoelectric effect.

where $h = 6.63 \times 10^{-34}$ joule/sec (known as *Planck's constant*). The important thing to recognize here is the discovery that energy comes in multiples of this number ($h$) and not in a continuous flow.

At this point we should note the relationship between wavelength and frequency. All electromagnetic disturbances travel through empty space at the same speed ($c$), approximately 300,000 km/sec. If a wave travels 10 m while going through one cycle, its wavelength ($\lambda$) is 10 m. Then how many waves could occur in one second? This is equivalent to asking the question: How many wavelengths are contained in the distance of 300,000 km (300,000,000 m)? The answer is:

$$\frac{300,000,000 \text{ m}}{10 \text{ m}} = 30,000,000 \text{ cycles per sec}$$

We have found the frequency ($f$) of the disturbance, and we see that these factors are related:

$$f = \frac{c}{\lambda} \text{ or } \lambda = \frac{c}{f} \text{ or } f\lambda = c$$

where $\lambda$ is the wavelength and $c$ is the speed of light.

In modern terminology the unit of frequency (cycles per second) is specified as *hertz*, after Heinrich Hertz, who in 1887 showed that electromagnetic waves do exist. Furthermore, the familiar symbols $kc$ (kilocycle per sec, 1000 cycles per second) and $Mc$ (megacycle per sec, 1,000,000 cycles per sec) are now called $kH$ (kilohertz) and $MH$ (megahertz), respectively.

Niels Bohr was particularly interested in understanding the spectrum of the hydrogen atom in relation to the structure of the atom itself. The theories of classical physics would have predicted that a charged particle (the electron), if accelerated in an orbit around the proton, would radiate electromagnetic energy and thereby loose energy continuously. If this were true, then according to Planck's $E = hf$, the frequency of light would change continuously and many colors would result—the spectrum would be continuous (contain all colors). On the contrary, the spectrum of the hydrogen contains only specific lines; hence Bohr concluded that classical physics must be inadequate to explain the atom in this respect. He asserted that the electron can exist only in very specific energy states, and only when the electron is excited to a higher energy state may it then radiate energy upon a downward transition, thus quantizing its output. His real inspiration came when he speculated that his quantized effect might show up in the angular momentum of the electron. He assigned the following values to each successive energy state:

$$\frac{h}{2\pi}, \frac{2h}{2\pi}, \frac{3h}{2\pi}, \ldots$$

where $h$ = Planck's constant.

Then he computed the radii of the orbitals that agreed with this assumption and the frequencies of emission that would result. The emitted frequencies were in good agreement with the observed spectrum; hence the concept of the quantum had succeeded in explaining another fundamental fact of nature.

By 1924, French physicist Louis V. deBroglie (1892-    ) had shown that the quantum nature of the hydrogen atom (and its spectrum) could be explained by simply assuming that the orbitals of the electrons must be of such a size that an integral number of wavelengths will fit around them. This is to say that their circumference ($2\pi r$) must equal a whole number ($n$) times the wavelength ($\lambda$),

$$2\pi r = n\lambda$$

in order to provide a stable orbit (see Figure 10.11).

It seems that the work of Einstein, Bohr, and deBroglie has led us from the

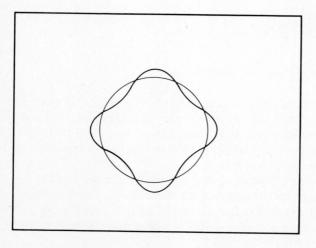

**Figure 10.11.** de Broglie waves associated with electron orbitals.

realization that light has both wave and particle properties to the realization that electrons, usually thought of as particles, have wave properties as well. In a much more general sense, deBroglie expressed the dual (particle and wave) nature of all matter in motion. Can a moving bowling ball be considered to have a wave nature? Yes, according to deBroglie; a bowling ball of mass 3 kg and velocity 2.21 m/sec would have a wavelength of $10^{-27}$ m, far too short to be noticed, but still an important factor in theoretical considerations. The deBroglie wavelengths ($\lambda$) are defined by:

$$\lambda = \frac{h}{mv}$$

where

$h = $ Planck's constant

$m = $ mass

$v = $ velocity

The wave nature of the electron is very clearly shown when we compare its diffraction pattern with that of X-rays, typically thought of as waves (see Figure 10.12).

**Figure 10.12.** Diffraction patterns: (a) electron diffraction created by passing a beam of electrons through beryllium, a crystal. (RCA Laboratories, Princeton, New Jersey); (b) X-ray diffraction, created by directing X-rays through a polycrystalline aluminum. (Mrs. H. M. Read, Bell Telephone Laboratories, Murray Hill, New Jersey).

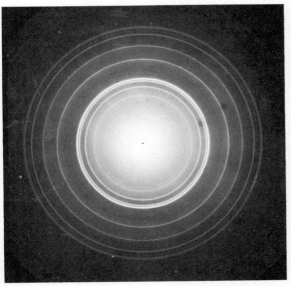

(a)

(b)

# FORCES WITHIN THE NUCLEUS

Rutherford discovered that the nucleus of the atom is very compact, but not until 1932 was it known that the nucleus was composed of more than one kind of particle. The neutron was discovered at that time. Thus at least two kinds of particles exist, the proton carrying a positive charge and the neutron having no charge, but each having roughly the same mass, about 1800 times that of the electron.

One of the most fundamental questions with which nuclear scientists are still struggling is: What holds the nucleus together? If we apply known laws of electrostatics and gravity, we would conclude that the repulsive electrical forces between protons (like charges repel) are so strong as compared to gravitational forces that the nucleus should surely fly apart. Since nuclei do not normally fly apart, we must conclude that a force that is stronger than the electrical force must bind the nuclear parts together, at least when they are close together as in a compact nucleus. Current theories cluster about the concept that protons and neutrons are continually exchanging particles, which may be thought of as substructures of the protons and neutrons themselves. We do know that many additional particles exist within the nucleus—over 30 have been identified from tracks in a cloud chamber when atoms are bombarded by fast-moving protons or neutrons (see Figure 10.13).

We will not pursue the substructure of the nucleus further; however, let's see

**Figure 10.13.** The tracks in this hydrogen bubble chamber photograph reveal the multiplicity of particles that exist in the nucleus of the atom. The tracks of neutral particles, shown as dashed lines on the right, are inferred from the tracks of the charged particles that result from their decay. (Brookhaven National Laboratory)

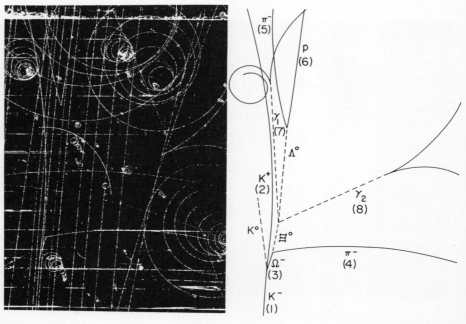

how the strong attractive force (nuclear binding force) interacts with the repulsive electrical force among the protons to produce either stable or unstable nuclei.

In the nucleus of any given atom, say that of neon, which contains 10 protons, we may find differing numbers of neutrons: 90.9% will have 10 neutrons (total weight approximately 20), 0.3% will have 11 neutrons (total weight approximately 21), and 8.8% will have 12 neutrons (total weight approximately 22). These are referred to as three different *isotopes* of the same element. The number of protons (atomic number) determines the element; the number of neutrons determines the isotope of that element. The total of the number of protons and neutrons is called the *atomic weight* and is written as a superscript: neon 20 is written $^{20}_{10}Ne$, neon 21 is written $^{21}_{10}Ne$, and neon 22 is written $^{22}_{10}Ne$. The subscript indicates the atomic number. If we were to average the weight of these three isotopes of neon, we would obtain the atomic weight shown in the Periodic Table (see front endpapers).

$$.909 \times 20 = 18.180$$
$$.003 \times 21 = .063$$
$$.088 \times 22 = 1.936$$
$$\text{Average} \quad 20.179$$

Note that the slight discrepancy between this number and that of the Periodic Table (20.183) results from the fact that individual isotopes do not have exact integral weights; i.e., neon 20 has a weight of 19.99244 atomic mass units.

As we look at the isotopes of elements of larger and larger atomic numbers (see Table 10.1), we see that the number of neutrons increases faster than the number of protons. This is shown by the fact that the plot of heavier elements (and their isotopes) lies entirely above the line $N = Z$, where $Z$ is the number of protons and $N$ is the number of neutrons, i.e., tin 119, Sn, has 50 protons and 69 neutrons; mercury 201, Hg, has 80 protons and 121 neutrons. This seems to demonstrate a possible role of neutrons. As more and more protons are accumulated in a nucleus, the repulsive electrical force goes up, and as a result the nucleus would have a greater tendency to fly apart. The addition of neutrons tends to neutralize this effect by averaging out the electrical force over more particles; hence the neutrons seem to add stability to the nucleus. However, we do see certain combinations that are unstable (radioactive) (see squares in Table 10.1).

## RADIOACTIVITY

Atoms that have unstable nuclei have very specific patterns of decay and rates of decay. This rate is specified by the concept of a half-life. Visualize a sample of uranium 238 ($^{238}_{92}U$), which contains 8 billion atoms. If you could observe this sample over a period of 4.5 billion years, you would find only 4 billion atoms of uranium remaining, the balance having spontaneously ejected nuclear parts to

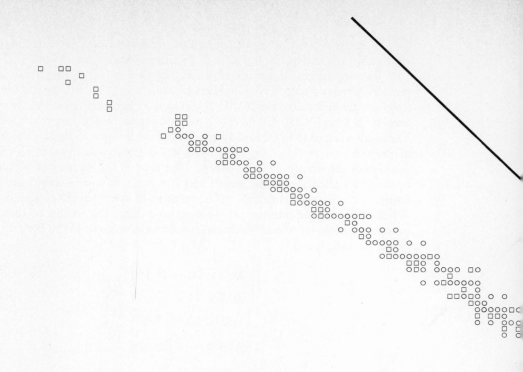

Table 10.1

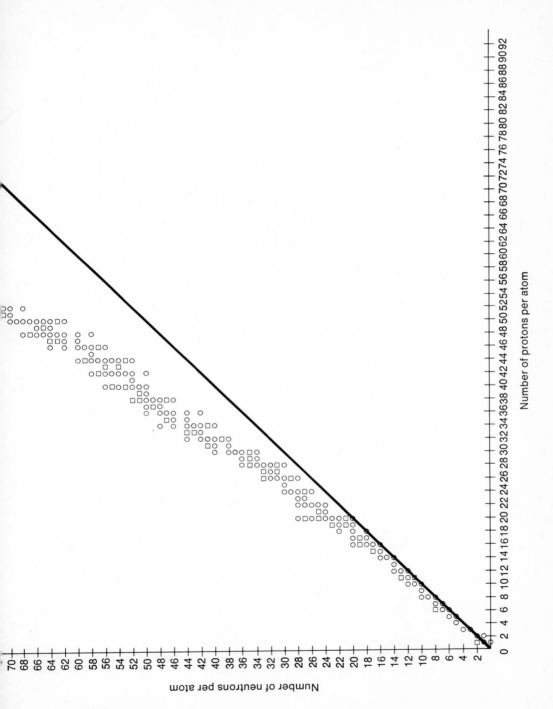

**Table 10.2**

| Time | Atoms remaining |
|---|---|
| $4.5 \times 10^9$ yr | 8 billion atoms → 4 billion atoms |
| $4.5 \times 10^9$ yr | 4 billion atoms → 2 billion atoms |
| $4.5 \times 10^9$ yr | 2 billion atoms → 1 billion atoms |
| $4.5 \times 10^9$ yr | 1 billion atoms → $\frac{1}{2}$ billion atoms |
| $4.5 \times 10^9$ yr | $\frac{1}{2}$ billion atoms → $\frac{1}{4}$ billion atoms |
| $4.5 \times 10^9$ yr | $\frac{1}{4}$ billion atoms → $\frac{1}{8}$ billion atoms |

become another element, as explained in the next section. Now, as though you were starting a new experiment with the 4 billion atoms, 4.5 billion years later you will see only 2 billion atoms of uranium. Thus, within each period of 4.5 billion years the supply of uranium is reduced by one-half. Therefore, the half-life of uranium is 4.5 billion years. This concept is drawn out over several periods in Table 10.2.

By contrast with this very long half-life, consider carbon 14, which has a half-life of 5700 years; or copper 64, which has a half-life of 12.8 hours. Would you expect copper 64 to occur (on the earth) naturally in large amounts today? No, for it would have had so many half-life periods in the 4.5 billion years of the earth's history that no measurable amount would remain. The copper we do find naturally consists of copper 63 and copper 65, both stable isotopes.

## METHOD OF DECAY

Radioactive elements decay by ejecting alpha particles or beta particles and sometimes gamma rays. Alpha ($\alpha$) particles were discovered to be fast-moving helium nuclei (by Rutherford), consisting of two protons and two neutrons. When an $\alpha$-particle is ejected, a new atom is created with its atomic number reduced by two and its atomic mass reduced by four.

Beta ($\beta$) particles are the fast-moving electrons similar to those with which Thomson experimented, but we are now thinking of electrons that are ejected from the nucleus of the atom. Scientists do not picture the nucleus as normally containing electrons, but rather think of them as being created at the time of ejection. They visualize a neutron as possessing both positive and negative charges equivalent to that of an electron. If an electron is created to carry off a negative charge from a neutron, that neutron is left positively charged; hence it becomes a proton—the atomic number is increased by one without reducing the atomic weight in any appreciable amount since the mass of an electron is only about 1/1800th that of a neutron.

$$^{238}_{92}U \xrightarrow{\alpha} {}^{234}_{90}Th \xrightarrow{\beta} {}^{234}_{91}Pa \xrightarrow{\beta} {}^{234}_{92}U \xrightarrow{\alpha} {}^{230}_{90}Th \xrightarrow{\alpha} {}^{226}_{88}Ra \xrightarrow{\alpha}$$

$$\longrightarrow {}^{222}_{86}Rn \xrightarrow{\alpha} {}^{218}_{84}Po \xrightarrow{\alpha} {}^{214}_{82}Pb \xrightarrow{\beta} {}^{214}_{83}Bi \xrightarrow{\beta} {}^{214}_{84}Po \xrightarrow{\alpha}$$

$$\longrightarrow {}^{210}_{83}Pb \xrightarrow{\beta} {}^{210}_{83}Bi \xrightarrow{\beta} {}^{210}_{84}Po \xrightarrow{\alpha} {}^{206}_{82}Pb$$

**Figure 10.14.** The decay series of uranium, leading to the production of lead-206.

Figure 10.14 shows the decay steps for uranium 238. First, an $\alpha$-particle is ejected, changing $^{238}_{92}U$ to thorium 234 ($^{234}_{90}Th$), which, according to its own half-life of 24 days, ejects a $\beta$-particle from its nucleus to become protactinium 234 ($^{234}_{91}Pa$), etc. You can see that after a number of such events, the stable element lead 206 ($^{206}_{82}Pb$) is formed. Although each step in this sequence has its own half-life, the half-life of uranium, because it is longer than any other in the series, dictates the half-life of the entire series. Because this half-life is known, the decay of uranium into lead becomes a tool for discovering the age of rocks of the earth, the moon, and other samples we may obtain in the future. The following sample is oversimplified but conveys the basic procedure for radioactive dating.

Suppose an experimenter found equal numbers of uranium 238 and lead 206 atoms in the crystalline structure of a moon rock. Could he then estimate its age? Since lead 206 is believed to originate almost exclusively in this decay process (we speak of lead 206 as the daughter product of uranium 238), it is reasonable to suppose that when the moon rock crystallized, no lead 206 was present. Therefore, we might expect the decay record, as shown in Table 10.3.

**Table 10.3**

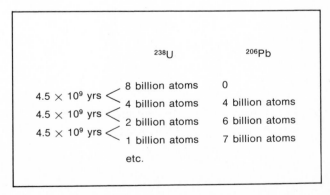

Table 10.4

|  | POTASSIUM 40 | ARGON 40 |
|---|---|---|
|  | 8 billion atoms | 0 |
| 1 billion yrs | 4 billion atoms | 4 billion atoms |
| 1 billion yrs | 2 billion atoms | 6 billion atoms |
| 1 billion yrs | 1 billion atoms | 7 billion atoms |
| 1 billion yrs | ½ billion atoms | 7½ billion atoms |

After 4 billion yrs, $\dfrac{\frac{1}{2}}{7\frac{1}{2}} = \dfrac{1 \text{ part potassium}}{15 \text{ parts argon}}$

At what point in time do equal numbers of atoms exist? Only after one half-life; hence the age of the rock is estimated to be 4.5 billion years.

As a verification of the uranium-lead method just described, potassium-argon ratios are also checked. Potassium 40 decays by beta emission to become argon 40 with a half-life of 1 billion years, as shown in Table 10.4. What ratio between these isotopes could you expect after 4 billion years?

One further example of radioactive dating is worthy of our attention, for it can be used with the remains of once living material such as plants or the bones of animals. The isotope of carbon that has the atomic weight of 14 is radioactive and decays with a half-life of 5700 years. When a plant or animal is alive, it absorbs a portion of carbon 14 from the atmosphere, and the ratio between this isotope and the more abundant carbon 12 isotope is known. When the plant or animal dies, the carbon 14 decays, changing the ratio. By the ratio ($^{14}$C : $^{12}$C) found in the bone of an animal, the period of time since its death may be estimated. Because of the rather short half-life of carbon 14, this method is limited to about 40,000 years, or approximately seven half-lives.

## ATOMIC ENERGY

The study of radioactivity ultimately led to a source of energy of enormous magnitude: nuclear energy. The structure of a nucleus may be altered by bombardment using either protons or neutrons as "bullets." Neutrons have the advantage of not being repelled by the electrical nature of the nucleus. When they are directed toward a given nucleus, the neutron may be captured to form a heavier isotope, or it may trigger a splitting off of a proton.

$$^{14}_{7}\text{N} + ^{1}_{0}\text{n} \longrightarrow ^{14}_{6}\text{C} + ^{1}_{1}\text{H}$$

This process occurs in the atmosphere with cosmic rays providing the accelerated neutron ($^1_0$n) and producing the radioactive isotope carbon 14. Alternately, the reaction may produce an $\alpha$-particle:

$$^{14}_{7}N + ^{1}_{0}n \longrightarrow ^{11}_{5}B + ^{4}_{2}He$$

Compared to these reactions, which split off only small particles, a neutron captured by the heavy element uranium 238 results in it splitting into two parts of more nearly equal masses:

$$^{238}_{92}U + ^{1}_{0}n \longrightarrow ^{145}_{56}Ba + ^{94}_{36}Kr$$

or in the case of uranium 235:

$$^{235}_{92}U + ^{1}_{0}n \longrightarrow ^{144}_{56}Ba + ^{90}_{36}Kr + 2\,^{1}_{0}n$$

The remarkable thing about such fission reactions, which split the nucleus more nearly "down the middle," is the tremendous amount of energy released—in the order of 200 million electron volts per reaction. To appreciate just how much energy this represents, 500 g (a little over one pound) of uranium 235 is capable of producing energy equivalent to that of two million pounds of coal. To achieve the full energy potential in a fission reaction, the process must not end with the first reaction but must be self-sustaining. This does happen in the case of uranium 235 because two or three neutrons are available from each reaction to trigger similar reactions in neighboring atoms. You can see the pyramid effect if each reaction triggers two or three more. Within a fraction of a second, billions of reactions occur and we have an explosion on our hands. However, several additional factors may tend to prevent or at least control such a reaction.

First, let's consider the fact that uranium 238 does not fission as readily as uranium 235; rather it tends to absorb neutrons and then proceeds to decay with only mild energy output. Uranium 238 is far more abundant than uranium 235, occurring in the ratio of 99.3% and 0.7%, respectively. A sample with this mix would not even be able to sustain a chain reaction. By contrast, the pure form of uranium 235 fissions so easily that it is difficult to control, especially if its mass is greater than that described as a critical mass. We can see what is meant by the term *critical mass* by considering a small lump of U-235, in which neutrons escape to the outside of the lump and are lost to further reactions [see Figure 10.15 (a)]. Then, as we think of the lump growing, it is evident that smaller and smaller percentages of the neutrons escape the lump [see Figure 10.15 (b)]. The fact that more neutrons stay within the lump allows for the chain reaction to develop more completely, in fact, to produce a bomb. The atom bomb was constructed in such a way that two subcritical masses of U-235 were placed at either end of a cylinder. At the time of detonation, these two masses were literally fired at each other, resulting in the sudden occurrence of a mass above the critical level. The resulting energy release surpassed that of the largest bombs by a factor of a million.

Can we turn such a reaction to useful, peacetime use? Yes, but it is essential that the reaction be under control at all times.

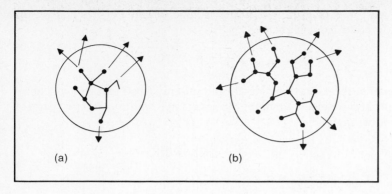

**Figure 10.15.** Uranium-235: (a) small lump; (b) larger lump, showing a smaller percentage of escaping neutrons.

   First, let us consider a refinement that makes the fission reaction more efficient. When U-235 splits, it releases fast neutrons; and because of their speed, they may pass through neighboring atoms without triggering another reaction. It would be advantageous to slow the neutron, thus increasing its chance of triggering a reaction. Several elements may be used to accomplish this slowing by collision with its own atoms. A pure form of carbon called *graphite* can be used as such a moderator. Modern fission reactors are constructed by interspersing fissionable material among graphite bricks (see Figure 10.16).
   One additional factor is needed to control the reaction. Control is accomplished by inserting cadmium rods into the reactor. Such rods absorb neutrons and prevent a runaway reaction. The insertion and withdrawal of these neutron absorbers must be automated to respond to temperatures generated in the reactor. There is a delicate balance between too much absorption, causing the reaction to cease, and too little absorption, which allows a runaway reaction and results in the melting of the reactor structure, subsequently contaminating the environment with radioactive products.
   One of the most serious problems that is and will continue to be of concern is the fact that all fission reactions produce radioactive byproducts with long half-lives, which, if released to the atmosphere or oceans, would have disastrous results. The best solution today involves storage of these byproducts in natural or man-made caves beneath the surface of the earth. Scientists do not always agree as to the severity of these dangers. Certainly they are united in the search for safer sources of nuclear energy. One such source involves the lightest elements like hydrogen (and its several isotopes) in a process called *fusion*.

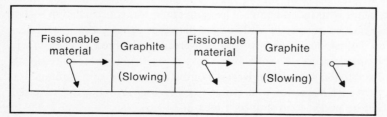

**Figure 10.16.** A fission reactor design, showing the slowing effect of the graphite.

178

# THERMONUCLEAR FUSION

The astronomer is well-acquainted with the fact that light elements may be fused, and in so doing, tremendous amounts of energy are released.

This is the basic nuclear reaction that is taking place in the sun and stars. One such cycle, the proton-proton cycle, is shown below:

$$_1^1H + {_1^1H} \longrightarrow {_1^2H} + {_0^1e}$$

$$_1^2H + {_1^1H} \longrightarrow {_2^3He}$$

$$_2^3He + {_2^3He} \longrightarrow {_2^4He} + 2{_1^1H}$$

This symbolic representation does not reveal the source of energy. If you could weigh the individual atoms involved in this reaction, you might be surprised to find that the mass of the helium ($_2^4$He) atom is lighter than the four hydrogen ($_1^1$H) atoms that entered into its creation. If this is true, then it will be useful to define a standard atomic mass unit (*amu*) as being one-twelfth the mass of a carbon 12 atom. On this basis, the hydrogen atom has a mass of 1.008 amu; hence four hydrogen atoms have a total of 4.032 amu. Yet when formed in the helium atom, the mass is only 4.003 amu. What happened to the difference of 0.029 amu (4.032 − 4.003 = 0.029)? It was converted to energy. Einstein recognized the equivalence of mass and energy in the expression $E = mc^2$ [energy = mass × (speed of light)$^2$]. Specified in the *cgs* system (cm-gram-sec) the speed of light (*c*) is equal to $3 \times 10^{10}$ cm/sec; therefore, $c^2 = 9 \times 10^{20}$ (900,000,000,000,000,000,000). Any loss of mass (in grams) is multiplied by this number ($c^2$) to yield the equivalent energy in ergs. If 70 kg of hydrogen could be made to fuse to form helium, the mass loss would be approximately 500 g. This mass loss would produce $4.5 \times 10^{23}$ ergs, the equivalent of energy represented by several billion tons of coal and with no radioactive byproducts.

Fusion certainly appears to be a logical approach to energy shortages, and in fact it is under investigation; however, the conditions under which fusion takes place are very difficult to create. These conditions include a pressure 3 billion times that which is exerted by the earth's atmosphere and a temperature measured in millions of degrees Kelvin (15,000,000° K minimum). No known container can withstand such pressure and temperature; however, these conditions do exist near the core of the sun, and fusion is the sun's primary source of energy. The sun converts approximately $4\frac{1}{2}$ million tons of its mass into energy every second by fusing hydrogen into helium. Hotter, more massive stars are thought to fuse elements on a step-by-step basis, up to iron.

Why should iron be a stopping point for this fusion process? The answer to this question will help us to understand both fusion and fission. If we could inspect the masses of individual protons and neutrons in the first 26 elements (hydrogen through iron), we would see a general downward trend; hence fusion of lighter elements into heavier throughout this series should result in the conversion of mass to energy (see Figure 10.17). Beginning with iron and proceeding through the heaviest elements, the mass of individual protons and neutrons increases; hence

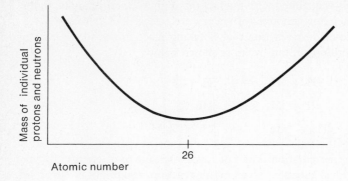

**Figure 10.17.** A plot of the changing amount of material in the protons and neutrons of atoms as their atomic number changes.

exceedingly large amounts of energy would be required to cause fusion to occur. On the other hand, if heavy elements can be split into two significantly lighter elements, their individual protons and neutrons are also lighter. The mass that is apparently lost is converted to energy.

When we see this pattern of atomic masses, we are also seeing a pattern of binding energies. This topic will be discussed in the next chapter.

## QUESTIONS

1. Why is it impossible to know whether any given model of the atom is correct?
2. True or false: Radioactive elements must be stimulated by light or x-rays in order to emit their typical radiation.
3. If the radiations of a radioactive element are allowed to flow between charged plates, what kind of particle will be attracted toward the negative plate and what kind of radiation will not be affected by the charged plate?
4. What property of an electron flow is demonstrated by the fact that a cross placed in their path casts a sharp shadow on the screen of the cathode ray tube?
5. True or false: A modern TV set utilizes the fact that a flow of electrons may be bent (and hence controlled) by both an electric field and a magnetic field.
6. What evidence do we have for thinking of the nucleus of the atom as a small, highly concentrated region rather than as a positively charged pudding?
7. Why is it dangerous to conceive of the atom as a miniature solar system?
8. Express in your own words the meaning of the term "quanta" with regard to the quantum theory.
9. Find the frequency of blue light that has a wavelength of $4.5 \times 10^{-7}$ meters.
10. How does the mass of the electron compare to the mass of the proton?

11. True or false: It is a simple matter to explain why the nucleus of an atom stays together, in spite of the fact that we know that like charges repel.
12. Give the number of protons and neutrons in each of the following: (a) $^{16}_{8}O$, (b) $^{17}_{8}O$, (c) $^{56}_{26}Fe$, (d) $^{206}_{82}Pb$, (e) $^{207}_{82}Pb$.
13. If the half-life of a radioactive element were 250,000 years, what fraction of a given quantity could be expected to remain after 500,000 years?
14. Radioactive elements decay by ejecting alpha particles, beta particles, and gamma rays. Describe each of these particles or rays.
15. Discuss the difference between atomic fission and atomic fusion, giving the advantages and disadvantages of each. Which of these processes is used in atomic energy plants today?

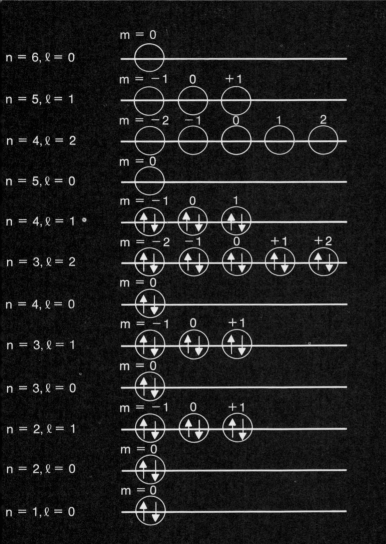

# the periodic nature of elements

The phrase *nuclear reaction* refers to a change in the nucleus of an atom. The phrase *chemical reaction* refers to a change in the structure of only the electrons. Here again our understanding of the structure of the electrons does not come from direct observation but is inferred from large-scale reactions, the combining and separating of atoms. For instance, when an electric current is passed through water (with a small amount of acid added), the water decomposes to form hydrogen (H) and oxygen (O) gas. Since twice as much hydrogen is formed as compared to oxygen, chemists feel reasonable in assuming that two hydrogen atoms bond with each oxygen, a ratio of 2:1. This ratio is expressed in the symbol for water, $H_2O$. When equal numbers of sodium (Na) and chlorine (Cl) atoms are brought together, they react to form common table salt (NaCl). Since no part of either gas is left over, we assume that sodium and chlorine combine in a ratio of 1:1, expressed as NaCl (sodium chloride). Whenever we say that two (or more) elements *react*, we mean that a new substance is formed, generally with physical properties different from those of the original constituents.

## UNITS OF MATTER

In the last paragraph, we use the terms *atom, element,* and *substance,* and we will be using terms such as *mixture, molecule,* and *compound.* We have also seen that the atom has a structure composed of *protons, neutrons, electrons,* and numerous other subatomic parts. But let's start with the atom in defining the more general terms. We think of an element as being composed of atoms, all of which are alike; that is, they all have the same number of protons. Iron (Fe), nickel (Ni), copper (Cu), and carbon (C) are examples of over one hundred different kinds of atoms known today. By contrast, water ($H_2O$) and sodium chloride (NaCl) are not elements but compounds composed of two different kinds of atoms bound together in definite ratios. The basic unit of a compound is called a *molecule* or *ionic pair,* but we will make this distinction later in this chapter. As long as a substance is composed of atoms bonded in a definite ratio, then that substance is called *pure.* By contrast, many substances do not possess such a definite ratio of elements; they are called *mixtures.* For instance, table salt dissolved in water to make a salt solution is a mixture, for any portion of such a solution contains an indefinite number of atoms

185

the periodic nature of elements

or molecules. Birdseed is a good example of a mixture, for it contains different kinds of seeds, and the number of each kind in a small sample is not predictable. The most general term that includes all terms just defined is *substance*—that is, anything that occupies space and has mass. We may summarize these terms in outline form:

Substance
    Mixture
        Solutions
        Birdseed, etc.
    Pure Substance
        Elements
            Atoms, all alike
        Compounds
            Atoms of different kinds
            Molecules ($H_2O$)
            Ionic pairs (NaCl)

## ATOMIC WEIGHTS

By the midnineteenth century, many of the elements were known, together with the atomic weight of one or more of their isotopes. Remember (from Chapter 10) that different isotopes of a given element merely refer to differing numbers of neutrons, i.e., carbon 12 (6 protons and 6 neutrons) and carbon 14 (6 protons and 8 neutrons). Because carbon 12 is far more abundant than carbon 14 or any other isotope of carbon, when we average the weights of these two isotopes we get 12.011 atomic mass units. The first 54 elements have been listed in Table 11.1 in the order of average atomic weights; Note that three elements—argon (18), cobalt (27), and tellurium (52)—are out of order according to their atomic numbers (counting protons only).

## FAMILY GROUPS OF ELEMENTS

About this time, the Russian chemist Dmitri Mendeleev (1834–1907) noticed a surprising repetition in chemical properties of the various elements. To illustrate what this means, consider a series of experiments in which two elements are brought together to see if they react. You can tell when a reaction occurs because a new substance is formed that has a different appearance than either element that entered into the reaction. Often a reaction can be encouraged by heating the elements. The following is a partial listing of elements that do react, along with an indication of ratios of atoms involved:

| LiBr | $Li_2S$ | $Li_3O$ | LiOH |
| NaBr | $Na_2S$ | $Na_3O$ | NaOH |
| KBr | $K_2S$ | $K_3O$ | KOH |

**Table 11.1** Table of elements

| ELEMENT | SYMBOL | ATOMIC WEIGHT | ATOMIC NUMBER |
|---|---|---|---|
| Hydrogen | H | 1.008 | 1 |
| Helium | He | 4.003 | 2 |
| Lithium | Li | 6.939 | 3 |
| Beryllium | Be | 9.012 | 4 |
| Boron | B | 10.811 | 5 |
| Carbon | C | 12.011 | 6 |
| Nitrogen | N | 14.007 | 7 |
| Oxygen | O | 15.999 | 8 |
| Fluorine | F | 18.998 | 9 |
| Neon | Ne | 20.182 | 10 |
| Sodium | Na | 22.990 | 11 |
| Magnesium | Mg | 24.312 | 12 |
| Aluminum | Al | 26.982 | 13 |
| Silicon | Si | 28.086 | 14 |
| Phosphorus | P | 30.974 | 15 |
| Sulfur | S | 32.064 | 16 |
| Chlorine | Cl | 35.453 | 17 |
| Potassium | K | 39.092 | 19 |
| Argon | Ar | 39.948 | 18 |
| Calcium | Ca | 40.08 | 20 |
| Scandium | Sc | 44.956 | 21 |
| Titanium | Ti | 47.90 | 22 |
| Vanadium | V | 50.942 | 23 |
| Chromium | Cr | 51.996 | 24 |
| Manganese | Mn | 54.938 | 25 |
| Iron | Fe | 55.847 | 26 |
| Nickel | Ni | 58.71 | 28 |
| Cobalt | Co | 58.933 | 27 |
| Copper | Cu | 63.54 | 29 |
| Zinc | Zn | 65.37 | 30 |
| Gallium | Ga | 69.72 | 31 |
| Germanium | Ge | 72.59 | 32 |
| Arsenic | As | 74.922 | 33 |
| Selenium | Se | 78.96 | 34 |
| Bromine | Br | 79.909 | 35 |
| Krypton | Kr | 83.80 | 36 |
| Rubidium | Rb | 85.47 | 37 |
| Strontium | Sr | 87.62 | 38 |
| Yttrium | Y | 88.905 | 39 |

**Table 11.1** Table of elements (*Cont.*)

| ELEMENTS | SYMBOL | ATOMIC WEIGHT | ATOMIC NUMBER |
|---|---|---|---|
| Zirconium | Zr | 91.22 | 40 |
| Niobium | Nb | 92.906 | 41 |
| Molybdenum | Mo | 95.94 | 42 |
| Technetium | Tc | 98 | 43 |
| Ruthenium | Ru | 101.07 | 44 |
| Rhodium | Rh | 102.905 | 45 |
| Palladium | Pd | 106.4 | 46 |
| Silver | Ag | 107.870 | 47 |
| Cadmium | Cd | 112.40 | 48 |
| Indium | In | 114.82 | 49 |
| Tin | Sn | 118.69 | 50 |
| Antimony | Sb | 121.75 | 51 |
| Iodine | I | 126.904 | 53 |
| Tellurium | Te | 127.60 | 52 |
| Xenon | Xe | 131.30 | 54 |

$$RbBr \quad Rb_2S \quad Rb_3O \quad RbOH$$
$$CsBr \quad Cs_2S \quad Cs_3O \quad CsOH$$

This tabulation seems to indicate very clearly that lithium (Li), sodium (Na), potassium (K), rubidium (Rb), and cesium (Cs) all react in a very similar manner; hence they can be thought of as a family group of elements. Although there are many other tabulations involving other family groups of elements, we will reserve them to a later time to prove the point that Mendeleev made. He worked with atomic weights, but the relationship that he saw is more obvious when we concentrate on atomic numbers. Notice the following pattern:

| ELEMENT | DIFFERENCE IN ATOMIC NUMBERS |
|---|---|
| Lithium (3) | |
| | 8 |
| Sodium (11) | |
| | 8 |
| Potassium (19) | |
| | 18 |
| Rubidium (37) | |
| | 18 |
| Cesium (55) | |

This repetition of chemical properties suggests that if the elements were arranged in rows, beginning with these elements, other similarities might show up.

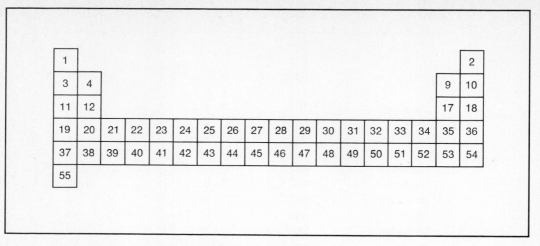

**Figure 11.1.** The periodic table of elements.

Let's start with potassium and arrange 18 elements in a row; then follow with rubidium leading another row of 18 (see Figure 11.1). We ask a fundamental question: Where do elements with atomic numbers of 4, 5, 6, 7, 8, 9, 10, 12, 13, 14, 15, 16, 17, and 18 logically fit to make groups of elements fall in vertical columns as did the elements with atomic numbers 3, 11, 19, 37, and 55? When experimenting, the elements with atomic numbers 2, 10, 18, 36, and 54 did not react with any other elements; hence they might logically form a single vertical column, namely the last column. Next, we might ask if elements with atomic numbers 9 and 17 have chemical properties like those with numbers 35 and 53. Fluorine (9), chlorine (17), bromine (35), and iodine (53) all combine with sodium in a 1:1 ratio; hence this seems to suggest their common group trait. It is a similar line of reasoning and evidence that shows other groups and completes what is called the Periodic Table, as shown in the front endpapers. Groups of elements are identified by Roman numerals I, II, III, IV, V, VI, VII, and VIII, with some subdivisions labeled $A$ and $B$. We have demonstrated why groups I, VII, and VIII contain their respective members. See if you can match the following reactions to a given family group:

| | | |
|---|---|---|
| BeO | $BeCl_2$ | |
| MgO | $MgCl_2$ | |
| CaO | $CaCl_2$ | Group _____ |
| SrO | $SrCl_2$ | (Be, Mg, Ca, Sr, Br) |
| BrO | $BaCl_2$ | |

188

189

the periodic
nature
of elements

CO₂    CF₃
SiO₂   SiF₃
GeO₂   GeF₃         Group _____
SnO₂   SnF₃         (C, Si, Ge, Sn, Pb)
PbO₂   PbF₃

(Answers are on page 202.) These groups are rather obvious, but some of the other subdivisions are not as obvious now as they will be after we develop the theory of electron structure. First, let us see one further clue that our grouping of elements is correct.

## METALS VERSUS NONMETALS

Perform three tests on each element in the list:

1. Is it ductile (can be easily formed into a new shape by pounding), or is it brittle (shatters when hit by hammer)?
2. Is it shiny (when freshly cut or polished), or is it dull?
3. Will it conduct an electrical current?

Generally speaking, if an element is ductile, shiny when cut, and passes an electrical current, it will be classified as a metal. If it fails these tests, it's a nonmetal. There are borderline cases that do not allow a clear-cut distinction, i.e.,

**Figure 11.2.** Metals and nonmetals, a natural division in the periodic table of elements.

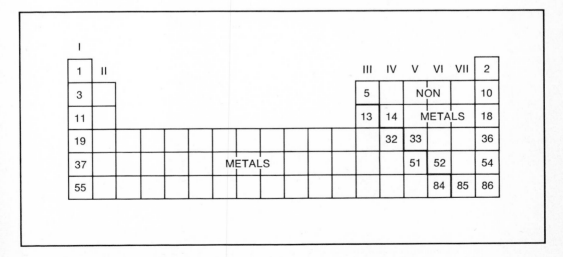

carbon that is brittle and dull but conducts an electrical current; so several elements like this one are rather arbitrarily assigned to either a metal or nonmetal category. The interesting thing is that as we go through the list of elements in order, we get the distinct impression that new rows on the Periodic Table should be started with elements 3, 11, 19, 37, 55. Why? Because these test out as metals, whereas several just before each of these tested as nonmetals. For example, 7, 8, 9, 10 are all nonmetals, and 11, 12, 13 are metals; so it seems logical to start a new row at the point of change. This line of reasoning also indicates that argon (a nonmetal) should come before potassium (a metal), even though argon has a larger atomic weight than potassium (see Table 11.1). When this general plan is followed, all metals fall to the left in the Periodic Table, and nonmetals fall to the right. The boundary between the two groups shows that there are far more metals than nonmetals (see Figure 11.2).

## ATOMIC ENERGY LEVELS

When we think of the Bohr model of the hydrogen atom, developed in Chapter 10, we recall that the spectrum of that element suggested the presence of rather distinct energy levels between which electrons could "jump." If we are dealing only with the first 20 elements, we could visualize the energy levels of any atom as a series of shell-like structures, the first of which can hold up to two electrons; the second can hold up to eight electrons; and the third, eight more electrons. This model would explain why lithium, sodium, and potassium behave similarly from a chemical viewpoint—they all have one electron in their outer energy level (see Figure 11.3).

Following this model, group II elements have two electrons in their outer energy level; group III elements have three electrons in their outer energy level; etc., for each group. Group VIII is pictured in Figure 11.4 to show that the outer energy levels are full.

This may help us to realize why group VIII members are called *inert gases* and do not react with other elements—their outer energy levels are already full. This is also the basis for the "octet" rule: each row of elements ends with eight

**Figure 11.3.** The electron pattern of elements that have similar chemical properties: Group I.

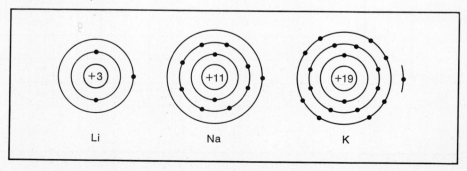

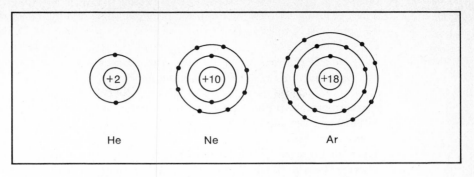

**Figure 11.4.** The electron pattern of elements that have similar chemical properties: Group II.

electrons in its outer energy level. In this state, the atom shows no tendency to give up electrons nor to absorb or share electrons; hence it does not readily react with other elements.

If we tried to pursue this model beyond element #20, we would find it inadequate; therefore, we will develop a more complete model called the *quantum mechanical model*.

## QUANTUM MECHANICAL MODEL

The spectrum of an excited gas, generated in a magnetic field, shows spectral lines that are split into multiple lines. This is called the *Zeeman effect,* after the Dutch physicist Peter Zeeman (1865–1943), and it reveals a substructure within each energy level of the atom. For the sake of simplicity, energy levels can be viewed as shell-like structures surrounding the nucleus of the atom at specified distances and representing the most probable locations for electrons. We will refer to these shell-like energy levels as *electron orbitals* and their substructures as *suborbitals,* and we will superimpose the modern quantum mechanical view of the atom on this visual picture.

The state of any electron may be described by four quantum numbers: $n$, the quantum number corresponding to the size of the orbital (energy level); $\ell$, the orbital quantum number that tells us something about the shape of the orbital; $m$, the magnetic quantum number that tells us something of the angle of the electron's spin axis; and $\sigma$ (sigma), the spin direction of the electron. The following restrictions are placed on these four numbers: $n$ can take on integral values 1, 2, 3, 4, . . . ; but $\ell$ is limited to integral values 0, 1, 2, 3, . . . up to $(n - 1)$ only. That is, when $n = 1$, $\ell = 0$; when $n = 2$, $\ell = 0$ or 1; when $n = 3$, $\ell = 0$ or 1 or 2, etc. The magnetic quantum number $(m)$ is restricted to integral values from $-\ell$ to $+\ell$; i.e., when $\ell = 2$, then $m$ can equal $-2, -1, 0, 1$, or 2. But when $\ell = 0$, then $m$ can only equal 0. Finally, the spin quantum number $(\sigma)$ can have only two values $+\frac{1}{2}$ (spin up, ↑) and $-\frac{1}{2}$ (spin down, ↓).

We are going to start filling orbitals and suborbitals following these restrictions but also guided by the exclusion principle suggested by the Austrian–American physicist Wolfgang Pauli (1900–1958):

> *No two electrons can occupy the same quantum state—no two electrons may have the same set of quantum numbers.*

Furthermore, we presume that the lowest energy states should be filled first; but within any given suborbital, $\sigma$ numbers of $+\frac{1}{2}(\uparrow)$ will be assigned ahead of $-\frac{1}{2}(\downarrow)$. Figure 11.5 shows various quantum states, and each circle can hold two electrons, one with up spin ($\uparrow$) and one with down spin ($\downarrow$). Note that the suborbital described by $n = 3$, $l = 2$ represents a higher energy than $n = 4$, $l = 0$; hence the latter will fill first.

The spin arrows have been filled in to show krypton (36 electrons). If you

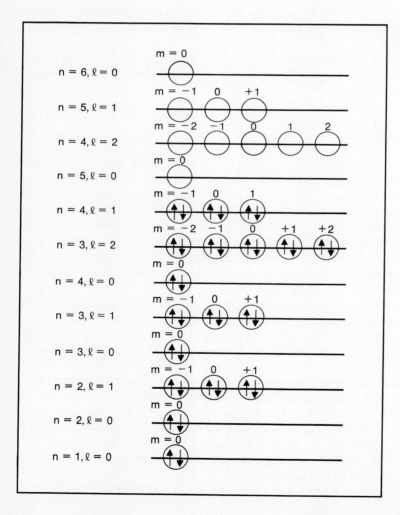

**Figure 11.5.** The quantum states of the 36 electrons of a krypton atom.

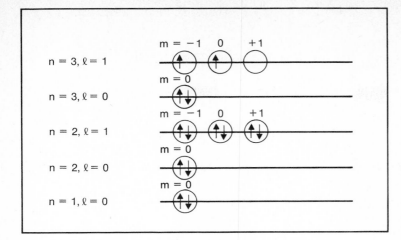

**Figure 11.6.** The electron configuration of silicon (14 electrons).

total the number of electrons in the fourth orbital, you will find 8, demonstrating the octet rule for row four in the Periodic Table.

We have already stated the fact that the group numbers indicate directly the number of electrons in the outer orbital. For instance, we may show the electron configuration for silicon (14 electrons), a group IV element (see Figure 11.6).

See if you can confirm that oxygen has 6 electrons in its outer orbital, chlorine has 7, potassium has 1, and magnesium has 2. The number of electrons in the outer orbital of an atom is believed to determine its chemical properties—its reaction and bonding characteristics.

## VALENCE

We have already seen that an atom is most stable when it completes its outer orbital. A group I element can achieve this by giving up its single outer electron, leaving it with a $+1$ charge. Hence we say the valence of a group I element is $+1$. A group II element must lose two electrons to leave it with a complete octet, making it a $+2$ ion; therefore, we assign a valence of $+2$ to this group. Group III elements carry out this same pattern and have $+3$ valence numbers.

If we consider the inert gases of group VIII, which have complete outer shells, we see that a valence of zero (or no valence number at all) is appropriate. Group VII elements need 1 electron to complete their outer orbital, which if received would make them $-1$ in charge; hence this group is assigned a $-1$ valence number. Likewise, group VI elements carry a valence of $-2$, and group V elements carry a valence of $-3$. It should be noted that in these latter cases (nonmetals) other valences are possible when the elements appear in certain combinations. Neglecting these exceptions, we can say in general that the valence of an atom is numerically equal to the charge that atom would carry if its outer orbital were complete.

Valence numbers allow us to predict the way certain reactions should occur.

For instance, a single magnesium (+2 valence) atom will react with a single oxygen (−2 valence) atom to form magnesium oxide:

$$2\,Mg + O_2 \longrightarrow 2\,MgO$$

On the other hand, two potassium (+1 valence) atoms are required to satisfy each oxygen (−2 valence) atom:

$$4\,K + O_2 \longrightarrow 2\,K_2O$$

Note: If the symbols are replaced by their valence numbers, the sum is zero on either side.

$$4(+1) + 2(-2) = 0$$

## CHEMICAL BONDS

Consider the sodium atom as shown in Figure 11.7. Every electron experiences an electrostatic attraction toward the positive nucleus; however, the single electron in the outer orbital experiences a shielding effect from the other electrons that repel it. Thus, the single outer electron in this atom, or others like it, is very loosely held. It is easily dislodged by an electric current or by the presence of another element that would become more stable if it could gain an electron. Remembering the stability of the group VIII elements, apparently due to full octets of electrons in their outer orbitals, we move to the left one group on the periodic table to find elements that need one electron to complete an octet. Chlorine is such an element. As you already know, when sodium and chlorine atoms are brought together, a reaction occurs to produce NaCl. Now we see what is actually taking place. The rather loosely bound electron of the sodium atom experiences a greater force

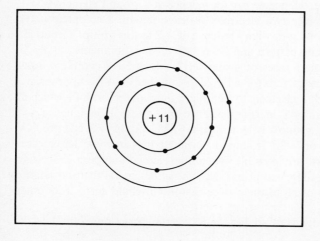

**Figure 11.7.** The electron pattern of sodium.

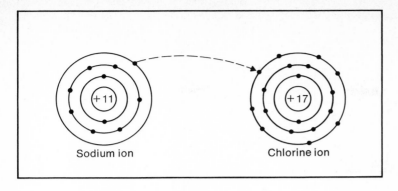

**Figure 11.8.** Ionic bonding of sodium and chlorine.

toward the chlorine atom and transfers to it, providing its octet completion (see Figure 11.8). The loss of one electron by the sodium atom leaves it charged positively (a $+1$ ion), and the gain of one electron by the chlorine atom gives it a negative charge (a $-1$ ion). Two such ions are held together because of their unlike charge. This describes an ionic bond. We do not consider this kind of bond as producing a molecule but rather an ionic pair. The crystalline form of NaCl can be pictured as an arrangement of positive and negative ions in alternate positions (see Figure 11.9).

Following this example, typically we might expect elements of groups I or II (with loosely bound electrons) to form ions with elements of groups VI or VII (which need one or two electrons to complete their octet). Some examples include: CaO, KCl, $MgCl_2$, and $Na_2S$ (the last two examples are called *ion aggregates* because more than two atoms are involved).

Because elements of groups I and II give up their outer electron so easily and, therefore, react more violently, they are called the *active metals*. Likewise, those elements that lack a single electron to complete their octet (group VII elements) are called the *halogens* (salt formers). Typically when an active metal reacts with a halogen to form an ion, the product is called a *salt*.

**Figure 11.9.** The sodium chloride crystal.

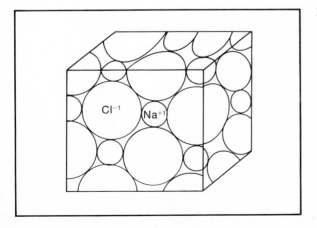

**Table 11.2** Ionization energies of selected atoms

| H 13.6 | | | | | | | He 24.6 |
|---|---|---|---|---|---|---|---|
| Li 5.4 | Be 9.3 | B 8.3 | C 11.3 | N 14.5 | O 13.6 | F 17.4 | Ne 21.6 |
| Na 5.1 | Mg 7.6 | Al 6.0 | Si 8.1 | P 11.0 | S 10.4 | Cl 13.0 | Ar 18.0 |
| K 4.3 | Ca 6.1 | Ga 6.1 | Ge 8.1 | As 10 | Se 9.8 | Br 11.8 | Kr 14.0 |
| Rb 4.2 | Sr 5.7 | In 5.8 | Sn 7.3 | Sb 8.6 | Te 9.0 | I 10.4 | Xe 12.1 |
| Cs 3.9 | Ba 5.2 | Tl 6.1 | Pb 7.4 | Bi 8 | Po | At | Rn 10.7 |

A very useful way to specify how active elements are in forming ions is to specify the energy required to remove a single electron from the outer orbital (see Table 11.2).

In Table 11.2 notice the trend toward lower ionization energies as you go down a given column. This is due to the increased size of the atom, resulting in weaker electrostatic attraction of the outer electrons. Notice also the increase in ionization energies as you go to the right across any row. This is due to the fact that as one moves to the right, the positive charge of the nucleus increases with very little change in the size of the atom.

## COVALENT BONDING

Many of the elements that are normally gaseous occur in molecular form—that is, two or more atoms are bonded together. Oxygen, for example, normally occurs as pairs of oxygen atoms—a fact we would be hard pressed to explain as being an ionic bond, for in no sense can oxygen atoms make themselves alternately positive and negative ions. Let us look to another form of bonding. Oxygen atoms need two electrons to complete an octet; hence if two oxygen atoms could simply share two electrons, each could behave as though their octets were complete. This is called *covalent bonding* and is represented simplistically in Figure 11.10. Notice that if you ignore where the electrons came from, each atom seems to have eight electrons in its outer orbital. Since the completed octets equate to more stability, these atoms "hang" together. The bonding tendency is more clearly seen if we visualize the electron clouds of each atom as being distorted with greater density between the two atoms. This infers that an electron is more likely to be found in that central region; hence the distribution of charge is as shown in Figure 11.11.

The attraction between regions of unlike sign value provides the bonding tendency. The halogens typically form into atomic pairs (molecules) by covalent

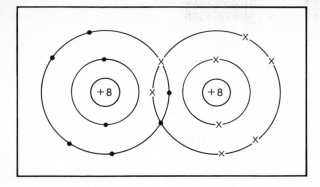

**Figure 11.10.** The principle of covalent bonding.

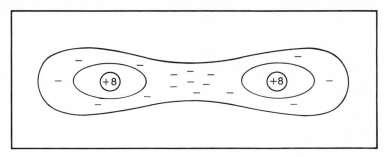

**Figure 11.11.** The electron distribution within a covalent bond.

bonding, sharing one electron. Two hydrogen atoms can share their electrons to provide complete outer orbitals.

## POLAR COVALENT BONDS

It is often true that in covalent bonds one atom attracts the electrons more than the other, thus producing a molecule with electrical properties—slightly negative at one end and slightly positive at the other. One of the most important polar molecules, in terms of life as we know it, is the water molecule ($H_2O$) (see Figure 11.12). Because of the nature of the oxygen atom, the two hydrogen atoms bond at an angle of 105°. This V-shaped configuration, together with the fact that the oxygen atom attracts electrons more than the hydrogen atoms, creates a polar situation that can be treated as a dipole (see Figure 11.13). This property tends to strengthen the attraction between adjacent water molecules and thus holds water in a liquid state at higher temperatures than would be expected. For instance, $H_2S$, $H_2Se$, and $H_2Te$ all exist as gases at room temperature even though they are similar to water ($H_2O$). Water remains a liquid at room temperature because of the strong polar nature of its molecules.

It is possible for molecules to bond in a form so as to cancel out their polar natures. For example, note the chemical form of molecules used to coat cooking utensils to keep them from sticking (see Figure 11.14). The fluorine (F) atoms attract the electrons more strongly than the carbons (C), so the arrows (vectors)

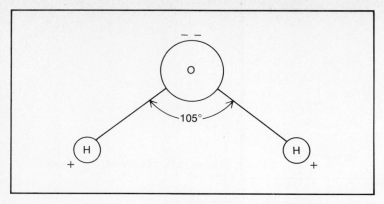

**Figure 11.12.** A polar molecule (water).

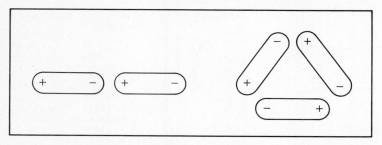

**Figure 11.13.** The attractive force between polar molecules.

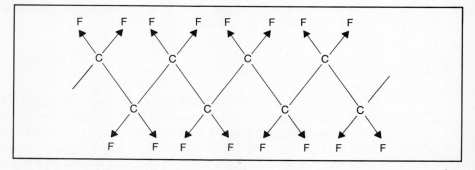

**Figure 11.14.** The model for the nonsticking coatings used on cooking utensils.

point toward the negative direction of each polar bond. Note, however, that for each vector in one direction there is another vector pointing in the opposite direction. The net effect is to neutralize these electrical forces and produce a nonpolar chain. This lack of electrical forces produces the nonsticking surface.

By contrast, consider the chain shown in Figure 11.15. Chlorine (Cl) attracts electrons more than carbon (C), and carbon attracts electrons more than hydrogen (H). Here the vectors complement each other, producing a strong polar character in this chain. A household product that utilizes this polar character is plastic wrap, which is sold under various brand names. Plastic wrap will stick to itself or to a dish, etc., because of this molecular structure. This will give you an idea of what is

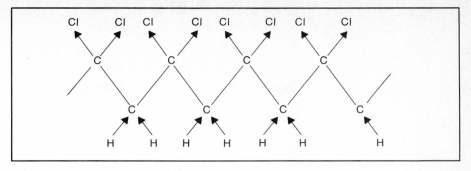

**Figure 11.15.** A model of a polar chain i.e., plastic wrap, that clings to other substances.

meant by molecular engineering. By combining certain atoms in a specific array, chemists can create products that will have desired characteristics.

In this section on polar bonds we have raised two additional questions: (1) Why does the water molecule form with an angle of 105° between its hydrogen components? (2) Why does the oxygen atom attract electrons more than the hydrogen atom? To answer the first question we need to refine our thinking about electron orbitals in the atom. We have pictured the oxygen atom as shown in Figure 11.16. Now we visualize the suborbital structure by quantum numbers (see Figure 11.17). Note the two unpaired electrons in the $2p$ level. We further refine the most probable orbitals by picturing any $\ell = 0$ orbital as spherical (s) and any $\ell = 1$ orbital as one of the $p$ orbitals shown in Figure 11.18. As you can see, the axes of these orbitals are separated by an angle of 90°, and each of the two unpaired electrons can be thought of as "residing" somewhere in the $2p_y$ and $2p_z$ orbitals. By covalent bonding with two hydrogen atoms the octet rule is satisfied, yielding a configuration like that shown in Figure 11.19.

However, in this molecule the oxygen atom attracts the electrons more strongly than the hydrogen atoms, and the hydrogen nuclei are virtually exposed as

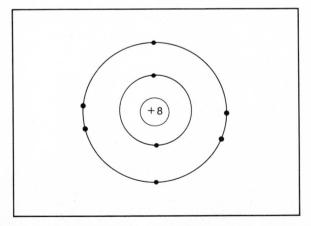

**Figure 11.16.** The oxygen atom.

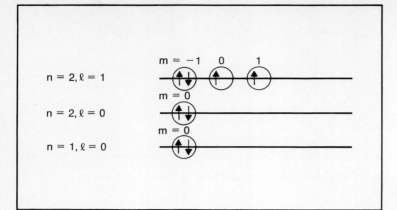

**Figure 11.17.** The electron configuration of the oxygen atom.

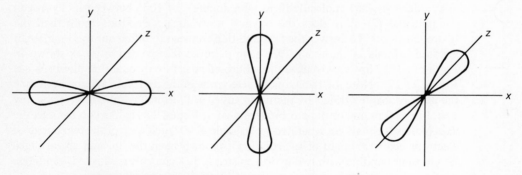

**Figure 11.18.** The $p$-orbitals.

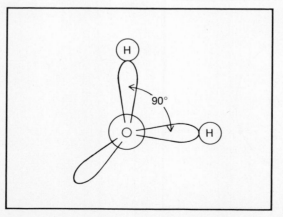

**Figure 11.19.** The expected form of the water molecule, based on $p$-orbitals.

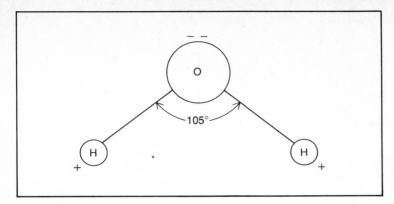

**Figure 11.20.** The actual form of the water molecule, resulting from the repulsive force between electrons.

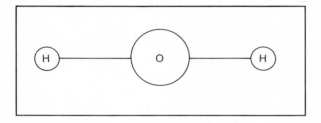

**Figure 11.21.** The expected form of the water molecule if $p$-orbitals were not mutually perpendicular.

bare protons. Two such like charges tend to repel each other, thus opening up the angle between them to 105° (see Figure 11.20).

The quantum mechanical picture of the $p$ orbitals explains why a water molecule does not appear as in Figure 11.21, for the unpaired electrons do not exist in directly opposite orbitals. If they did, then water would not have a polar nature, and its characteristics as a solvent would be quite different. This explanation of the form of a water molecule will help you understand some of the crystalline forms in solids.

ELECTRONEGATIVITY

The electronegativity of an atom is a measure of how strongly it attracts electrons. As we have seen, the fact that electrons are more strongly attracted to some atoms than to others gives molecules a polar nature. We might think of the electronegativity of an atom as being proportional to the energy required to break a bond formed by two atoms of the same element. For instance, the $O_2$ bond requires more energy to break than an $H_2$ bond, hence the higher electronegativity of oxygen as compared to hydrogen. Table 11.3 is a partial listing of electronegativities. Note the similarity of pattern with that of ionization energies (cf. Table 11.2).

In Table 11.3 the electronegativity decreases as you go down any family column, but it increases as you go toward the heavier elements in any given row.

**Table 11.3** Electronegativities of selected atoms

| H 2.1 | | | | | | | He |
|---|---|---|---|---|---|---|---|
| Li 1.0 | Be 1.5 | B 2.0 | C 2.5 | N 3.0 | O 3.5 | F 4.0 | Ne |
| Na 0.9 | Mg 1.2 | Al 1.5 | Si 1.8 | P 2.1 | S 2.5 | Cl 3.0 | Ar |
| K 0.8 | Ca 1.0 | Ga 1.6 | Ge 1.8 | As 2.0 | Se 2.4 | Br 2.8 | Kr |
| Rb 0.8 | Sr 1.0 | In 1.7 | Sn 1.8 | Sb 1.9 | Te 2.1 | I 2.5 | Xe |
| Cs 0.7 | Ba 0.9 | Tl 1.8 | Pb 1.8 | Bi 1.9 | Po 2.0 | At 2.2 | Rn |

The electronegativity of the inert gases is zero; they have no tendency to attract electrons because of full octets in the outer orbitals. From this table you will be able to predict the polar nature of many covalent bonds.

In the next chapter we will see how the polar nature of molecules and the intermolecular bonds that result play against other forces to determine the state of matter: gas, liquid, or solid.

Answers to questions on page 188: The elements Be, Mg, Ca, Sr, and Br form group II, and the elements C, Si, Ge, Sn, and Pb form group IV.

## QUESTIONS

1. Identify the following as either mixture, element, or compound: (a) carbon, (b) carbon dioxide, (c) salt solution, (d) salt (NaCl), (e) water ($H_2O$).
2. True or false: The order of the elements is the same regardless of whether we count protons alone or both protons and neutrons.
3. What is the most convincing evidence that the chemical nature of various elements are similar and that these similarities are repeated in a periodic fashion?
4. Group VII elements lack only one electron of completing their outer energy level. With what group will they logically form ionic bonds?
5. What is the difference between elements that are clearly metals and those that are clearly nonmetals?
6. How many electrons could have the following quantum numbers in the same atom? ($n = 2$, $l = 1$, $m = 0$)
7. How many valence (+1) atoms would you expect to bond with a (−2) atom?
8. Why do molecules form in definite proportions?

9. True or false: Group VIII elements are unusual in that they do not readily react with other elements.
10. Why are group I and group II elements so active (highly reactive) with so many other elements?
11. What factors in the structure of an atom tend to dictate how easily electrons are removed?
12. In covalent bonding, atoms share electrons. Why does this make the atoms "stick" together?
13. Water is a polar molecule. Can you explain what this means and why it is polar?
14. Why should the electronegativity of an inert gas be zero?

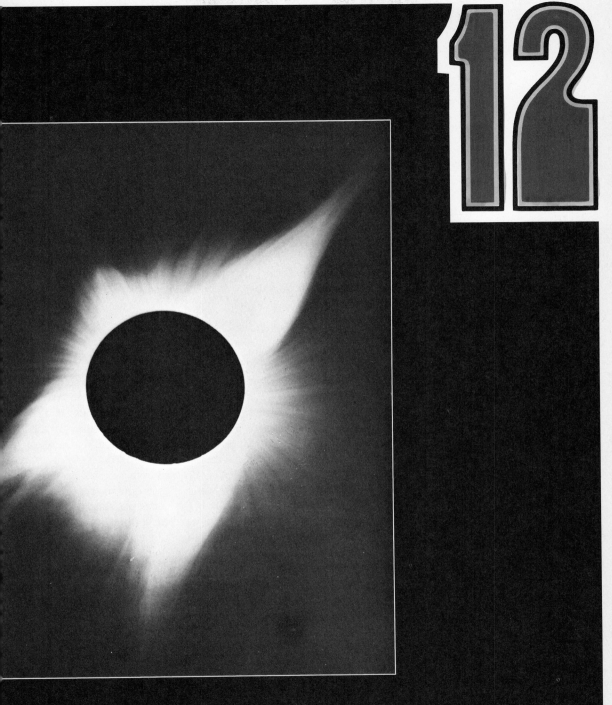

# 12

## states of matter

Matter may exist in any one of four states: solid, liquid, gas, or plasma. We are familiar with three of these states in water: ice (a solid), liquid water, and steam (a gas). An example of a plasma—the fourth state—is that of a fluorescent light in which ions of a gas conduct an electrical current. Any given element has different properties for each state in which it may exist; yet we can draw some similarities for all gases, for all liquids, and for all solids.

GASES—THEIR PROPERTIES

Air is a mixture of gases: 78% nitrogen, 21% oxygen, the remainder of 1% being composed of carbon dioxide ($CO_2$), argon, and traces of a few other elements. If you have used a bicycle pump, you are most likely aware that when you start a downward stroke of the piston, no air enters the tire until about one-half of the stroke is completed (see Figure 12.1). What is happening to the air inside the cylinder during this first half-stroke? It is being compressed into a smaller volume. This is characteristic of most gasses—they are highly compressible. Now if you release the handle of the pump, it will spring upward to its original position. Why? Because all gases exert a pressure on the walls of their container, including the piston. That pressure can be expressed in $lb/in.^2$ (pounds per square inch) or $dynes/cm^2$ (dynes per square centimeter). If we were to build a fire under the pump, the piston would experience a greater force and would try to raise the handle farther. These reactions can be experienced without seeing a single molecule of air. How can we explain what is actually taking place on the submicroscopic level?

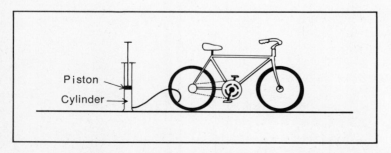

**Figure 12.1.** During the first half-stroke of a bicycle pump, the air in the cylinder is being compressed.

## KINETIC-MOLECULAR THEORY

The theory that seems to do the best job is called the *kinetic-molecular theory:*

1. A gas is composed of molecules that in themselves are too small to be seen.
2. The volume of empty space surrounding these molecules is much larger than the volume of the molecules themselves.
3. The molecules of a gas typically do not exert either electrostatic or magnetic forces on each other but move about independently.
4. The molecules are in motion, and their average kinetic energy is proportional to the temperature of the gas, on the Kelvin scale. The higher the temperature, the faster their motion on the average. Evidence of their rapid motion is the fact that the odor, say of ammonia, is rapidly dispersed throughout a room when a bottle of this substance is opened.
5. The molecules travel in straight lines until they collide; and when they do collide, they may transfer energy, but the net energy of the system of molecules remains unchanged so long as the temperature is held constant. This says that their collisions are perfectly elastic. When molecules bounce off each other or off the walls of the container, no energy is lost.

A good theory should be consistent with observed facts, should not violate know physical laws, and should predict results in experiments that can be performed. Let's test the kinetic-molecular theory in relation to the way gases behave under changes in volume, pressure, and temperature.

## GAS LAWS

To see the relationship between volume, pressure, and temperature of a gas, let's visualize an experiment in four parts. The first part involves a cylinder of gas with a movable piston (see Figure 12.2). A thermometer and a pressure gauge are attached so as to record temperature and pressure, respectively. Selected molecules are shown with arrows (vectors) to indicate their direction of motion and speed. The longer the arrow, the faster the motion. Let's start with the volume, pressure, and temperature as shown in Figure 12.2(a). In Figure 12.2(b) the piston is lowered, reducing the volume to four-fifths the original volume; the temperature is maintained as in (a). Note that the pressure has increased from 20 to 25 dynes/cm². This change from (a) to (b) illustrates Boyle's Law of Gases:

*The pressure* (P) *of a gas at constant temperature* (T) *is inversely proportional to its volume* (V).

$$P_1 V_1 = P_2 V_2 \text{ (where temperature is constant)}$$

We may explain the increase in pressure as due to more molecules hitting each square centimeter of the piston (and the walls).

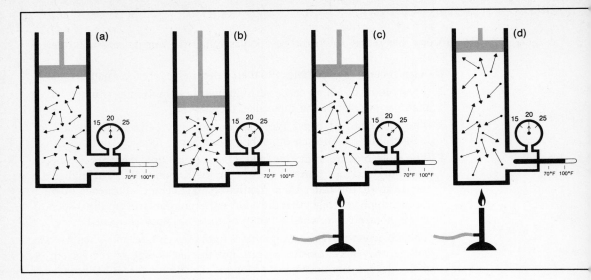

**Figure 12.2.** Gas laws: (a) The basis for comparison in each of the following cases; (b) as volume decreases, pressure increases (temperature remains the same); (c) as temperature increases, pressure increases (volume remains the same); (d) as temperature increases, volume increases (pressure remains the same).

Now compare the third experiment, Figure 12.2(c), with that of the first, Figure 12.2(a). The volume is the same (note the position of the piston), but now the gas is heated, producing higher average velocities in the molecules. This means that the molecules will strike the piston (and the walls) with greater force, hence an increase in pressure. This relationship was expressed by the French physicist Jacques A. Charles (1746-1823) as follows:

*At constant volume, the pressure of a gas is proportional to its temperature (Kelvin).*

$$\frac{P_1}{P_2} = \frac{T_1}{T_2} \text{ (where volume is constant)}$$

The higher the temperature, the higher the pressure in a confined volume.

The fourth experiment, shown in Figure 12.2(d), when compared to Figure 12.2(a), shows that in spite of a higher temperature the pressure may be held constant by increasing the volume (note the higher position of the piston). This suggests a third relationship:

*When the pressure of a gas is held constant, its volume (V) is proportional to its temperature (T), measured on the Kelvin scale.*

$$\frac{V_1}{V_2} = \frac{T_1}{T_2} \text{ (where pressure is constant)}$$

The increase in temperature causes the molecules to hit the walls and piston harder, but the increase in volume reduces the number of hits per square centimeter in a given time. These two factors balance each other, allowing pressure to remain constant.

All three quantities can be combined in one ideal gas law:

$$\frac{P_1 V_1}{T_1} = \frac{P_2 V_2}{T_2}$$

The usefulness of this law can be illustrated by solving a simple problem (see below). Find the final pressure of a gas in which the initial factors are shown along with final volume and temperature as measured.

*Initial factors*  
$P_1 = 300$ dynes/cm$^2$  
$V_1 = 200$ cm$^3$  
$T_1 = 400°$ K  

*Final factors*  
$P_2 = unknown$  
$V_2 = 150$ cm$^3$  
$T_2 = 350°$ K  

$$\frac{300 \text{ dynes/cm}^2 \times 200 \text{ cm}^3}{400° \text{ K}} = \frac{P_2 \times 150 \text{ cm}^3}{350° \text{ K}}$$

$$\frac{300 \text{ dynes/cm}^2 \times 200 \text{ cm}^3 \times 350° \text{ K}}{150 \text{ cm}^3 \times 400° \text{ K}} = P_2$$

$$350 \text{ dynes/cm}^2 = P_2$$

This calculation is based on an assumption that we are dealing with an ideal gas, and in practice a given gas may have slightly different characteristics. However, we have a starting point, and this law together with the statements in the kinetic-molecular theory go a long way toward describing the phenomena of the various experiments. We have already observed that the molecules of a gas show no form or arrangement. However, if we should cool the gas sufficiently, the kinetic energy of the molecules will decrease so as to allow the intermolecular forces to dictate, at least to a limited degree, the arrangement of those molecules. At this point we have a liquid.

## THE LIQUID STATE

One of the most distinctive characteristics of a liquid is its lack of compressibility. One can heat a liquid and a small amount of expansion will result, or cool a liquid and find a small amount of contraction; but most liquids do not respond to pressure changes by changing their volume. This suggests that when a gas condenses to a liquid state, most of the empty spaces between the molecules are filled with molecules: hence the liquid cannot be compressed further. This fact makes liquids very useful in systems such as a hydraulic jack (see Figure 12.3).

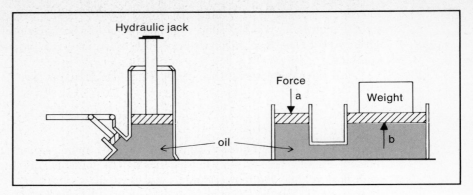

**Figure 12.3.** The hydraulic jack. The force per cm² exerted at "a" is equal to the lifting force per cm² acting at "b."

Suppose a force of 500 newtons (nt) is applied to the piston at (a) as in Figure 12.3, and its area is 10 cm². This amounts to a pressure of 50 nt/cm² (500 nt/10 cm² = 50 nt/cm²). If the cavity is filled with a noncompressible liquid, e.g., oil, then this same pressure will be transmitted to every square centimeter of the container, including every square centimeter of the piston (b). If this piston has an area of 100 cm², then it will experience a force of 5000 nt (100 cm² × 50 nt/cm² = 5000 nt). Thus a weight ten times as heavy as the force exerted can be lifted by such a device.

We can accurately picture the individual molecules of a liquid as being in continual motion. This can be proven by dusting the surface of a liquid very lightly and then side-lighting the surface. The fine dust particles will be seen to vibrate and move about as a direct result of the motions of the invisible molecules that support the dust grains. This is known as Brownian motion, named after a Scottish botanist, Robert Brown (1773-1858), who first performed such an experiment in 1827.

The evaporation of an exposed liquid further demonstrates molecular motion. Some of the molecules attain sufficient velocity so that they are able to escape the surface tension of the liquid and pass into the air above. Of course this represents a loss of kinetic energy on the part of the liquid and normally results in a cooling effect. You have probably experienced this cooling effect when you climb out of a swimming pool and allow the water that clings to your body to evaporate. In the case of a dish of water, left to evaporate, the heat lost is continually replaced from the surroundings, and molecules continue to attain velocities that permit their escape.

While in a liquid state, intermolecular forces are stronger than the forces due to kinetic energy of the molecules, and therefore the liquid "holds" together in a given volume. However, the intermolecular forces are not strong enough to hold the liquid in a rigid form; hence it flows under the influence of gravity. We might picture the flow of a liquid as a continuous rearrangement of intermolecular bonds. Should the liquid be cooled further, however, the kinetic energy of the molecules will be reduced to the point that the intermolecular forces are very dominant and a solid crystalline structure forms.

## THE SOLID STATE

Let's first consider water molecules being cooled to form ice. The very structure of the water molecule (see Figure 12.4) suggests that intermolecular bonds will occur at certain angles. Picture a three-dimensional arrangement that would allow a hydrogen (+) atom of one molecule to be near an oxygen (−) atom of another molecule, and yet no two hydrogens or no two oxygens adjacent to each other, in other words, an arrangement that follows the rules of charged particles—like charges repel and unlike charges attract. Furthermore, in a crystalline solid, the geometric arrangement of molecules should allow for repetition of form in all directions. The crystalline form of water (ice) illustrates these principles very clearly. From one point of view the crystal takes on a hexagonal form, the oxygen

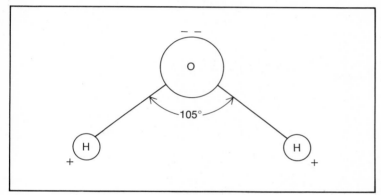

**Figure 12.4.** The polar nature of the water molecule.

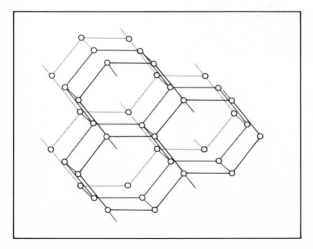

**Figure 12.5.** The crystalline form of ice, as seen from one perspective.

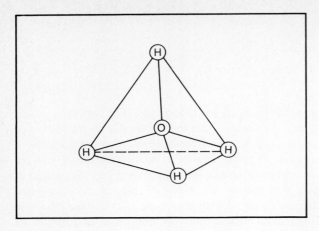

**Figure 12.6.** The crystalline form of ice, as seen from a perspective that reveals its tetrahedron form.

atoms forming the vertices of a six-sided polygon and the hydrogen atoms providing the bonding forces along the sides (see Figure 12.5). From still another view each oxygen atom has four hydrogen atoms as its nearest neighbors, forming a tetrahedron (see Figure 12.6). These solid forms can be repeated side by side in a three-dimensional space. There are only five regular solids in the sense that all

**Figure 12.7.** The Platonic solids: (a) tetrahedron, (b) cube, (c) octahedron, (d) icosahedron, (e) dodecahedron.

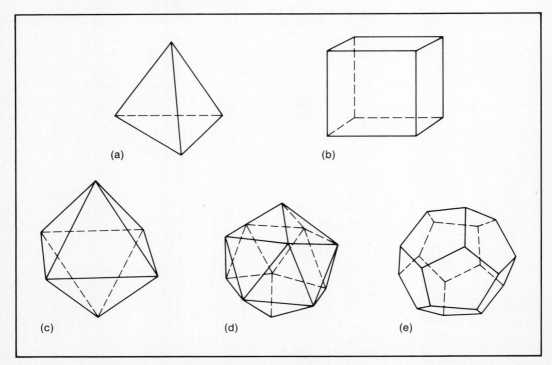

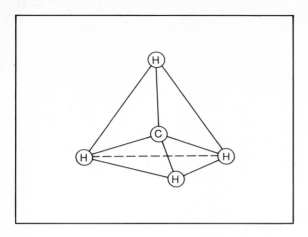

**Figure 12.8.** The methane molecule.

faces are equal-sided and all angles are equal. These solids were discovered by Pythagoras in the sixth century B.C. and received so much attention by the Greek philosopher Plato that they are still called the *Platonic solids*. As illustrated in Figure 12.7, they include the tetrahedron, the cube, the octahedron, the icosahedron, and the dodecahedron. The methane molecule ($CH_4$) illustrates the tetrahedron, placing the hydrogen atoms as far apart as they can be placed in three-dimensional space (see Figure 12.8). Methane is a primary constituent of natural gas and is created by the decay of organic material.

We have already seen a cubic form in the arrangement of sodium chloride pairs (NaCl). Each positive sodium ion has six negative chlorine ions as its nearest neighbors (see Figure 12.9). Similarly, each chlorine ion has six sodium ions for its nearest neighbors. Only the cubic form makes this possible.

One of the hardest crystalline forms is that of the diamond, the result of carbon atoms bonding covalently in such a way that each atom has four other atoms

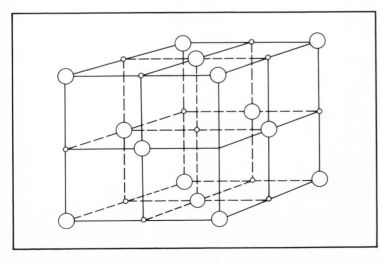

**Figure 12.9.** The cubic form of sodium chloride (NaCl). Atoms are shown proportionately small to show arrangement more clearly.

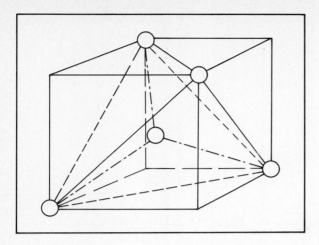

**Figure 12.10.** A body-centered cubic form of carbon—a diamond crystal.

bonded to it in a tetrahedral form, although its form is also described as cubic. Figure 12.10 shows this relationship. Every other corner of the cube contains a carbon atom—a body-centered cubic form. Carbon atoms take on this form only under extreme pressure; hence we would not expect to find natural diamonds forming near the surface of the earth. A more common crystal of carbon is called *graphite,* one of the softer crystals that can be used as a dry lubricant. How can carbon form diamonds that are so hard on one hand and graphite that is so soft on the other hand? The answer to this question lies entirely within the crystalline structure. Diamonds acquire their hardness from their tetrahedral form, whereas graphite is soft because it forms in layers of hexagonal configurations (see Figure 12.11). The bonds holding the sheetlike layers together are rather weak and can be easily separated by a sharp blade.

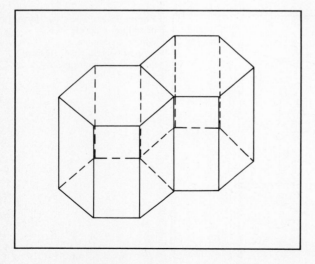

**Figure 12.11.** The sheetlike crystalline form of carbon—graphite.

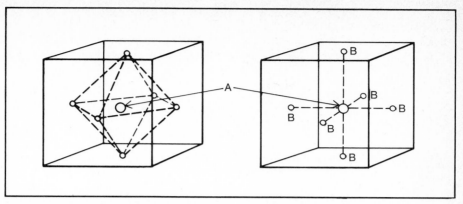

**Figure 12.12.** A variation on the cubic form.

## OTHER CRYSTALLINE FORMS

The octahedron can be inscribed in a cube as shown in Figure 12.12. You can see the equivalency of these two figures; however, the face-centered cube shown in Figure 12.12 (b) is a little easier to visualize. This crystalline form lends itself to the bonding of a central atom (A) with six other atoms spaced so that they are as

**Figure 12.13.** Basic crystalline shapes.

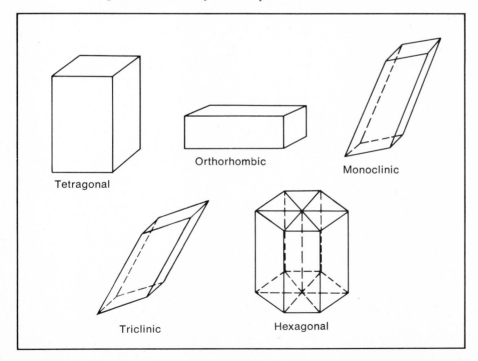

far removed from each other as is possible. Numerous crystals of the form $AB_6$ take on this arrangement.

We will not pursue crystalline structures beyond an illustration of the five basic types (see Figure 12.13), which are simply variations on the themes already presented.

## MOLECULAR MOTION IN SOLIDS

When we think of a liquid being cooled to become a solid, we might be tempted to think that the motions of the molecules cease at the freezing temperature. This is far from true. A solid tends to bind atoms into certain arrangements as we have just seen; however, their motions continue within certain boundaries near the corners or center of forms such as shown in Figure 12.14.

Let's return to the ice crystal to recognize a very important and unusual property. When water freezes, the crystalline structure, shown in Figure 12.5 represents a more open structure of molecules than in the liquid form. This is to say that a given number of water molecules occupy more space as a solid than as a liquid; hence the solid is less dense than the liquid form of water. We see several negative aspects of the expansion of water as it freezes, i.e., the bursting of water pipes exposed to freezing temperatures or the bursting of an engine block that is unprotected by antifreeze. On the other hand, we see a very beneficial aspect in the fact that ice floats in water. Can you imagine the consequences if this were not true? Ice forming on lakes in the winter would sink to the bottom, killing plant and animal life as the entire lake would eventually freeze. However, because of the expanded volume of the ice, it floats on the top of the lake, providing some insulation and thereby moderating the temperature of the water beneath. Plants and fish therefore survive the winter unharmed. Water is unusual in this respect, as most liquids become more dense when assuming a solid state.

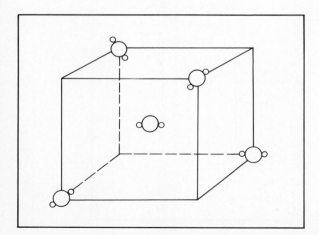

**Figure 12.14.** Molecular motion in a solid.

## CHANGE OF STATE

There are two basic factors that cause a substance to change state: temperature and pressure. In our immediate discussion we will hold pressure constant at one atmosphere (approximately 15 lb/in. or 10 nt/cm$^2$) and note the effect of a change in temperature. Suppose that we start an experiment by placing 1000 cm$^3$ of ice, at $-10°C$, in a beaker. We light a Bunsen burner beneath the beaker and assume that it will supply heat at a steady rate throughout the experiment. We record the temperature change as time goes by and plot the results as a graph. Figure 12.15 shows a typical outcome. Note that the energy supplied by the flame caused a steady increase in temperature to 0°C. Approximately 0.5 calorie of heat is required to raise each gram of ice 1°C. When the temperature of 0°C was reached, the temperature remained unchanged for an interval of time from (b) to (c) (see Figure 12.15). During this time the energy was utilized in breaking the intermolecular bonds of the solid to produce a liquid state, and no increase of temperature was seen. The energy necessary to change the state from solid to liquid is called the *heat of fusion* and amounts to 80 cal/g. As a liquid, at 0°C, continued heating produced increased temperature at the rate of one degree per gram for each calorie supplied until the temperature reached 100°C, the boiling point of water. Again a change of state was imminent, and we see the large amount of energy that is necessary to free the molecules from their intermolecular bonds. This is called the *heat of vaporization* and amounts to 540 cal/g. When all the liquid water has been converted to steam, its temperature could be raised further by continued heating. Although the freezing and boiling points of other elements may be quite different, the basic concepts presented here are similar.

Now let's consider another factor that influences the state of matter—namely, pressure. We know that under one atmosphere of pressure, as experienced at sea level, water boils at 100°C. At this temperature potatoes will cook (by boiling) in one half-hour; but suppose you hiked up a high mountain, you would find that

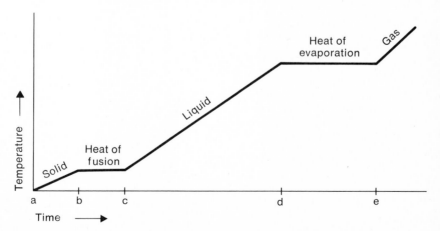

**Figure 12.15.** Graph showing the change of state as heat is supplied to a block of ice.

potatoes would require a full hour or more to cook there. Why? Because the atmospheric pressure is reduced at this higher elevation. The boiling point of water is also reduced to 90°C, and therefore the potatoes take much longer to cook. If you carried this same experiment to Mars, where the atmospheric pressure is almost nil, you would find the water vaporizing at a temperature below the normal freezing point of water (on earth). Thus an ice cube (on Mars) would turn directly into a gas with no intermediate liquid state. This is called *sublimation.* We are familiar with frozen carbon dioxide (dry ice), which sublimates directly from the solid to the gaseous state. The solid $CO_2$ was formed originally by subjecting the gaseous form to very high pressure. So we see that changes in pressure affect both the freezing and the boiling points of substances.

## PLASMAS

The three states of matter we have been discussing—gas, liquid, and solid—are certainly the most obvious states in our everyday experience; however, a fourth state is far more common when we consider the universe as a whole. This fourth state is called a *plasma*—a gas that will conduct electricity. Normally a gas is almost a perfect insulator; however, if that gas is ionized such that the electrons of the atoms are separated from their respective nuclei (protons and neutrons), these ions can conduct electricity. The fluorescent tube we use for lighting is an example of a plasma that glows under the ionizing influence of an electric current. If a magnet is brought near such a glowing tube, the light (glow) will appear distorted to one side of the tube. This illustrates a very important characteristic of a plasma—it can be influenced (controlled) by a magnetic field. Furthermore, a plasma may be used to reflect electromagnetic waves. The ionosphere of the earth consists of plasma layers ranging between 80 and 320 km in elevation. Radio communication is achieved around the curvature of the earth by reflecting the longer radio signals off these ionized (plasma) layers (see Figure 12.16).

One of the most dramatic illustrations of the interaction between a plasma and a magnetic field is that shown by the plumes of the sun's corona during a total

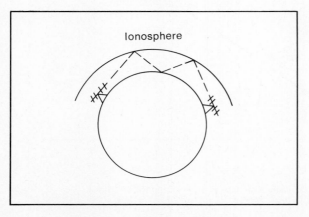

**Figure 12.16.** The earth's ionosphere permits radio communication around the curvature of the earth.

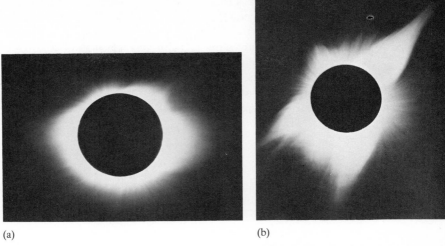

(a)   (b)

**Figure 12.17.** The corona of the sun: (a) at sunspot maximum (Lick Observatory); (b) at sunspot minimum. (Yerkes Observatory)

solar eclipse. Figure 12.17 shows that these plumes are not always formed in the same way. When the sun is very active, having many spots and a particularly strong magnetic field, a rather uniform coronal structure develops, as in Figure 12.17(a). When quiet (few spots), the sun's weaker magnetic field produces only

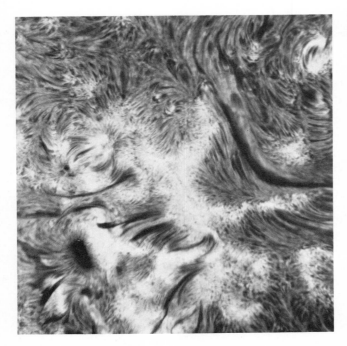

**Figure 12.18.** A strong magnetic field is suggested in this red-light view of the sun. (Hale Observatories)

limited plumes, as in Figure 12.17(b). Even the visible surface of the sun (the photosphere) is a plasma that is controlled by strong magnetic fields, as suggested by the photograph taken in the red light of hydrogen (see Figure 12.18). The sun is a star, and it is safe to say that stars, glowing nebulae between the stars, and the galaxies as a whole are almost exclusively plasmas, that is, ionized gases that possess much higher energy values than normal gases. Only in very limited, cool regions of the universe, like our own planet, does matter exist in other than plasma form. In the next chapter, we will see how this "other than plasma" type material interacts in chemical reactions.

## QUESTIONS

1. When the hydraulic brake system of a car gets air in the lines, the brakes become spongy, whereas the normal liquid provides a solid feel to the brakes. Why is this true?
2. What is the source of the pressure that an enclosed gas exerts on the walls of a container?
3. What can one expect to happen if a closed container of gas is heated?
4. Why should a decrease in the volume of a gas produce an increase in pressure, assuming temperature is held constant?
5. What molecular changes take place when a liquid is cooled to become a solid?
6. How does evaporation demonstrate the fact that molecules are in motion?
7. When water freezes, it occupies a greater volume than as a liquid. Why is this so?
8. What are the possible consequences of the fact that water expands upon freezing?
9. Why are diamonds harder than graphite even though they are both composed of the same element—carbon?
10. True or false: All molecular motion stops when a liquid freezes.
11. What is the maximum temperature that liquid water may attain at sea level before it turns into steam?
12. Why does water boil at a lower temperature in high elevations?
13. True or false: A fluorescent lamp represents a plasma.

# 13

chemical energy

Chemical energy refers to the energy stored in the bonds between atoms, and to a lesser degree to the intermolecular forces that tend to hold molecules in a certain arrangement. We may think of this chemical energy as potential energy, and may compare it to that of a bowling ball perched atop a slide as illustrated in Figure 13.1. In relation to the bottom of the slide, the ball is $h$ units above; hence its potential energy (P.E.) is equal to its weight ($w$) times its height ($h$) (P.E. = $wh$). The ball seeks its lowest possible energy in its neighborhood, and so its present stable position $A$ represents the lowest energy in that neighborhood. However, if someone gave it additional energy by lifting it to position $B$, then it could roll downward, releasing energy and assuming a lower energy state at $C$.

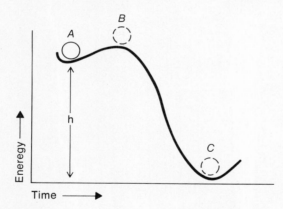

**Figure 13.1.** An analogy of chemical energy: a bowling ball on a slide.

## EXOTHERMIC OR ENDOTHERMIC REACTIONS

The energy stored in the bonds of a molecule may be pictured the same way, as shown in Figure 13.2. Point $A$ on the graph represents the energy of a certain compound that is stable at a given temperature and pressure. However, if the compound is heated, thus raising the energy to $B$, then a reaction may occur that releases chemical energy by rearranging bonds—destroying some and forming others. A very simple illustration of such a reaction is the burning of paper. A piece of paper left at room temperature is quite stable in its own form and does not burn spontaneously. However, if a flame is brought too near, it will supply the

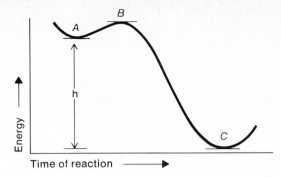

**Figure 13.2.** Chemical energy.

necessary activation energy, causing a reaction (burning) that releases energy. Once a reaction begins, it may provide its own activation energy to continue the reaction. When the change in bonds produces a net release of energy, we speak of it as an *exothermic reaction*. One kilogram of coal, when burned, releases approximately 7000 kcal of energy. Natural gas (methane), when burned, produces new bonds that represent lower energy states than prior burning, and hence release energy in the order of 12,000 kcal/kg. The chemical formula for this reaction is:

$$CH_4 + 2\,O_2 \longrightarrow CO_2 + 2\,H_2O + \text{energy}$$

This is also to suggest that $CO_2$ and $H_2O$ molecules are more stable (have lower energy) than $CH_4$ and $O_2$. Table 13.1 lists numerous reactants that are used in everyday experience as sources of energy. Note, however, that the chemical

**Table 13.1** Heat of combustion

| SUBSTANCE | REACTION | HEAT OF COMBUSTION |
|---|---|---|
| Hydrogen | $2\,H_2 + O_2 \longrightarrow 2\,H_2O + \text{energy}$ | 3,200 kcal/kg |
| Propane | $C_3H_8 + 5\,O_2 \longrightarrow 3\,CO_2 + 4\,H_2O + \text{energy}$ | 12,000 kcal/kg |
| Anthracite coal | $C_{75}H_{24} + 81\,O_2 \longrightarrow 75\,CO_2 + 12\,H_2O + \text{energy}$ | 7000 kcal/kg |
| Domestic oil | $C_{12}H_{24} + 18\,O_2 \longrightarrow 12\,CO_2 + 12\,H_2O + \text{energy}$ | 11,000 kcal/kg |
| Pine wood | $C_6H_{12}O_6 + 6\,O_2 \longrightarrow CO_2 + 6\,H_2O + \text{energy}$ | 5000 kcal/kg |

formulas show only the primary participants in the reactions and neglect many of the elements that are present.

We might logically ask why these substances possess such high chemical energies—How did such bonds form? They must have been formed under circumstances in which they could absorb energy. When a net increase in chemical energy occurs in a reaction, it is called *endothermic*. The sun is a source of energy for such reactions. Through a process of photosynthesis, in the presence of carbon dioxide and water, glucose and oxygen molecules are formed with bond energies higher than those of carbon dioxide and water.

$$6\,CO_2 + 6\,H_2O + \text{energy} \longrightarrow \underset{\text{(glucose)}}{C_6H_{12}O_6} + 6\,O_2$$

The glucose is utilized in the cellulose structure of plants and in the starch or carbohydrates of foods. To recover this energy we burn wood or eat food, gradually oxidizing the carbohydrates to provide body heat and muscle energy.

The decomposition of water into hydrogen and oxygen gas is another endothermic process. Energy in the form of an electric current is supplied to the water, breaking the bonds to allow hydrogen to collect at one terminal and oxygen at the other.

$$2\,H_2O + \text{energy} \longrightarrow 2\,H_2 + O_2$$

The oceans of the earth represent a very large supply of hydrogen. If methods could be devised to break the bonds of the water molecules without the use of fossil fuel sources, perhaps by utilizing solar energy, hydrogen could be collected, compressed into a liquid state, and eventually burned in an "engine." Such a burning is called an *oxidation reaction;* however, the term *oxidation* can be used over a much broader range of application.

## OXIDATION-REDUCTION REACTIONS

The term *oxidation* can be applied to any reaction that involves the loss of electrons. The very nature of oxygen, lacking two electrons in its outer orbital, makes it an electron "grabber." A piece of iron rusts because oxygen, coming in close contact with the iron atom, "grabs" electrons from the iron to form an ionic bond, making iron oxide. There are numerous other nonmetals that can act in a manner similar to oxygen, and therefore we should think of them as oxidizing agents as well. In the following examples, Cl, Br, I, F, S, and N are oxidizing agents, in a sense adding themselves to another element:

$$2\,Fe + 3\,Cl_2 \longrightarrow 2\,FeCl_3$$
$$2\,Fe + 3\,Br_2 \longrightarrow 2\,FeBr_3$$
$$2\,Fe + 3\,I_2 \longrightarrow 2\,FeI_3$$

$$Mg + F_2 \longrightarrow MgF_2$$
$$Cu + S \longrightarrow CuS$$
$$3\,Mg + N_2 \longrightarrow Mg_3N_2$$

By contrast certain reactions remove oxygen or its counterpart from a compound, and this is called *reducing the compound*. For example, when iron ore is burned with charcoal, oxygen is removed from the ore. Charcoal is the reducing agent.

$$\underset{\text{(iron ore)}}{2\,Fe_2O_3} + \underset{\text{(charcoal)}}{3\,C} \longrightarrow \underset{\text{(iron)}}{4\,Fe} + \underset{\text{(carbon dioxide)}}{3\,CO_2}$$

Note, however, that the charcoal becomes oxidized at the same time as the iron is reduced; hence we can usually assume that both oxidation and reduction occur in any reaction. Additional examples of oxidation-reduction reactions follow. Can you pick out the oxidizing agent and the reducing agent in each reaction? (Answers are found at end of chapter.)

1. $CuO + H_2 \longrightarrow Cu + H_2O$
   (Copper is reduced and hydrogen is oxidized.)
2. $Fe_2O_3 + 2\,Al \longrightarrow 2\,Fe + Al_2O_3$
   (Iron is reduced and aluminum is oxidized.)
3. $CuCl_2 + H_2 \longrightarrow Cu + 2\,HCl$
   (Copper is reduced and hydrogen is oxidized.)

## BALANCING EQUATIONS

In the three chemical equations listed above note that the number of each kind of atom on the left side of the equation always equals the number of that kind of atom on the right side. This is a statement that atoms are neither created nor destroyed by a chemical reaction—they are merely rearranged. Can you see that the following equation is out of balance? What coefficients are needed to make it balance?

$$?\,Na + ?\,Cl_2 \longrightarrow ?\,NaCl$$

Since there are two chlorines on the left, we must expect two chlorines on the right; hence:

$$?\,Na + Cl_2 \longrightarrow 2\,NaCl$$

Now there are two sodiums (Na) on the right; therefore we need two on the left and we have a balanced equation.

$$2\,Na + Cl_2 \longrightarrow 2\,NaCl$$

Try to balance each of the following: (Check your answers with those found at end of chapter.)

1. $?\ H_2O_2 \longrightarrow ?\ H_2O + ?\ O_2$
2. $?\ CuO + ?\ H_2 \longrightarrow ?\ Cu + ?\ H_2O$
3. $?\ H_2 + ?\ CO \longrightarrow ?\ CH_3OH$
4. $?\ KClO_3 \longrightarrow ?\ KCl + ?\ O_2$
5. $?\ Al + ?\ NaOH + ?\ H_2O \longrightarrow ?\ NaAlO_2 + ?\ H_2$
6. $?\ C_2H_5OH + ?\ O_2 \longrightarrow ?\ H_2O + ?\ CO_2$

Example #6 predicts that one molecule of alchohol ($CH_5OH$) will combine with three molecules of oxygen to produce three molecules of water and two molecules of carbon dioxide. However, it is not convenient to handle single molecules of a given substance; so chemists have defined a convenient unit that maintains the proportions in a given chemical equation.

## GRAM-FORMULA-WEIGHT

The unit is called a *mole*, which stands for the *gram-formula-weight* of any compound. The experimental evidence of the Italian scientist Amadeo Avogardro (1776–1856) (and others) indicates that the approximate weight of $6.0238 \times 10^{23}$ atoms (or molecules) of any element (or compound) has the same numerical value in grams as their atomic weight (or formula weight). For example, the atomic weight of carbon 12 is 12; hence, Avogardro's work suggests that $6.0238 \times 10^{23}$ carbon 12 atoms will weigh 12 grams. Likewise, 1 gram of hydrogen contains Avogadro's number of atoms because each hydrogen atom is approximately one-twelfth the mass of the carbon 12 atom (see Figure 13.3).

When several atoms are bonded to form a molecule, like that of alcohol ($C_2H_5OH$), we must take the sum of all the atomic weights to achieve the weight in grams of $6.0238 \times 10^{23}$ molecules.

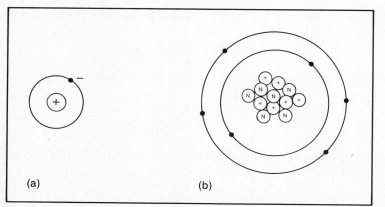

**Figure 13.3.** A model of (a) the hydrogen atom, (b) the carbon atom.

$$\begin{array}{cccc} C_2 & H_5 & O & H \\ 2\,(12) + 5\,(1) & + 1\,(16) & + 1\,(1) & = 46 \end{array}$$

We see that 46 g of $C_2H_5OH$ will contain Avogadro's number of molecules and will represent 1 mole of alcohol. Consider again #6 from page 226.

$$C_2H_5OH + 3\,O_2 \longrightarrow 3\,H_2O + 2\,CO_2$$
$$46\text{ g} + 3\,(32\text{ g}) = 3\,(2 + 16)\text{ g} + 2\,(12 + 32)\text{ g}$$
$$46\text{ g} + 96\text{ g} = 54\text{ g} + 88\text{ g}$$
$$142\text{ g} = 142\text{ g}$$

This says that we may expect 46 g of alchohol and 96 g of oxygen to react to produce 54 g of water and 88 g of carbon dioxide. You can see that this still preserves the proper ratios of reactants and products but allows us to use very practical units of each. Of course, one could take any desired fractional amount of each quantity and still expect a complete and balanced reaction because the proportionate numbers of atoms (and/or molecules) are involved.

Now we turn our attention to a fundamental division among compounds—that of acids and bases.

## ACIDS AND BASES

Perhaps we would most easily recognize an acid or a base if we tasted such a solution, although this is not a recommended procedure for unknowns. The sour taste of a lemon is characteristic of an acid. Chemists would explain that taste and define an acid by recognizing the presence of hydrogen-plus ions in the solution. Where do such ions come from? When certain hydrogen compounds, like HCl (hydrogen chloride gas), are dissolved in water, the following reaction takes place:

$$HCl + H_2O \rightleftarrows H_3O^+ + Cl^-$$

The extra hydrogen ion that is now attached to the water molecule (producing $H_3O^+$) came from the hydrogen chloride molecule. The HCl is a proton donor, hence an acid. We could express such a reaction more simply by omitting the water in the equation; for example:

$$HCl \rightleftarrows H^+ + Cl^-$$

This form emphasizes the fact that a proton ($H^+$) has been donated by the hydrogen chloride, and it is this positive ion that gives an acid its characteristic properties. It should not be inferred that the hydrogen ion exists as an entity itself but rather in association with a water molecule. As such, it is called a *hydronium ion* ($H_3O^+$).

**Table 13.2** pH—The strength of an acid

| CONCENTRATION OF H⁺ IONS | pH | INTERPRETATION |
| --- | --- | --- |
| $10^2$ moles/liter | $-2$ | Very acidic |
| $10^{-1}$ | 1 | Acidic |
| $10^{-4}$ | 4 | Acidic |
| $10^{-6}$ | 6 | Slightly acidic |
| $10^{-7}$ | 7 | Neutral |
| $10^{-8}$ | 8 | Slightly basic |
| $10^{-10}$ | 10 | Basic |
| $10^{-12}$ | 12 | Basic |
| $10^{-15}$ | 15 | Very basic |

The double arrow indicates that the reaction goes in both directions simultaneously. However, only relatively few hydronium ions recombine with the chlorine ions to form HCl; hence only a short reverse arrow is shown. This is characteristic of a strong acid—the reaction moves far to the right. Sulfuric acid is another strong acid that reacts with water to form ions as follows:

$$H_2SO_4 + H_2O \rightleftarrows H_3O^+ + HSO_4^-$$

The hydrogen sulfate ion further dissociates into:

$$HSO_4^- + H_2O \rightleftarrows H_3O^+ + SO_4^{-2}$$

By contrast, acetic acid is relatively weak, and the reaction does not proceed to the right very far, meaning that only a relatively few ions are formed, and hence the short forward arrow.

$$CH_3CO_2H + H_2O \leftrightarrows H_3O^+ + CH_3CO_2^-$$

To quantify the strength of an acid, we define a term called *pH* as being the concentration of $H^+$ ions in a solution (see Table 13.2).
You will note that when the concentration of $H^+$ ions falls to $10^{-7}$ (1 in 10,000,000), the solution is termed *neutral*, and when it falls below that level, like $10^{-8}$, it moves into another concept, that of a base. Let's see how a base is defined.

## BASES

A basic solution may be recognized by a bitter, brackish taste, due to the presence of the hydroxide ion ($OH^-$). Ammonia ($NH_3$), for instance, is such a base; it reacts with water to form a positive and a negative ion.

$$NH_3 + H_2O \rightleftharpoons NH_4 + OH^-$$

This is a weak base and only relatively few ions form. The larger portion of $NH_3$ remains associated.

An example of a very strong base is sodium hydroxide (also known as caustic soda), which reacts with water as follows:

$$NaOh + H_2O \rightleftharpoons Na^+ + OH^-$$

In a strong base, the percentage of $OH^-$ ions formed is high, and again we say that the reaction moves far to the right. A common washing product called *borax* ($Na_2B_4O_7$) when dissolved in water produces a base as follows:

$$Na_2B_4O_7 + 2H_2O \longrightarrow H_2B_4O_7 + 2OH^- + 2Na^+$$

## ACID-BASE REACTIONS

In the example of acids and bases just preceding this, we have seen water act as either a base or an acid. Water serves as a base in the following reaction:

$$HCl + H_2O \rightleftharpoons H_3O^+ + Cl^-$$

Water serves as an acid in this reaction:

$$H_2O + NH_3 \rightleftharpoons NH_4^+ + OH^-$$

Both of these reactions and in fact all acid-base reactions can be summarized in the form:

$$\underset{\text{(acid)}}{HA} + \underset{\text{(base)}}{B} \rightleftharpoons \underset{\text{(new acid)}}{HB^+} + \underset{\text{(new base)}}{A^-}$$

In the following example we react an acid with a base in such a way as to neutralize both acidic and basic natures. The product is a salt NaCl.

$$\underset{\text{(acid)}}{H^+Cl^-} + \underset{\text{(base)}}{Na^+OH^-} \longrightarrow \underset{\text{(water)}}{H_2O} + \underset{\text{(neutral salt)}}{Na^+Cl^-}$$

We are familiar with products on the market called *antiacids* for the relief of an overacidic condition of the stomach. These products represent bases that tend to neutralize an acid, as shown in the following:

**230**
chemical energy

$$HCl + NaHCO_3 \longrightarrow H_2O + NaCl + CO_2$$
(acid)   (bicarbonate of soda)   (water)   (salt)   (gas)

As you can see, the $CO_2$ gas is one of the products of this reaction, and this gas is the source of the "burping" reaction that usually follows the use of an antiacid.

*Answer to questions on page 225.*

| Oxidizing agent | Reducing agent |
|---|---|
| 1. Oxygen | Hydrogen |
| 2. Oxygen | Aluminum |
| 3. Chlorine | Hydrogen |

*Answers to questions on page 226.*

1. $2 H_2O_2 \longrightarrow 2 H_2O + O_2$
2. $CuO + H_2 \longrightarrow Cu + H_2O$
3. $2 H_2 + CO \longrightarrow CH_3OH$
4. $2 KClO_3 \longrightarrow 2 KCl + 3 O_2$
5. $2 Al + 2 NaOH + 2 H_2O \longrightarrow 2 NaAlO_2 + 3 H_2$
6. $C_2H_5OH + 3 O_2 \longrightarrow 3 H_2O + 2 CO_2$

## QUESTIONS

1. What is the difference between chemical energy and kinetic energy?
2. What is the source of much of the energy stored in food stuffs (starches, for instance) and wood products (which can be burned)?
3. What is the difference between an exothermic and an endothermic reaction?
4. In oxidation-reduction reactions, the element that looses oxygen is said to be (a) oxidized or (b) reduced. Which?
5. State the guiding principle(s) used to balance equations.
6. One can expect Avogadro's number of iron atoms ($^{56}_{26}Fe$) to weigh how many grams?
7. A strong acid, when mixed with water, produces a fairly high concentration of what kind of ion?
8. How does a base differ from an acid?
9. When is a solution considered neutral?
10. Explain how an antiacid may relieve an acid stomach.
11. If your swimming pool tested out to have a pH of 10, what would you add to bring it to be more nearly neutral?

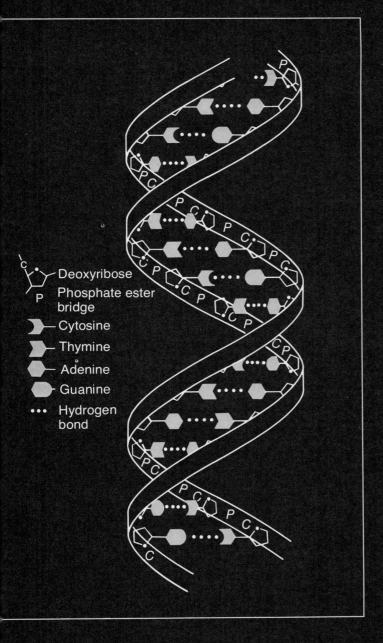

# 14

# chemistry of living organisms

Well over 2 million different compounds are known to exist, and a very large proportion of these molecules contain the element carbon. Furthermore, most of these carbon compounds are associated with living organisms; hence the name *organic* molecules. It was once thought that only living organisms could produce such molecules; however, in 1828, the German chemist Friedrich Wohler (1800-1882) produced urea ($NH_2CONH_2$), an organic molecule, from ammonia cyanate ($NH_4OCN$). Since that time over a million organic molecules have been produced in the laboratory. One of the most significant hopes for the future is the synthesis of viruses. If a given virus can be created in the laboratory, this is synonymous with saying that the chemist understands its chemical structure and may proceed to alter that structure so as to interrupt the reproductive ability of a natural virus, thus destroying it.

## CARBON, THE CENTRAL ELEMENT

A fundamental question of organic chemistry is, Why should carbon be the central element around which so many different compounds are formed? We observe that carbon is a group IV element in the Periodic Table, indicating that it has four electrons in its outer orbital. This fact opens up an almost limitless variety of structural arrangements, primarily through covalent bonding. Consider the three-dimensional structure of the carbon atom, the four valence orbitals being distributed in a tetrahedral form [see Figure 14.1(a)].

Figure 14.1(b) shows how the carbon atom can covalently bond with four hydrogen atoms to form methane ($CH_4$). The hydrogen-carbon bond represents one of the strongest possible bonds that can be formed (second only to the hydrogen-hydrogen bond). The third strongest bond is the carbon-carbon bond, a carbon atom bonding covalently with another carbon atom. Utilizing only these two bonds, C—C and C—H, we can form an almost limitless number of compounds known as *hydrocarbons*.

$$C_2H_6 \qquad H-\underset{\underset{H}{|}}{\overset{\overset{H}{|}}{C}}-\underset{\underset{H}{|}}{\overset{\overset{H}{|}}{C}}-H$$

(ethane)

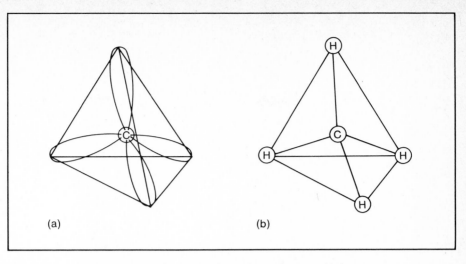

**Figure 14.1.** The carbon atom: (a) showing four valence electron orbitals; (b) showing its covalent bonding with four hydrogen atoms, yielding a methane molecule ($CH_4$).

$$C_3H_8 \qquad \begin{array}{c} H\ H\ H \\ | \ | \ | \\ H-C-C-C-H \\ | \ | \ | \\ H\ H\ H \end{array}$$
(propane)

$$C_4H_{10} \qquad \begin{array}{c} H\ H\ H\ H \\ | \ | \ | \ | \\ H-C-C-C-C-H \\ | \ | \ | \ | \\ H\ H\ H\ H \end{array}$$
(butane)

As shown here, these molecules appear to be flat chains, but remember that in reality the tetrahedral form is repeated, yielding the three-dimensional form shown in Figure 14.1. We encounter the hydrocarbon compound octane ($C_8H_{18}$) every time we buy gasoline, some cars requiring higher percentages of this molecule than others.

(octane)

## chemistry of living organisms

To create a completely different set of hydrocarbons, two carbon atoms may bond doubly (sharing two pairs of electrons), leaving a total of only four electrons to bond with hydrogen. Such a molecule is called *ethene* ($C_2H_4$), and the double bond is shown by a pair of bond lines (see display equation below).

$$\begin{array}{c} H \\ \phantom{H}\diagdown \\ \phantom{HH}C=C \\ \diagup \phantom{CC} \diagdown \\ H \phantom{CCCC} H \end{array}$$
(ethene)

This represents only one of an entire series of molecules that utilize this double bonding. The acetylene gas used by welders is an example of a further variation that utilizes a triple bond (see below).

$$H - C \equiv C - H$$
(acetylene)

To add to the great variety of organic molecules is the possibility that a given number of carbons (and hydrogens) may form in differing arrangements called *isomers*. A few are given below as illustrations of possible forms.

(isomers of hexane)

chemistry of living organisms

These structures are referred to as *isomers of hexane* (6 carbons). Each arrangement (and there are many more) gives the molecule slightly different physical and chemical properties, even though their chemical formula is the same ($C_6H_{14}$).

## CARBON RINGS

Carbon atoms may also bond in combinations that produce a ring, and a simple example of such a structure is represented by benzene ($C_6H_6$), a cleaning fluid.

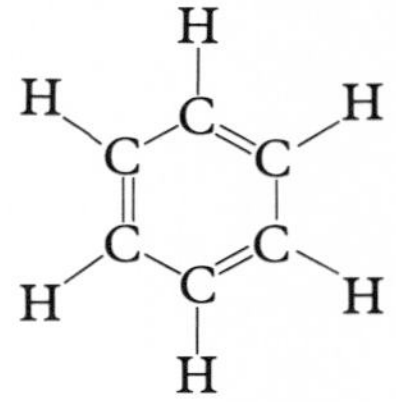

(benzene)

Perhaps one further example of the ring structure will show the virtually endless combinations that are possible. This represents the structure of naphthalene ($C_{10}H_8$).

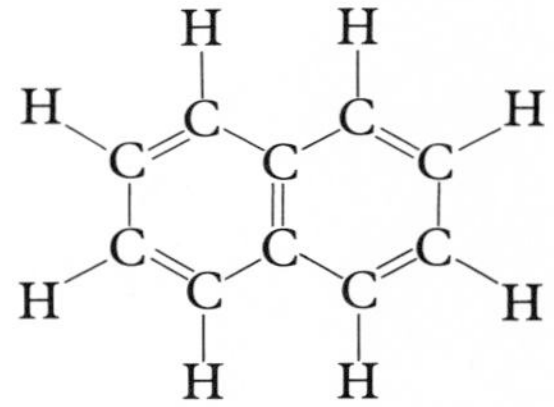

(naphthalene)

You can see that the symmetry of this molecule allows other rings to be added.

Virtually all organic compounds are hydrocarbons or derivatives of hydrocarbons. For instance, if in a methane molecule ($CH_4$) one of the hydrogen atoms is replaced by the hydroxyl radical (OH), we obtain the molecule methyl alcohol ($CH_3OH$). All alcohols represent variations on this theme; for example:

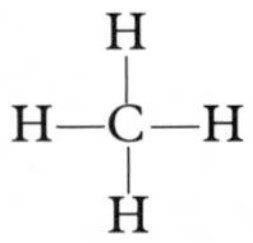

(methane) 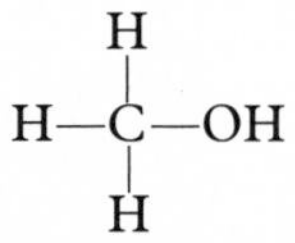    (methol alcohol)

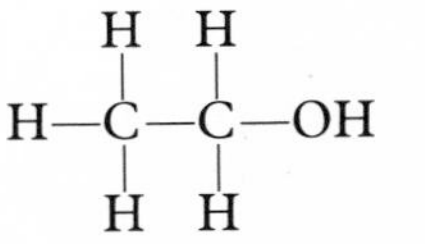

(ethyl alcohol)

## chemistry of living organisms

$$\begin{array}{c} \text{H} \quad \text{H} \quad \text{H} \\ \text{H}-\text{C}-\text{C}-\text{C}-\text{H} \\ \text{H} \quad \text{OH} \quad \text{H} \end{array}$$
(isopropyl alcohol)

$$\begin{array}{c} \text{H} \quad\quad \text{H} \quad\quad \text{H} \\ \text{H}-\text{C}-\quad\text{C}-\quad\text{C}-\text{OH} \\ \text{H} \quad \text{H}-\text{C}-\text{H} \quad \text{H} \\ \text{H} \end{array}$$
(isobutyl alcohol)

In the pages that follow you will see the molecular structure of many familiar substances. Note how the theme of a carbon chain or a carbon ring is played out.

## FOODS

The basic structure of food is that of the hydrocarbon or its derivative. By a process of photosynthesis, plants combine carbon dioxide from the air with water from the soil to produce a carbohydrate and oxygen. This process is activated by sunlight and utilizes chlorophyll as the catalyst to make it happen (see equation below).

$$6\,CO_2 + 6\,H_2O + \text{sunlight} \longrightarrow \underset{\text{(sugar)}}{C_6H_{12}O_6} + 6\,O_2$$

Thus the radiant energy of the sun is converted to chemical energy and is stored in the form of the carbohydrate until utilized by the body to supply energy to muscles or to other body functions.

## CARBOHYDRATES

Carbohydrates are so named because the carbon atom is bonded to hydrogen and oxygen, the latter occurring in the ratio of 2 to 1, as in water ($H_2O$). You will see this ratio in the equation given above, that is, in the structure that represents a sugar ($C_6H_{12}O_6$). Sugars occur in a variety of forms; for instance, glucose and fructose are shown here:

$$\begin{array}{c} \text{H}-\text{C}=\text{O} \\ \text{H}-\text{C}-\text{OH} \\ \text{OH}-\text{C}-\text{H} \\ \text{H}-\text{C}-\text{OH} \\ \text{H}-\text{C}-\text{OH} \\ \text{CH}_2\text{OH} \end{array} \qquad \begin{array}{c} \text{C}-\text{H}_2\text{OH} \\ \text{C}=\text{O} \\ \text{OH}-\text{C}-\text{H} \\ \text{H}-\text{C}-\text{OH} \\ \text{H}-\text{C}-\text{OH} \\ \text{CH}_2\text{OH} \end{array}$$
(glucose) \qquad\qquad (fructose)

These sugars may also occur in cyclic form, as shown below

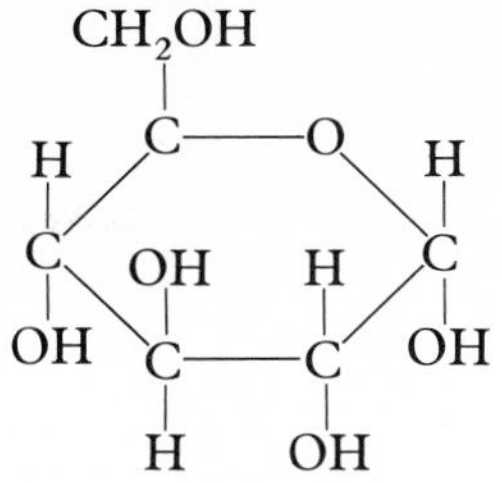

(glucose) 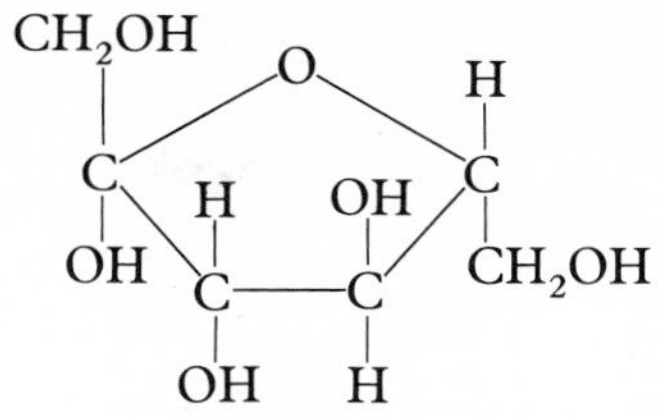   (fructose)

Although the equation for both forms is the same, their physical and chemical properties are different. Glucose is often called *blood sugar*, whereas fructose is called *fruit sugar*. Still another form of sugar is derived from sugar cane or from sugar beets, called *sucrose*, and this type is normally used as table sugar. Its structure is shown below. Note that sucrose is formed from glucose plus fructose.

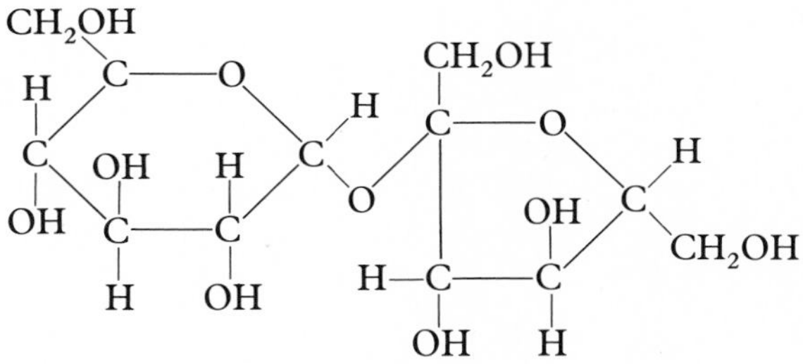

(sucrose)

Also included among the carbohydrates are the starches that a plant holds as its reserve food supply, the glycogens that form a reserve of carbohydrates in animals, and the cellulose found in the cell walls of plants. Cellulose is used to manufacture paper, cellophane, rayon, etc. Cellulose cannot be digested by man due to the lack of the necessary enzymes in his body. On the other hand, the cud-chewing cow (or the like) can digest cellular materials such as straw because this kind of animal has the necessary enzymes. We will define the enzyme more fully later in this chapter.

## LIPIDS (FATTY ACIDS)

Only a slightly different arrangement of carbon, hydrogen, and oxygen atoms give still another class of organic molecules—the lipids. A major subclass of the lipids are the fatty acids, including palmitic, stearic, oleic, linoleic, linolenic, and arachidonic acids. These fatty acids are the basic constituents of some very common foods:

| | |
|---|---|
| Butter | $C_3H_7COOH$ |
| Coconut oil | $C_{11}H_{23}COOH$ |
| Corn oil | $C_{17}H_{31}COOH$ |
| Olive oil | $C_{17}H_{33}COOH$ |
| Liver oil | $C_{19}H_{31}COOH$ |

The lipids are essential to health in that they are present in the membrane of every living cell, although they may also represent a threat to a person, contributing to the problems of excess weight and an overproduction of cholesterol by the body itself. To understand this potential danger, we must first consider the difference between saturated and unsaturated fatty acids. For instance, consider corn oil, which is largely composed of linolenic acid, shown below:

$$H-\underset{H}{\overset{H}{C}}-\underset{H}{\overset{H}{C}}-\overset{H}{C}=\overset{H}{C}-\underset{H}{\overset{H}{C}}-\overset{}{C}=\overset{}{C}-\underset{H}{\overset{H}{C}}-\overset{}{C}=\overset{}{C}\cdots \overset{O}{\overset{\|}{C}}-OH$$
(linolenic acid)

A molecular structure of this type that contains several double bonds (C=C) is called a *polyunsaturated fat*, and this particular fatty acid is a liquid at room temperature. However, the public would often prefer a solid like margarine, which can be spread like butter (see structure below).

$$H-\underset{H}{\overset{H}{C}}-\underset{H}{\overset{H}{C}}-\underset{H}{\overset{H}{C}}-\underset{H}{\overset{H}{C}}-\underset{H}{\overset{H}{C}}-\underset{H}{\overset{H}{C}}-\underset{H}{\overset{H}{C}}-\underset{H}{\overset{H}{C}}-\overset{H}{C}=\overset{}{C}\cdots \overset{O}{\overset{\|}{C}}-OH$$
(a saturated fat)

Therefore the manufacturer simply heats the oil in the presence of hydrogen (and a catalyst), and a structural change occurs in which perhaps two of the double (C=C) bonds are broken with the addition of hydrogen, yielding a more saturated fat. This process is called *hydrogenation*. The melting point is raised in the process, and hence the product is solid at room temperature. The more double (C=C) bonds that are broken, the more the consistency becomes hard like butter.

But what does this have to do with cholesterol? Evidence is pointing to the fact that polyunsaturates tend to inhibit the body's production of cholesterol. Excessive cholesterol tends to deposit plaque in the blood vessels and may be the cause of arteriosclerosis. Therefore the use of polyunsaturated fats, over the saturated variety, seems preferable.

Our understanding of the total effect of certain foods on the body is incomplete; hence students are encouraged to be aware of current research at all times. The following structural diagram will give you some idea of the complexity of the cholesterol molecule. It may be synthesized in the body by a series of 30+ steps in a very short period of time.

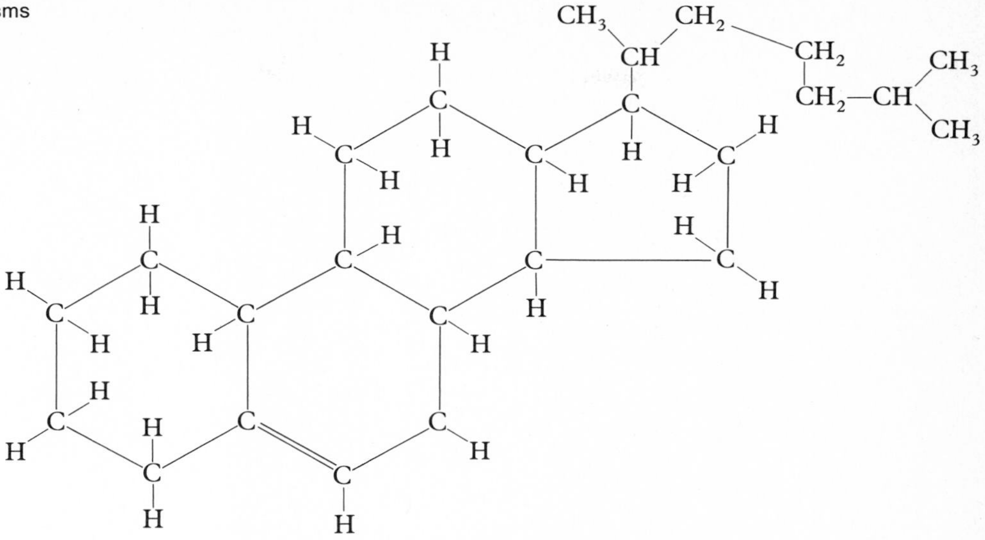

(cholesterol)

## SOAPS AND DETERGENTS

Soaps and detergents are also members of the lipid family; in fact, soap can be easily made from saturated animal fats. In order to understand how soap can loosen dirt and grease, we must recall the concepts of polar and nonpolar molecules, as presented on page 197. Recall that the water molecule is polar in that the oxygen attracts the electrons more strongly than the hydrogens, yielding the oxygen end negative (see Figure 14.2). By contrast, many hydrocarbons, like oil, grease, and gasoline, are nonpolar, as illustrated by the following structure:

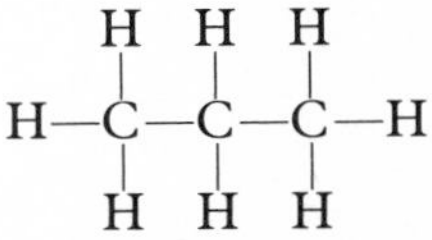

(nonpolar molecule)

Such nonpolar substances will not dissolve in polar solvents. For instance, running water over greasy hands has little effect; but if soap is added to the water, the grease will be dissolved, and with the grease any dirt will be washed away. How can soap serve this function? The chemical structure of soap may be described as a long hydrocarbon chain (nonpolar in nature), which can be thought of as the tail of the molecule, and a polar ($CO_2$) group which can be thought of as the head of the molecule.

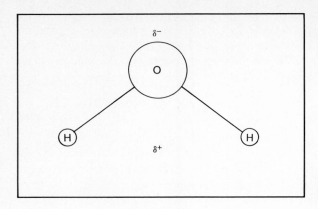

**Figure 14.2.** The polar form of a water molecule.

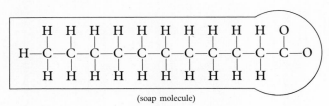

(soap molecule)

Because of its structure, a soap molecule may act as both a polar and a nonpolar molecule. The nonpolar tail dissolves grease and is thereby able to imbed itself in that grease. The polar head, meanwhile, is attracted electrically to the polar water molecules, which flow over the greasy surface. The soap molecules are pulled along by the water, pulling the grease and dirt with them.

As you are well aware, soap often leaves a scum behind, as in the case of the bathtub ring, especially when the water is hard—that is, when the water contains metallic ions like $Mg^{+2}$ or $Fe^{+2}$. The nature of the head of the soap molecule tends to make these ions precipitate—to fall out, even as we think of rain precipitating. The precipitated ions form the scum.

Molecules can be created that resemble the soap molecule but with a slightly different "head"—one that does not make the metallic ions precipitate out and therefore does not form scum. Typical of such a product, called a *detergent*, is the following form:

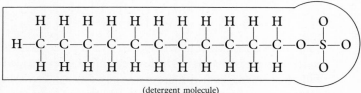

(detergent molecule)

Although detergents serve the cleansing function of a soap, without leaving a scum, they are not as easily disposed of in sewer treatment processes. Normally soaps can be broken down by the microorganisms present in treatment plants, and therefore soap is spoken of as *biodegradable*. However, certain early detergents were not biodegradable, and they created the problem of foaming in the treatment plant, in rivers, and in wells. The addition of certain enzymes to modern detergents has reduced this problem considerably.

## STEROIDS

The lipids include a subclass called the *steroids*, one of which we have already seen—cholesterol. Also included in this group are the sex hormones (estrone, testosterone, progesterone, and androsterone) and certain counterparts in the form of oral contraceptives. Vitamin $D_2$, cortisone, and digitoxin are also steroids.

## PROTEINS

Proteins represent the third major class of organic molecules (the first two were the carbohydrates and lipids). Protein molecules enter into many diversified body functions, including growth and repair, muscle action, nerve-building, fighting disease, etc. Protein molecules are very large, ranging in size from hundreds to thousands of atoms in a single molecule. Five elements compose most proteins: namely, carbon, hydrogen, oxygen, nitrogen, and sulfur. Consider the protein oxyhemoglobin, which carries oxygen in the blood. Its molecular formula, $(C_{783}H_{1166}O_{208}N_{203}S_2FE)_4$, indicates the presence of 9452 atoms in one molecule.

The building blocks of proteins are significantly simpler—they are the amino acids and have a general structure like that shown below:

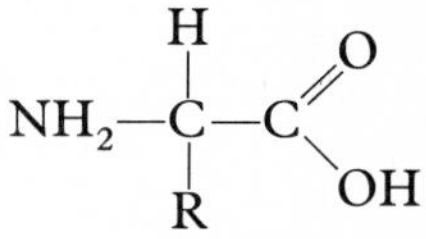

$$NH_2-\underset{R}{\overset{H}{C}}-C\underset{OH}{\overset{O}{\lessgtr}}$$

(amino acid)

They are characterized by the $NH_2$ (amino group) and the COOH (carboxyl group), where R may be simply a hydrogen atom or some side chain. For example, glutamic acid (considered by some to be a memory aid) has the form shown here:

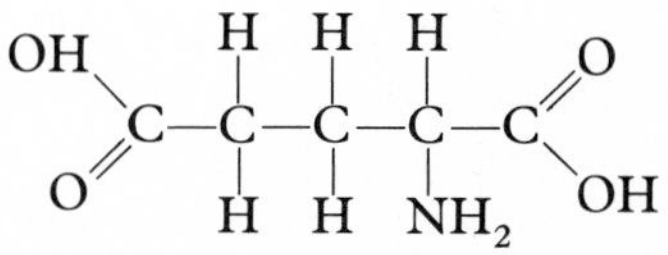

$$\underset{O}{\overset{OH}{\gtrless}}C-\underset{H}{\overset{H}{C}}-\underset{H}{\overset{H}{C}}-\underset{NH_2}{\overset{H}{C}}-C\underset{OH}{\overset{O}{\lessgtr}}$$

(glutamic acid)

Of the 26 amino acids known, some cannot be synthesized by the body; hence they must be present in the diet in the form of high-protein foods. Even though the

**Table 14.1**

| COMMON PROTEINS | SOURCES |
|---|---|
| Albumin | Egg white, milk, and blood |
| Albuminoids | Hair, horn, fingernails, and feathers |
| Chromoproteins | Hemoglobin, hair, and feathers |
| Globulins | Blood and milk |
| Glutelins | Rice and wheat |
| Glycoproteins | Saliva, tendons, and cartilage |
| Histones | Hemoglobin and thymus |
| Lipoproteins | Associated with lecithin and cholesterol, in egg yolks |
| Nucleoproteins | Nuclei of living cells |
| Phosphoproteins | Milk and egg yolks |
| Prolamine | Wheat and corn |
| Protamines | Fish sperm |

number of amino acids is small, an almost limitless variety of proteins are possible. The human body may utilize approximately 100,000 different proteins. Table 14.1 shows some common examples along with typical sources.

To become a given protein, the separate amino acids must be "put together" in an exact manner. Even a slight deviation will result in a different protein. How is it possible that a living cell can faithfully synthesize the proper protein at the proper time—say to make fingernails for the ends of the fingers and hair for the top of the head.

## DNA—THE CODE OF LIFE

In the nucleus of every cell is enough information to direct the production of any necessary protein and ultimately to reproduce an entire organism. This information is carried by a molecule called *DNA* (deoxyribonucleic acid), which may be envisioned as a double helix, as shown in Figure 14.3. When a certain protein is needed, its production may be triggered by an enzyme—a molecule that is in itself largely composed of protein. Enzymes direct most of the functions of living organisms. The enzyme, in effect, opens the double helix to allow a duplication of its message by a messenger RNA (ribonucleic acid) molecule. The RNA molecule may then leave the nucleus, enter the outer part of the cell, and there direct the production of a protein—meaning that it assembles the amino acids in the correct order to make the desired protein.

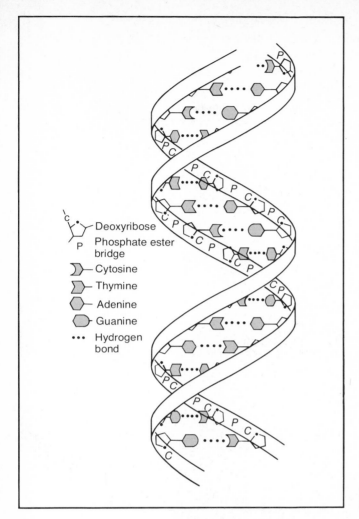

**Figure 14.3.** The DNA double helix molecule.

## QUESTIONS

1. What is the meaning of the word *organic* in relation to chemistry?
2. Why is carbon so well-suited to be the "base" for organic chemistry?
3. Illustrate several forms of bonding using carbon and hydrogen.
4. How is acetylene gas able to release energy when burned?
5. What role does the sun play in converting nonorganic molecules into organic molecules?
6. Two organic molecules may have the same chemical formula (i.e., $C_6H_{12}O_6$), yet have different physical properties (i.e., different melting points). How is this possible?

chemistry of living organisms

7. What role do lipids (fatty acids) play in living organisms?
8. Explain the influence of your diet on the formation of cholesterol. (Additional reading may be necessary.)
9. How can soap or detergent molecules attack grease particles, whereas plain water has little effect?
10. What advantage do detergents have (in general) over soaps?
11. How can only 26 amino acids form so many (over 100,000) proteins?
12. List food sources of protein.
13. What is the role of the enzyme in living cells? (Additional reading may be necessary.)

# 15

the dynamic earth

It is very tempting to think of the earth as permanent—unchanging, for haven't those mountains to the north of my hometown always been there? Hasn't the ocean shore always kept the same boundary? Even as we contemplate an answer to these questions, a muffled rumble is heard and the ground shakes beneath our feet. The earth responds, "No! I am not dead."

## EARTHQUAKES

When an earthquake occurs, do all parts of the earth shake at once? No; an earthquake is a localized event, for surely you have read about earthquakes that you have not felt yourself. We speak of the epicenter of an earthquake as being the location from which the disturbance seems to radiate. Could you find evidence that would tell you what happens when the earth shakes? If you traveled through an area that had recently experienced a severe earthquake, you might find two parts of a roadway displaced from each other or old stream beds having been shifted to one side or the other, requiring new beds to form (see Figure 15.1). These features bear witness to the fact that you are seeing a boundary between two parts of the earth that are moving in opposite directions. If you could fly above such an area, you might see the fault line extended as a faint marking across the landscape (see Figure 15.2).

If two portions of the earth's crust are trying to move in relation to each other, then why don't they move continually and smoothly? Why do they shift abruptly, causing an earthquake? The two portions of crust could be compared to two blocks covered with coarse sandpaper, held tightly together. One piece will not slide over the surface of the other readily, due to friction. However, if enough force in opposite directions is applied, a sudden shifting will occur (see Figure 15.3). If the surfaces of the sanding blocks could be lubricated, say with a thick grease, then they would slip more easily and, hence, could move often and smoothly. Some seismologists (persons who study earthquakes) are proposing the lubrication of faults with water to prevent the sudden shifts that can be so destructive; in other words, this would replace one large earthquake with 10 or 100 small ones. Such a course of action must be pursued with caution, for the initial infusion of water might unleash such a gigantic earth movement as to destroy many

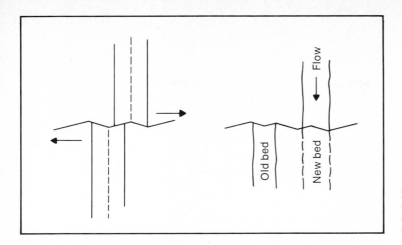

**Figure 15.1.** When severe earthquakes occur, roadbeds and riverbeds may be shifted significantly.

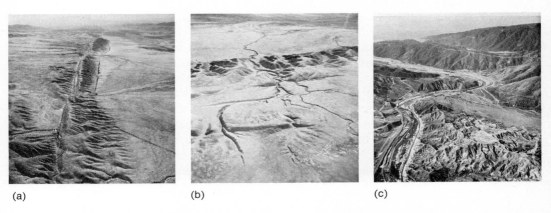

(a)  (b)  (c)

**Figure 15.2.** Three aerial views of the San Andreas fault in California: (a) the Carrizo Plains region, showing offsets in the fault line; (b) multiple fault traces and stream offsets; (c) Cajon Pass, main northeastern access route into the Los Angeles Basin. Valley alignment has been offset by one mile. (R. E. Wallace, U.S. Geological Survey)

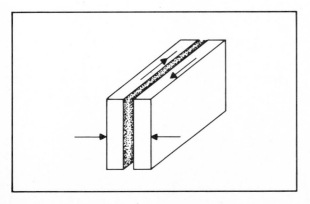

**Figure 15.3.** Sandpaper model of the earth's crustal plates.

cities. Currently attempts are being made to predict large earthquakes by sensing changes in the occurrence of smaller disturbances along a fault.

## SEISMOGRAPH

The instrument used to detect earthquakes is based on the concept of inertia, which we studied in Chapter 3—partially stated:

*An object at rest tends to remain at rest unless acted upon by an outside force...*

Thus, if a massive object is suspended in such a way that it need not move directly with the earth, that object can help us record the earthquake (see Figure 15.4).

When an earthquake occurs, two kinds of waves are generated: a pressure wave (P wave) in which the disturbance is in line with its direction of travel, and a sheer wave (S wave) in which the disturbance is at right angles to the direction of travel (see Figure 15.5).

By placing seismographs at numerous locations on the earth's surface, it is possible to record the nature and exact time of arrival for waves traveling from the epicenter of an earthquake. Such a record is illustrated in Figure 15.6. You may note that several different pulses arrived at times differing by as much as several minutes (assuming that there was only one main shock generated). Can you think of a reason for this delay? We might speculate that earthquake waves travel along the surface of the earth only and that different paths were followed [see Figure 15.7(a)], or that the waves travel through the earth in straight lines and may be reflected [see Figure 15.7(b)]. We might also compare the time of arrival at our

**Figure 15.4.** The seismograph.

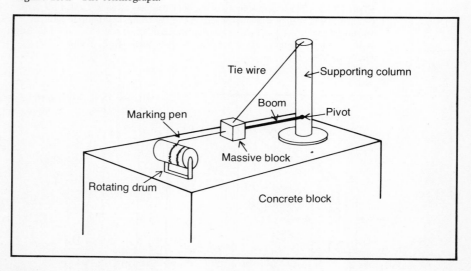

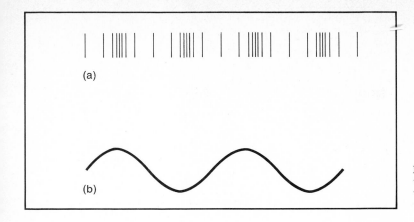

**Figure 15.5.** Earthquake waves: (a) pressure waves (P waves); (b) sheer waves (S waves).

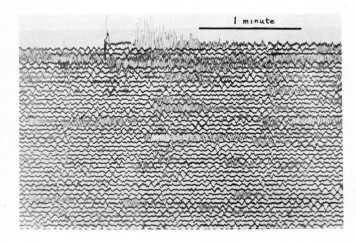

**Figure 15.6.** A seismogram, recorded on the northeast rim of the Kilauea caldera, Hawaii, showing an earthquake swarm. (G. A. Mac Donald, U.S. Geological Survey)

**Figure 15.7.** Possible models of earthquake wave propogation.

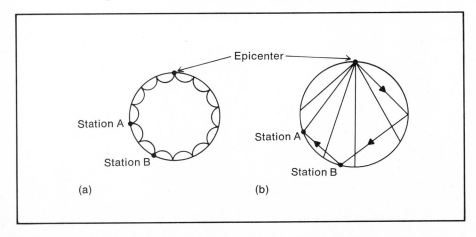

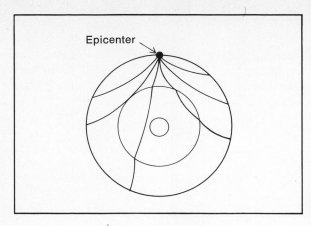

**Figure 15.8.** The earth's interior, as modeled from seismic studies.

station $A$, with the time of arrival at another station $B$, etc. When we do this, neither model seems correct, for the time delays are not proportional to the distance from the epicenter; that is, a station that is twice as far away does not experience twice the time delay. So we look for a better model. Thinking back to earlier experiments with light, we learned that light is bent (refracted) when passing from one material into another material in which its velocity changes; furthermore, the change in velocity often correlates with a change in density, as from air into water. In the case of earthquake waves, they may be bent as they pass through the earth if they encounter changes in density and corresponding changes in velocity. So we try to build up a model that will explain the time delays observed between stations and between different pulses at one station. It is the above procedure that has led to the current model depicted in Figure 15.8.

We see refracted and reflected wave paths that correspond accurately with time delays actually observed. Abrupt changes in velocity and direction of the wave occur at boundaries between the layers indicated (the crust, the mantle, the outer core, and the core) and result from sudden increases in density. Seismologists have observed that sheer waves are not transmitted through the layer labeled *outer core*, and it is this observation that has led to the identification of the outer core as a liquid. Experiments in the laboratory show that liquids will not transmit sheer waves.

Although an attempt, called the Mohole Project, has been made to drill through the earth's crust and thereby sample the mantle, the project was not successful before a combination of political and financial problems brought it to a premature end. Thus, we see that virtually all that is known about the interior of the earth has come from earthquake studies.

## VOLCANOS

Most dramatic of all evidences of a changing earth is the outpouring of molten rock from what seems to be the bowels of the earth itself. However, we must not be trapped into thinking of volcanic activity as having originated in the center of the earth. Volcanos are basically a crustal phenomenon with heat and molten materials surfacing from no deeper than the upper mantle. In fact, the molten materials that are spewed forth may be very similar to surface rock. To see how this conclusion was reached, let's look at the intimate relationship that exists between earthquakes and volcanos. A plot of earthquake centers and volcanos, as shown in Figure 15.9(a), reveals this close physical association.

In looking at this map, would you say that all localities are equally probable sites for earthquakes? No. In fact earthquake sites (and volcano sites) seem to outline 10 or 12 major sections of the earth's crust [see Figure 15.9(b)]. Remembering the illustration using sandpaper blocks, we might guess that these sections (plates) are moving in relation to each other, with adjoining boundaries slipping occasionally—creating an earthquake.

## PLATE TECTONICS

It is now known that large chunks of the earth's crust float on the more dense mantle below, even as ice floats in water, its density being only slightly less than the density of water. If the continental plates have been in motion over a period of millions of years or more, where would we look for evidence of that motion?

Have you ever noticed that the outline of the east coast of North and South America is very similar to the outline of the west coast of Europe and Africa? Perhaps these continental blocks were once very close together—that is to say, that the Atlantic Ocean did not even exist (see Figure 15.10). If the Atlantic opened up, perhaps the movement of the ocean floor has left scars. If we could empty the ocean, it would appear as in Figure 15.11. Sure enough, we see definite evidence of the sea floor spreading. Note the fault lines and the zig-zag appearance of features that were once aligned.

If we want to confirm the original "fit" of the continents, we might look for similarity of rock types, or see if there is a correspondence of fossil records on each shoreline. The commonality of plant and animal fossils that is actually found strongly suggests that the South American continent was once a part of the African continent, with similar connections to the north.

If parts of the earth's surface (called *plates*) are moving apart, it seems reasonable to expect relatively new material to well up within the fractures (cracks) that develop. When the material from the central region of the Atlantic Ocean floor is dated, it is found to be very young, and we will see how this dating is done very shortly. In fact, the upwelling process has built the mid-Atlantic Ridge to a height of approximately 3 km above the surrounding ocean floor (see Figure 15.11). All these observations lead us to the following model for explanation.

There appears to be a gradual increase in temperature as we move to greater depths in the earth, measuring over $1000°C$ at depths of 100 km and $2000°C$ at depths of 5000 km. Certain rocky materials become plastic (semimolten) at $1000°C$, so it is not difficult to imagine convection currents being generated in the upper mantle, which in turn may provide the motive force to spread the continents and supply the young material to fill in the fractures (see Figure 15.12).

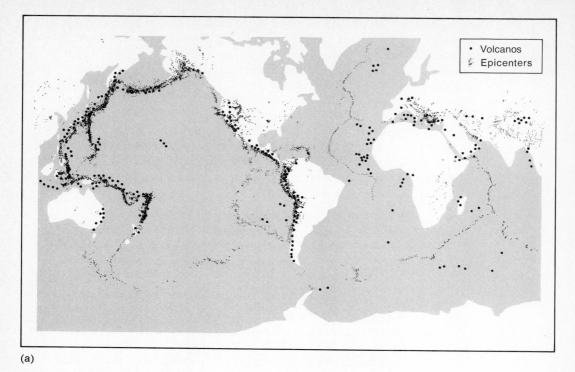

**Figure 15.9.** The distribution of earthquake epicenters and that of volcanos are shown (a) to coincide rather closely. The plot of these features also tend to outline the major plates of the earth (b).

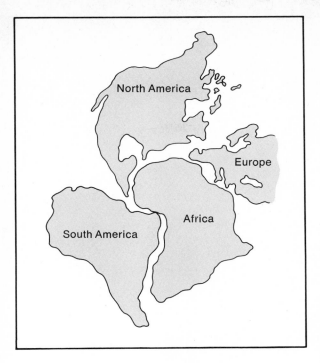

**Figure 15.10.** Millions of years ago, North and South America may have been in very close proximity to the African and European continents.

The rate at which plates are moving varies from about 1 cm/yr to 10 cm/yr, and estimates of the period over which the earth's plates have been moving generally range from 500 million to 2 billion years.

## COLLIDING PLATES AND SUBDUCTION

Obviously if some plates are moving apart, others must be colliding or at least grazing past each other and creating pressures that are periodically released as earthquakes but may also be responsible for other geologic changes as well. We have already noted the incidence of volcanos along plate boundaries, and inspection of Figure 15.13 will reveal that major mountain ranges occur near these boundaries; sometimes deep ocean trenches also develop nearby.

A concept that seems to provide a unified explanation for many of these geological activities is the following: As two plates butt against each other, a bowing upward of the land may result—a mountain-building process (see Figure 15.14). But eventually the pressure may be released by one plate slipping beneath the other, a process called *subduction*, or by other more involved changes. With the release of pressure, block faulting may occur, producing a series of mountain ranges and valleys as shown in Figures 15.15 and 15.16. Furthermore, the subduction process carries surface material downward toward the mantle where it is heated to a molten state. Then, because this molten material is less dense than the sur-

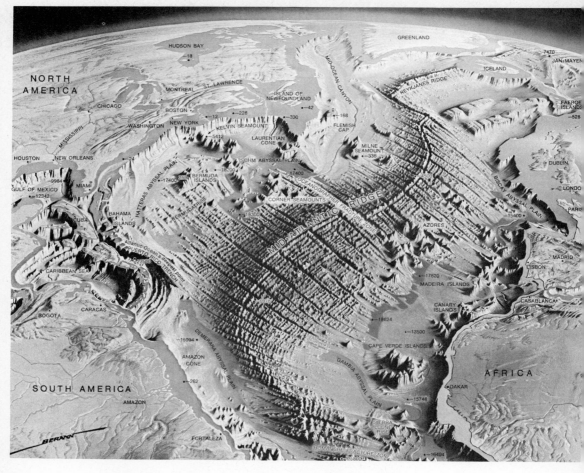

**Figure 15.11.** The Atlantic Ocean floor, showing evidence of the spreading that has occurred. (Alcoa)

**Figure 15.12.** The crustal plates of the earth, shown in cross-section. Convection currents in the mantle or lower crust may be responsible for the spreading of plates.

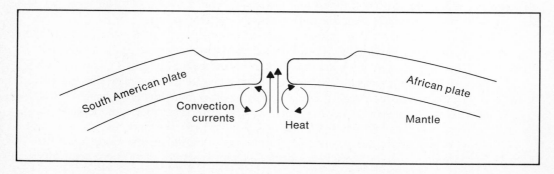

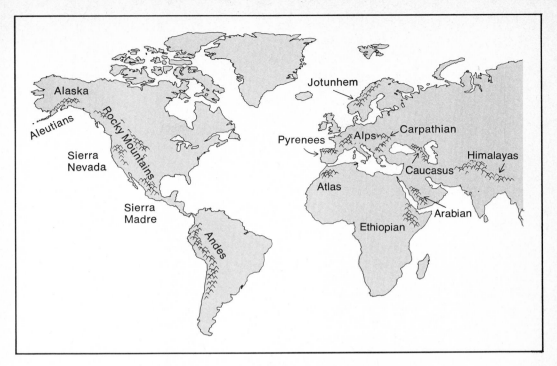

**Figure 15.13.** The borders of crustal plates are often regions of mountain building.

rounding material, it tends to rise (float) through crevices and vents back to the surface as a volcano. Volcanic eruptions represent still another mountain-building process. There are also many processes whereby mountains are torn down. These processes are capable of leveling the entire earth many times over in its long history.

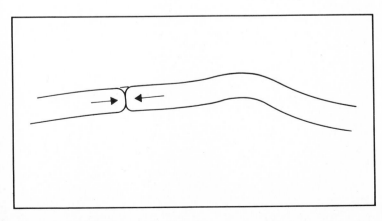

**Figure 15.14.** Pressure from colliding plates provides an uplifting force—a mountain building process.

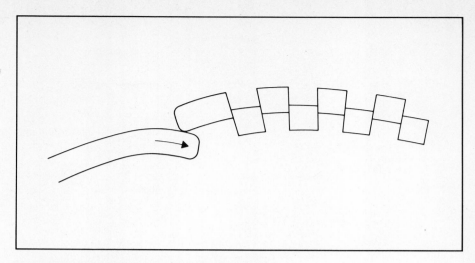

**Figure 15.15.** Block faulting—a mountain forming process.

**Figure 15.16.** This east-west profile through the Owens, Panamint, and Death Valleys clearly shows the effect of block faulting. (*After* a drawing by Ed Haase)

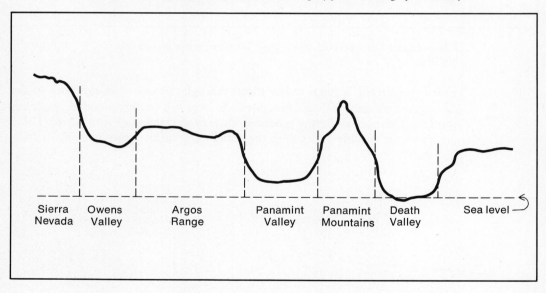

# EROSION

Perhaps the most obvious method whereby material is carried from higher to lower places is by running water. Rain falling on unprotected soil soon cuts small grooves that build into rivulets, streams, and full-fledged rivers. Rivers carry large quantities of mud and soil into the sea. Basically we are seeing the work of gravity, for it is this force that causes the rivers to run downhill.

Often associated with running water is a form of chemical erosion, for mild acids in the water attack the mineral composition of the rocks and tend to weaken their crystalline structure. Any surface exposed to the elements of weather experiences a similar breakdown in structure, making it susceptible to other forms of erosion. For example, a huge monument was brought from arid Egypt to humid New York, only to experience such severe *weathering* that inscriptions that had survived hundreds of years in Egypt were eradicated in only a few years in its new environment.

Glaciers form still another erosion factor. When snow survives a summer season, only to have more snow piled on top, the bottom layers are subjected to so much pressure that they turn to ice. The great mass of ice then experiences the downward force of gravity and tends to move to lower levels, carrying vast quantities of rock and soil with it. A glacier may easily scrape out an entire valley measuring a mile deep and a mile or more wide (see Figure 15.17). The evidence that certain valleys have been formed by this process lies in the typical U-shape as opposed to the V-shape of a valley cut by the action of a river. Glaciers also leave

**Figure 15.17.** U-shaped, glacier-formed valley. Red Mountain Pass, Ouray County, Colorado. (L. C. Huff, U.S. Geological Survey)

**Figure 15.18.** The polish and scratches evident in this rock clearly reveal the passage of a glacier over its surface. (W. C. Alden, U.S. Geological Survey)

telltale scratches on the harder rock surfaces over which they have passed (see Figure 15.18).

Through all these erosion factors, material is broken into finer grains, and much of it is carried into low basins or into the sea where it is deposited to form one of several rock types to be discussed in the next section.

## READING THE HISTORY OF THE EARTH IN THE ROCKS

Can you pick up a rock and tell how it was formed? Not even a highly trained geologist can recognize every given rock type on every occasion; however, certain characteristics may give us clues. Three basic types are shown in Figure 15.19. Which examples seem to suggest that they were created by successive layers of sediments being deposited one upon the other? Examples (a) and (b) show they are the sedimentary type, for you can see the stratified layers. Sometimes it is necessary to see specific examples from a distance before their stratified layers are evident, as in the walls of the Grand Canyon (see Figure 15.20).

Still another class of rock is shown in Figure 15.19(c) and (d). This type is called *igneous* (from the Latin word *igneus,* meaning "fire"). Rocks of the igneous type were once molten and then cooled to their present crystalline form. If they cooled slowly underground, allowing large crystals to grow, they are called *plutonic* (after Pluto, the Greek god of the underworld). If they cooled rapidly because of being cast forth on the surface of the earth, thus allowing only for small crystal growth, they are called *volcanic.* Generally speaking, lava flows are of this latter type, although ash and other byproducts of a volcanic eruption may produce sedimentary rock as well.

The third basic type of rock is called *metamorphic,* meaning rock that has been changed (metamorphosed). Rocks that were once on the surface of the earth and were of sedimentary or volcanic origin are often buried by erosion processes. The accumulation of sediment then exerts a pressure on these rocks, and they may

**Figure 15.19.** Rock types: (a) alternating layers of limestone and shale (sedimentary); (b) sandstone (sedimentary); (c) course grain, typical of granite (igneous, plutonic); (d) columnar joints in basalt (igneous, volcanic); (e) gneiss (metamorphic); (f) slate (metamorphic). [(a) G. K. Gilbert; (b) J. R. Stacy; (c) T. S. Lovering; (d) H. E. Malde; (e) W. R. Hansen; (f) H. E. Malde; U.S. Geological survey]

**Figure 15.20.** The Grand Canyon of the Colorado River. (N. W. Carkhuff, U.S. Geological Survey)

also undergo some heating but not enough to melt them. Under these conditions rocks may be deformed and their original structure changed. Usually the deformation is more evident on a larger scale rather than in a single rock (see Figure 15.19(e) and (f).

Table 15.1 below provides a listing of subcategories under the three main rock types. See how many you already recognize.

We may further define rocks by their mineral content.

**Table 15.1**

| SEDIMENTARY | IGNEOUS | METAMORPHIC |
|---|---|---|
| Shale | *Plutonic* | Marble |
| Sandstone | Granite | Slate |
| Limestone | Diorite | Schist |
| Coal | Gabbro | Gneiss |
|  | *Volcanic* | Quartzite |
|  | Basalt |  |
|  | Andesite |  |
|  | Rhyolite |  |
|  | Tuff |  |

**Figure 15.21.** Cubic form of a sodium chloride crystal.

## MINERALS AND CRYSTAL STRUCTURE

Minerals are substances found in nature that have a definite internal structure—that is, a definite arrangement of atoms. For instance, when ordinary table salt (NaCl) solidifies, crystals that have a cubic shape usually form. The sodium and chlorine atoms have a natural arrangement in which chemical bonds are essentially at right angles to one another (see Figure 15.21).

Quartz ($SiO_2$), if allowed to form its most natural crystalline form, will build plane faces that meet at 120° angles with variations in cross section (see Figure 15.22).

**Figure 15.22.** Quartz crystal. (W. T. Schaller, U.S. Geological Survey)

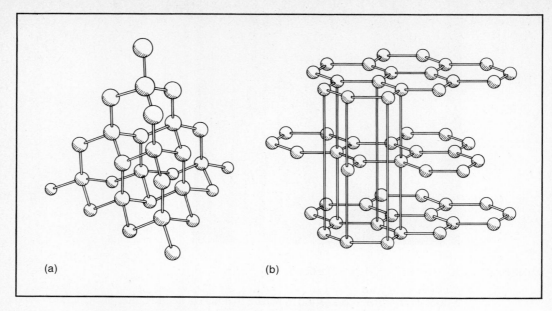

**Figure 15.23.** Crystals: (a) diamond; (b) graphite.

Variations in crystalline structure within the same chemical composition are readily apparent when we compare diamonds with graphite—both of which are composed of carbon. In the form of diamonds, carbon could be called the hardest substance in the world; but in the form of graphite, it is soft and is used as a lubricant. The difference between these two forms can be analyzed in terms of their crystalline structure (see Figure 15.23).

Consider where each of these crystalline forms is found. Diamonds come from materials once buried deep within the earth's crust, and graphite comes from surface material. Does this provide a clue as to how each is formed? Deep within the crust, pressures exist equivalent to a thousand atmospheres or more, and at such

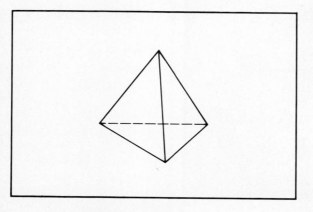

**Figure 15.24.** Tetrahedron.

**Figure 15.25.** The calcite crystal. (W. T. Schaller, U.S. Geological Survey)

pressures atoms are forced into the arrangement of smallest volume—somewhat like packing golf balls into the smallest possible box. You would not place them in layers with cardboard between them, but would allow them to arrange themselves to fill in all available spaces as efficiently as possible. Such an arrangement forms three-sided pyramids (tetrahedrons) rather than layers (planes). Figure 15.24 shows the basic form of the diamond is that of a tetrahedron. The calcite crystals shown in Figure 15.25 reveal by their shape the way in which the atoms that form this mineral bond. We have only touched on a few basic crystalline arrangements.

Elements that are most abundant in the mineral content of crustal rocks are shown in Table 15.2.

You can see that the above elements account for almost 99% of all elements by weight, leaving only 1% for the total weight of all other elements. Certain isotopes of the remaining elements are radioactive (unstable, breaking down into lighter components).

## RADIOACTIVE ELEMENTS

In Chapter 10 we noticed that any given element may occur in several forms, called *isotopes*. For instance, the element uranium has 92 protons, but may vary in its number of neutrons from 141 to 146, thus forming different isotopes. Let us consider one particular isotope of uranium, namely $^{238}_{92}U$ (146 neutrons). This atom is unstable, meaning that it periodically ejects protons, neutrons, and electrons, thus

Table 15.2

| ATOMIC NUMBER | ELEMENT | BY WEIGHT | BY VOLUME |
|---|---|---|---|
| 8 | Oxygen (O) | 46.6% | 91.00% |
| 14 | Silicon (Si) | 27.7 | 0.80 |
| 13 | Aluminum (Al) | 8.1 | 0.80 |
| 26 | Iron (Fe) | 5.0 | 0.70 |
| 20 | Calcium (Ca) | 3.6 | 1.50 |
| 11 | Sodium (Na) | 2.8 | 1.60 |
| 19 | Potassium (K) | 2.6 | 2.10 |
| 12 | Magnesium (Mg) | 2.1 | 0.60 |
| 22 | Titanium (Ti) | 0.4 | 0.03 |
| Total percent of earth's crust | | 98.9 | 99.13 |

reducing itself to lighter elements until it finally becomes a stable lead atom ($^{206}_{82}$Pb). The series of steps through which it disintegrates are shown in Figure 10.14, on page 175.

The fact that invisible "rays" were being emitted by uranium salts was first discovered by the French physicist Henri Becquerel in 1896, when he found that photographic film was exposed through its protective wrapping when the uranium salts were brought nearby. Later investigation revealed that in reality there are three kinds of rays emitted. By placing the radioactive material in a lead

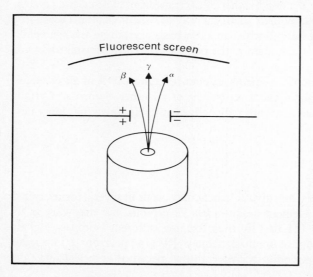

**Figure 15.26.** The detection of radioactive emissions.

container that is capable of capturing emissions, a desired beam may be created. If this beam is passed between charged plates, negative particles will be attracted toward the positive plate, positive particles will be attracted toward the negative plate, and uncharged particles (or radiation) will be unaffected. All three possible results were detected as shown in Figure 15.26.

The particles that reacted as if positively charged were called *alpha* ($\alpha$) particles; these are now believed to have the form of a helium nucleus (that is, two protons and two neutrons). The particles that reacted as if negatively charged were called *beta* ($\beta$) particles and are now thought to be electrons. The form of radiation that is unaffected by the electric field is the *gamma* ($\gamma$) *rays*, and these are now known to be that part of the electromagnetic spectrum that has extremely short wavelengths.

## AGE OF ROCKS

One of the very interesting properties of radioactive elements is the fact that their rate of decay is unaffected by heating, cooling, mixing to form compounds, or by any physical or chemical change. For instance, uranium 238 disintegrates at such a rate that half of any beginning quantity will convert itself into lead 206 in $4\frac{1}{2}$ billion years. This period is called its *half-life*. Lead (206) is spoken of as the *daughter product* because it is produced by the disintegration of uranium (238). Suppose we started with 8 billion atoms of uranium 238 and could watch it decay for $4\frac{1}{2}$ billion years. Only 4 billion atoms of uranium 238 would remain, 4 billion atoms having been converted to lead 206. In another $4\frac{1}{2}$ billion years the 4 billion atoms of uranium 238 would reduce to 2 billion atoms, and now 6 billion atoms of lead 206 would have been formed. During successive half-lives, the quantities of $^{238}$U would reduce to 1 billion atoms, $\frac{1}{2}$ billion atoms, $\frac{1}{4}$ billion atoms, $\frac{1}{8}$ billion atoms, etc., but never be completely exhausted. Of course, no one has ever witnessed one half-life of uranium 238. However, the rate of decay has been observed over many years, and these observations are extrapolated (extended into the future) to yield a half-life of $4\frac{1}{2}$ billion years (see Table 15.3).

**Table 15.3** Radioactive decay

| TIME (IN BILLIONS OF YEARS) | $^{238}$U (IN BILLIONS OF ATOMS) | $^{206}$PB |
|---|---|---|
| 0 | 8 | 0 |
| $4\frac{1}{2}$ | 4 | 4 |
| 9 | 2 | 6 |
| $13\frac{1}{2}$ | 1 | 7 |
| 18 | $\frac{1}{2}$ | $7\frac{1}{2}$ |

Table 15.4

| RADIOACTIVE ISOTOPE | DAUGHTER PRODUCT | TIME |
|---|---|---|
| $^{238}$U | $^{206}$Pb | $4\frac{1}{2}$ billion years |
| $^{40}$K | $^{40}$A | 1 billion years |
| $^{64}$Cu | $^{64}$Zn | 12.8 hours |
| $^{14}$C | $^{14}$N | 5600 years |

Using this knowledge, geologists can assess the age of uranium-bearing rocks by comparing the ratio of $^{238}$U to $^{206}$Pb atoms existing side by side in a given sample. Lead 206 is not the usual isotope of lead found in nature, but comes almost exclusively from uranium decay; hence if a sample has equal numbers of $^{238}$U and $^{206}$Pb, we might assume that rock to be $4\frac{1}{2}$ billion years old. Note that equal amounts occur only at the end of one half-life (see Table 15.3).

Each radioactive isotope has its own half-life, and of those listed you will see quite a wide variation in periods (see Table 15.4).

The age of any given rock sample that bears both $^{238}$U and $^{40}$K could be dated separately by these two elements; thus one is a check for the other. The oldest rocks found exposed on the surface of the earth are about $3\frac{1}{2}$ billion years old; whereas the oldest moon rocks are approximately $4\frac{1}{2}$ billion years old. In fact, the youngest moon rocks returned by Apollo flights date back to $3\frac{1}{2}$ billion years. Hence, we think of the moon as geologically dead since that time; whereas the earth is continually producing young rock as in the mid-Atlantic ridge. The aging of a given rock sample begins only when it is crystallized (solidified); therefore the earth could be as old as the moon and still not give evidence to its full age.

Later we will discuss the unified theory of the solar system and will conclude that the earth, moon, meteorites, sun, and remaining planets are approximately $4\frac{1}{2}$ billion years old. It appears to be a mere coincidence that the age of the solar system coincides with the half-life of uranium.

## DATING ORGANIC MATERIAL

Any radioactive isotope with a relatively long half-life can be used to date rocks, provided the rock contains that element. In the case of $^{14}$C, as shown in Table 15.4, the half-life is far too short to be used in dating the earth. However, $^{14}$C is useful in dating once-living organisms. When a plant or animal is alive, it absorbs $^{14}$C from the earth's atmosphere to a certain level in relation to $^{12}$C, which is the more usual isotope. When the animal dies, $^{14}$C begins to decay to form $^{14}$N, and the ratio of $^{14}C:^{12}C$ begins to diminish. The degree to which it has diminished tells the observer the time since death. For instance, the age of Stonehenge, the large array of boulders arranged on the Salisbury Plain of England, has been estimated by dating a deer antler thought to have been used as a digging tool by the Stonehenge builders.

One further technique can be used to complement those already discussed, in an attempt to tell the history of earth in its rocks. In many sedimentary rocks we find the fossil remains of animals (and plants) that experienced a rather sudden burying when these rocks were forming. If animals had always been the same, say over hundreds of millions of years, they would not form a clue as to when they were buried; however, evidence exists that animals change with time. Animals of certain types have been abundant at a certain time in the past; then have become extinct. The biologists see an orderly development among many species from the simple to more complex forms. This is born out by the fact that when sedimentary layers are essentially undisturbed, the older layers are at lower levels than the younger layers, and fossils appear to be more complex in the upper layers. When certain life forms are dated by radioactivity in surrounding rock, the fossils in turn become age-indicators.

## GEOLOGICAL ERAS

The time scale for the earth that has evolved from all of these methods suggests the following eras (see Table 15.5):

The Grand Canyon of Arizona is a remarkable exposé of these ages, for in its walls we see the history of this area from about 3 billion years ago (see Figure 15.27). Furthermore, the Grand Canyon shows evidence of numerous periods of uplift, faulting, erosion, and sedimentation. Reading its history from the lowest layers shown in Figure 15.27, we see precambrian igneous and metamorphic rock (a), with intrusion (b); precambrian sediments (c), which experienced uplift, leaving them set at an angle, later being eroded to level (d). During the Paleozoic and Mesozoic eras, layers of sediment were laid down as indicated by (e) and (f),

**Table 15.5**

| | |
|---|---|
| Cenozoic | The age of mammals began about 60 million years ago. |
| Mesozoic | The age of reptiles began 230 million years ago. |
| Paleozoic | The age of fishes and amphibians began 600 million years ago. |
| Precambrian | The age of virtually no fossilized remains began 3 billion years ago. |

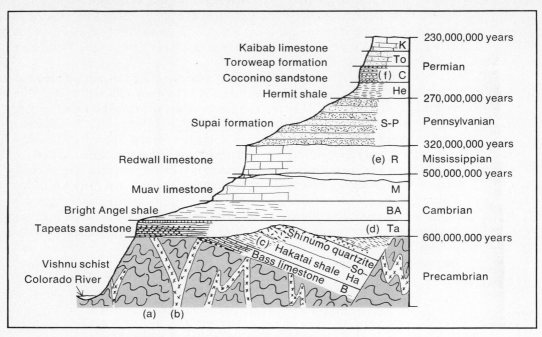

**Figure 15.27.** Time scale of the Grand Canyon.

respectively. A number of these features are shown and identified in the caption of Figure 15.28.

The basic assumptions under which the geologist works and which we have seen working for us in this chapter are summarized here:

*Uniformitarianism:* Processes that act to change the earth today are the same as the processes that have acted in the past. Hence the present is the key to understanding the past.

*Horizontality and Superposition:* Virtually all sediments are deposited in horizontal layers, and if undisturbed, we may assume that the older sedimentary rock lies beneath the younger layers.

*Organic Succession:* Based on our understanding of the changes that have occurred in plants and animals over millions of years and particularly our knowledge of when certain types of life became extinct, it is possible to use the fossil remains, found in sedimentary rock, to date that rock.

Certainly it is an understatement to say that we live on a dynamic earth—one that is ever-changing. A significant aspect of its dynamic character centers around its oceans and its atmosphere, which we will study in the next chapter.

**Figure 15.28.** The Grand Canyon, with formations identified: K, Kaibab limestone; To, Toroweap formation; C, Coconino sandstone; He, Hermit shale; S-P, Supai formation; R, Redwall limestone; M, Muav limestone; BA, Bright Angel shale; Ta, Tapeats sandstone; So, Shinumo quartzite; D, Dox sandstone; Ha, Hakatai shale; B, Bass Limestone. (N. W. Carkhuff, E. D. Mc Kee, U.S. Geological Survey)

## QUESTIONS

1. List the ways in which you could recognize an earthquake fault if you had an opportunity to fly over one.
2. Why are volcanos often associated with regions of high earthquake activity?
3. By what evidence do geologists conclude that the outer core of the earth is liquid?
4. What evidence exists to indicate that the Atlantic Ocean at one time did not exist, but North and South America joined the European and African continents?
5. What are some of the geologic changes that typically occur along the boundary of colliding plates?
6. List the major erosion factors that continually tend to level the mountains.
7. Valleys may be made by river erosion or by glacial movement. What is the difference in the shape of a valley that is created by each of these methods?

8. True or false: Rock that is subjected to high pressure, but not sufficient to melt the rock, is called igneous.
9. True or false: Granite is plutonic (igneous) rock, which means it cooled rapidly.
10. Graphite and diamonds are both carbon crystals. How does their crystalline structure differ and why?
11. Oxygen is the most abundant element in the crustal material of the earth. Why doesn't it intermix with the atmosphere?
12. Carbon 14 is a radioactive isotope. Why can't it be used to date the rocks of the earth?
13. Fossils of plants and animals are usually found in what kind of rock?
14. True or false: The principle of uniformitarianism assumes that the same geological processes occur today as in the past.
15. True or false: Within sedimentary layers the younger material is always on top.

# 16

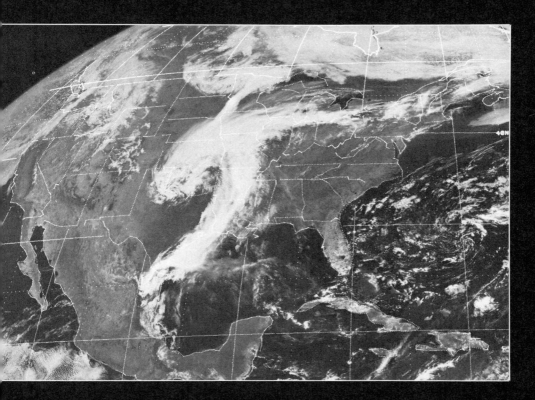

an ocean
of air and water

The earth is unique in the solar system with respect to both its atmospheric composition and its large quantity of surface water. Furthermore, the oceans and the atmosphere are so closely related that it is difficult to know where to begin our discussion. The current of atmosphere (winds) drives the sea currents, but in turn it is the ocean that controls the winds to a large degree.

## EARTH'S ATMOSPHERE

Let us begin by visualizing the earth's atmosphere as a series of layers of air. The boundary of each layer is not as well defined as those we visualized within the earth. Hence, let's think of these layers as merely a convenient way of talking about the atmosphere. The height of the layers may vary from season to season or even from day to day.

The layer that contacts the earth's surface is called the *troposphere*, sometimes referred to as the "sphere of weather," for it is mostly within this layer that weather disturbances (clouds, rain, snow, winds, etc.) occur. We will discuss weather phenomena later in this chapter. The troposphere is about 16 km deep. Above the troposphere lies the *stratosphere*, which extends to approximately 50 km in elevation. Contained within this layer is the ozone layer, which effectively absorbs much of the ultraviolet energy of the sun. It is the ultraviolet radiation of the sun that produces suntans and/or sunburns; thus the ozone layer protectively screens us from an excess of this portion of the sun's radiation. How is this accomplished? The first step in the formation of ozone is the disassociation of ordinary oxygen ($O_2$) gas into atomic oxygen ($O\cdot$). This disassociation is produced by the absorption of an ultraviolet photon, its energy being converted to chemical energy. The atomic oxygen is highly reactive and combines with $O_2$ to form $O_3$. This bonding process releases harmless infrared radiation. Ozone itself continually undergoes disassociation by further absorption of ultraviolet photons, thus preventing this potentially harmful form of radiation from reaching the surface of the earth.

We are currently concerned about the release of certain carbon compounds that contain fluorine and chlorine, commonly used to propel hairspray and deodorants. When these substances reach the ozone layer, they cause the $O_3$ to disassociate into $O_2$ and $O\cdot$, without an accompanying absorption of ultraviolet energy, thus allowing larger amounts of the energy to reach the earth's surface.

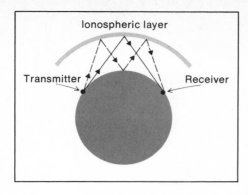

**Figure 16.1.** The ionosphere reflects radio signals longer than 15 m.

The immediate effect is to increase the incidence of skin cancer, and the ultimate effect is to destroy certain living forms (unless artificial protection is provided).

Extending from 50 to 80 km elevation is the *mesosphere* and above that the *thermosphere,* which has an indefinite limit of over 320 km. Within these two layers we will find sublayers composed of ions, atoms in which electrons have been lost or gained (hence, charged atoms). Because of the charged nature of these *ionospheric* layers; they quite effectively reflect radiowaves that are longer than 15 m. When radio communications are desired between points on the earth's surface, points that are separated by 160 km or more, it is impossible to depend on straight-line radiation due to the curvature of the earth (see Figure 16.1). Therefore, one must depend on the ability of the ionospheric layers to reflect radio signals. Sometimes a double or triple bounce off these layers has permitted communications halfway around the earth. Wavelengths shorter than 15 m typically penetrate the layers; however, during times when the sun is very active (as

**Figure 16.2.** Temperature changes within the layers of the earth's atmosphere.

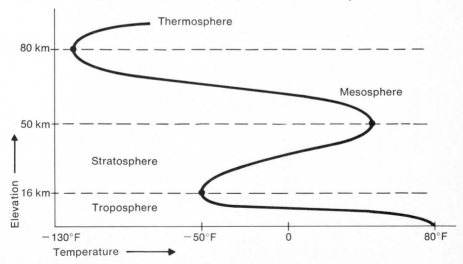

indicated by lots of sunspots), the ionospheric layers may be so well formed as to reflect a 10 m (or slightly shorter) wave. Fortunately for the radio astronomer, these layers do not prevent him from receiving radio energy from stars, nebulae, and galaxies in the order of 1 m wavelength or less.

One of the primary justifications for considering the atmosphere in layered form is seen in Figure 16.2. This is a plot of the average temperature at various elevations.

## A FUNCTION OF TEMPERATURE

It is clear from this graph that the air temperature decreases as we move to higher elevations in the troposphere. For instance, most transcontinental flights by commercial jets reach an elevation of approximately 11 km, and readings of outside temperatures average about $-40°C$ ($-40°F$). At approximately 16 km, the cooling trend is reversed and warming occurs with increased heights in the stratosphere up to about the 50 km level. The trend is once again decreasing temperatures with increasing height in the mesosphere. The thermosphere represents just what its name implies—a sphere of increasing temperature—reaching approximately $315°C$ ($600°F$) at the 320 km elevation. We know that temperature is a measure of the average velocity of molecules, and this is true in this case as well. However, the heating effectiveness of the atmosphere at this great height is vastly limited by its low density. Let's look at the way density changes with height.

At the surface of the earth the average density of the atmosphere is $0.00129$ g/cm$^3$ (about 1/800th as dense as water). As we go higher into the atmosphere, we experience a thinning out of the air ( a decrease in density). By the time we reach the 320 km level, the air density is so low that the heating effect of the air on an object is negligible. The temperature of such an object would be due to its own absorption of the sun's radiant energy.

## ATMOSPHERIC PRESSURE

Rather than dealing with air densities, let's look to a method of measuring a correlated quantity: that of atmospheric pressure. We saw in our study of the earth's interior that gravity is responsible for producing increased pressure (and density) as we move toward the center of the earth. Similarly, each molecule of air experiences an attractive force of gravity toward the center of the earth. This force causes a packing together (increased density) near the surface of the earth and a pressure equivalent to the weight of air above a given area. Consider a single square cm of area at sea level. The weight of air above that square cm is approximately 1,000,000 dynes, and so we say that the atmospheric pressure at sea level is approximately $10^6$ dynes/cm$^2$ (15 lb/in.$^2$). The amazing thing is this: We only need to climb to an elevation of 9 km (the height of Mt. Everest) to find the atmospheric pressure reduced to one-third of normal sea level pressure. Hence, two-thirds of the atmospheric molecules lie below 9 km. This pattern of decreasing

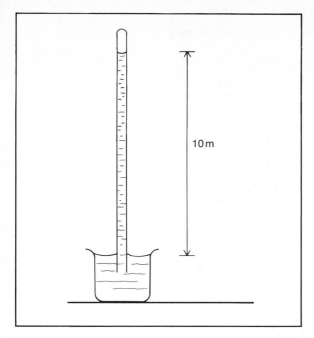

**Figure 16.3.** The barometer.

pressure continues with approximately 90% of the earth's atmosphere lying below 16 km.

In the early seventeenth century the Italian physicist Evangelista Torricelli (1608–1647), devised an experiment that measured atmospheric pressure directly. He inverted a tube, closed at one end and filled with water, in a tub of water without allowing any air to enter the tube. (see Figure 16.3). The water in the tube stood at a certain height above the water in the tub, leaving a vacuum at the top. Torricelli concluded that the water was supported in the tube by the pressure of the atmosphere. By calculating the weight of water in the column and the cross-sectional area of the tube, he was able to determine the pressure exerted by the atmosphere. This could be considered the first barometer. Today we perform the same experiment, except we fill the tube full of mercury that has a density of 13.6 g/cm$^3$ (13.6 times as dense as water). As a consequence, the column of mercury normally stands at a height of 760 mm (approximately 30 in.); when a normal reading is observed, we say that the atmospheric pressure is 1000 millibars. Even at sea level the pressure may vary from time to time; hence 1012 millibars would indicate a pressure slightly higher than normal, and 985 millibars would indicate a pressure slightly lower than normal. Later in this chapter we will see how such variations affect our weather.

If we could carry such a barometer to the height of Mt. Everest, the column of mercury would stand only about one-third as high. However, the mercury barometer is not a convenient form to carry to such a height, so the aneroid barometer was devised. It consists of a thin-walled, sealed can, which is easily deformed by changing outside pressures. Such changes are recorded by the needle and linkage

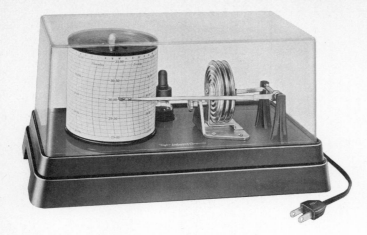

**Figure 16.4.** The aneroid barometer. (Taylor Instruments)

within the can (see Figure 16.4). You can easily see the application of the aneroid barometer as an indicator of altitude in an aircraft. The higher the plane flies, the lower the atmospheric pressure; hence the needle need only be calibrated to read altitude directly. However, due to atmospheric changes, the altimeter of an aircraft must be recalibrated upon each take off, using the known altitude of the airport, and the pilot must also be aware of possible changes while in the air. Therefore he never trusts the altimeter absolutely.

A UNIQUE ATMOSPHERE

The earth has an atmosphere that is unique in the entire solar system, as you will see when we look at the other planets in Chapter 18, for it is the only planet with a nitrogen- and oxygen-rich atmosphere. Nitrogen accounts for slightly over 77% by volume and oxygen 21%, leaving only about 2% for all other components. Argon constitutes almost 1%, and it is thought to derive largely from the radioactive decay of potassium 40. The earth's atmosphere also contains minute percentages of carbon dioxide ($CO_2$) and water vapor ($H_2O$). The $CO_2$ absorbs infrared radiation from the surface of the earth and thus maintains the earth surface at a temperature of approximately 27°C (80°F) above that which would prevail without the $CO_2$ content. This is referred to as the *greenhouse effect*. The percentage of water vapor in the earth's atmosphere varies significantly from season to season and from location to location and depends largely on temperature.

HUMIDITY

If we specified the actual content of water vapor in the earth's atmosphere at a given instant, this would be termed a measure of *absolute humidity*. However, it seems more useful to specify the relative humidity. At a given temperature air is capable of holding just so much water vapor. If we warm the air, it can hold more. If we cool the air, it will hold less. When the actual moisture content is compared to the maximum the air could hold, at a specified temperature, we call this the *relative humidity*. How would you interpret a weather report that stated the temperature at 32°C (90°F) and the relative humidity at 80%? This would mean that the air contains 80% of the maximum water vapor it can hold at 32°C (90°F). In desert regions the relative humidity often falls to 5% or less.

Where does the moisture in the air come from? A major source is the evaporation of oceans and lakes, but significant amounts also come from fields of growing plants and trees. The reason the ocean, lakes, plants, and trees release the moisture is due to heating by the sun. Approximately one-half of the solar energy reaching the earth produces evaporation in these areas.

## SOLAR HEATING

The earth is heated most effectively when the sun's energy strikes the surface of the earth at right angles. Imagine a "bundle" of rays from the sun that measures one square meter in cross section. If those rays strike the earth from a direction perpendicular to the surface, then all the energy will fall on one square meter of surface. However, if that same "bundle" of rays strikes at an angle of less than 90°, the rays will fall on an area greater than one square meter, with the effect of less energy per square meter—meaning a net reduction in heating efficiency (see Figure 16.5). This illustration helps us to realize that a given mass of air will be heated to the degree that the sun is directly overhead (perpendicular to the earth's

**Figure 16.5.** The efficiency of solar heating depends upon the angle at which the sun's rays strike the earth.

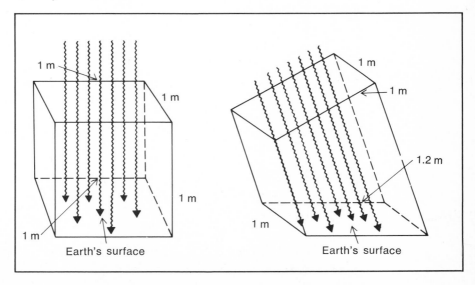

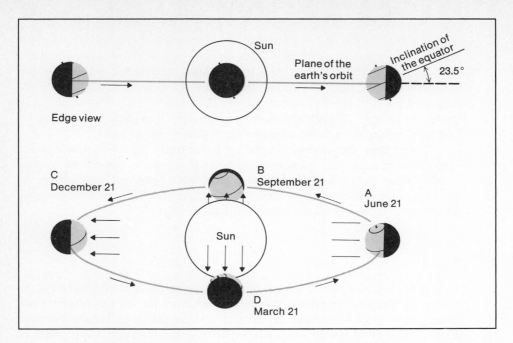

**Figure 16.6.** The seasons, a consequence of the earth's tilt of axis and revolution.

surface). Because the earth is rotating and because its axis of rotation is tilted $23\frac{1}{2}°$ away from the normal, we experience different angles of radiation during the day and during the changing seasons (see Figure 16.6).

Along the equatorial band between the Tropic of Cancer and the Tropic of Capricorn, the earth receives the most efficient heating throughout the year. Air is warmed in this region and thus can hold large percentages of water vapor. Likewise, evaporation of the oceans is abundant, often saturating the air with as much water vapor as it can hold, producing the tropical-type weather associated with these latitudes.

By contrast, the polar regions experience very low efficiency in solar heating due to the large angle between the sun's rays and the perpendicular to the earth's surface. As you can see, the north polar region receives no solar radiation in late December, and the south polar region receives no solar radiation in late June. Obviously, air in these regions would be very cold. The difference in air temperatures from place to place on the earth's surface is responsible for the winds that blow, and in fact is responsible for almost every phenomenon we call weather. Solar energy thus controls the flow of air and ultimately the flow of ocean currents.

## CURRENTS OF AIR (PREVAILING WINDS)

Recalling the gas laws from Chapter 12, we know that if a mass of air is heated, it will tend to expand. Since the same mass now occupies greater space, the density

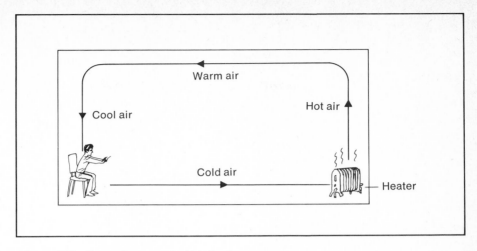

**Figure 16.7.** Convection currents aid in heating a room.

has decreased, and like ice floating on water, the less dense air tends to float (rise) upward on the more dense (cooler) air around it. This leaves a low pressure (void) into which the cooler surrounding air will tend to move. Horizontal movement of air is called a *wind*. A circulation pattern has been established as in a room that has a heater only on one side (see Figure 16.7). You feel the heat on the other side only because of the convection current thus established, and before the warm air reaches you, it has already cooled significantly. You can see that the equator is a region of hot air tending to rise, and the polar area is a region of cold air that tends to flow in to take the place of rising air. Thus the earth's atmosphere might be expected to be two large convection currents circulating as indicated in Figure 16.8.

This would probably be true if the earth were standing still (not rotating); but because of the earth's rotation, the circulation is broken into several subsystems.

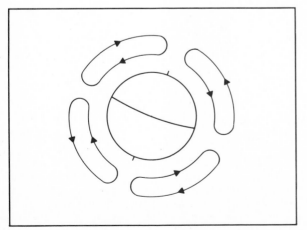

**Figure 16.8.** Convection currents in the earth's atmosphere if the earth were not rotating.

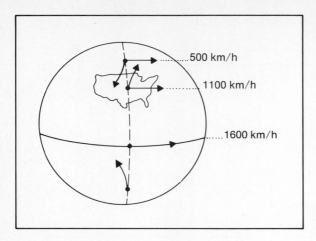

**Figure 16.9.** The Coriolis effect, because of the earth's rotation.

Furthermore, the flow of air is not just north or south but is deviated by what is called the *Coriolis effect*, as explained below. The earth turns as a rigid body, all parts rotating in just under 24 hours. A point on the equator travels at approximately 1000 mph in an eastward direction due to the rotation of the earth, whereas a point in the central United States travels at approximately 700 mph, a point in northern Canada at 300 mph, and a point at the north pole at 0 mph. Suppose a rocket was fired due south from the northern Canadian site. Because its eastward component of motion was only 300 mph, it would seem to lag behind the rotation of the earth as it flew south, for locations within the United States would be moving

**Figure 16.10.** The prevailing winds.

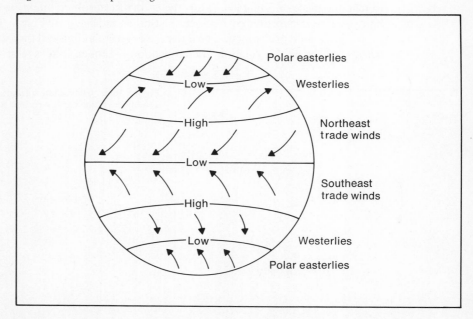

280

faster toward the east than the rocket (see Figure 16.9). If you were flying aboard the rocket, you would say, "I am veering to the right." Now suppose a rocket was fired due north from the central United States. It would have an eastward component of 700 mph, but as it flew northward the land would be moving more slowly. This time the rocket would get ahead of the earth's rotation. A passenger would still say, "I am veering to the right." Note that the opposite would be true in the Southern Hemisphere—rockets headed north or south would veer to their left in relation to the spinning earth beneath.

Applying this to winds, we see the same result. Study Figure 16.10 to see how the Coriolis effect works to produce winds that are either prevailing easterlies (out of the east) or prevailing westerlies (out of the west) at different latitudes. Note that all arrows veer to the right in the north and to the left in the south.

## OCEAN CURRENTS

When wind blows over bodies of water, waves are produced, and through such action currents are established and maintained. By comparing the pattern of wind currents (Figure 16.10) with that of ocean currents (see Figure 16.11), the relationship between the two becomes apparent. The Coriolis effect applies to flows of water as it does to winds. Note the general pattern of clockwise rotation in the Northern Hemisphere, and counterclockwise in the Southern Hemisphere. Among the interesting consequences of ocean circulation is the fact that water warmed by direct solar radiation in the equatorial current just north of South America passes up the east coast of North America and across the Atlantic to warm (to moderate) the climate of the British Isles and to a lesser degree that of Norway. Likewise, we see cold arctic waters flowing southward from the Labrador to cool northeastern Canada.

Typically the oceans form landlocked basins forming closed systems; however, a consistent westward flow is experienced around Antarctica because of the lack of land masses (see Figure 16.12).

## TIDES

Superimposed on the currents we have been discussing is a fluctuation of the ocean level that we call *tides*. Imagine yourself sitting at the seashore when the tide is in (high tide). How long will you have to wait for the tide to go out (low tide), perhaps to expose some tide pools or rocks that you want to explore? The answer is a little more than 6 hours. At this rate we could experience two high tides and two low tides in almost any 24-hour period. But what causes tides and why do high tides occur twice daily rather than only once daily?

Let us perform another mental experiment, one that you could easily verify if you spent a few evenings by the seashore. On the first night you might observe the moon to be passing high overhead at 8 P.M. and the high tide would follow shortly afterward. On the second night you would observe the moon to pass overhead at

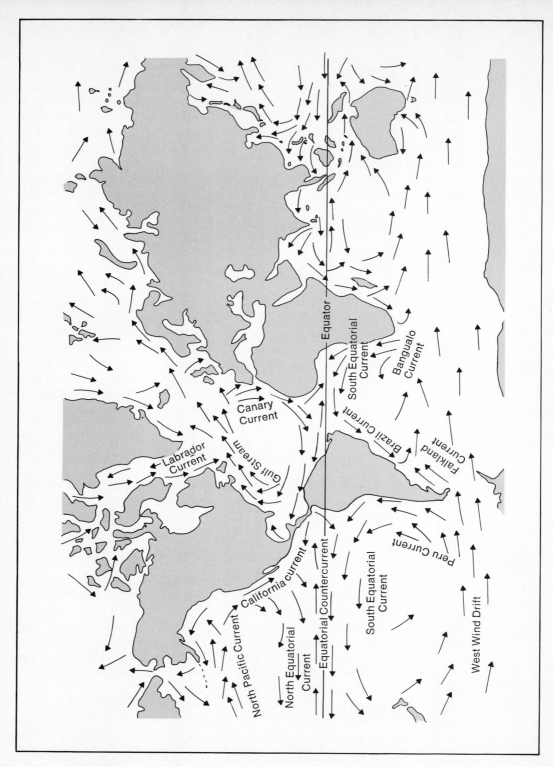

Figure 16.11. Ocean currents.

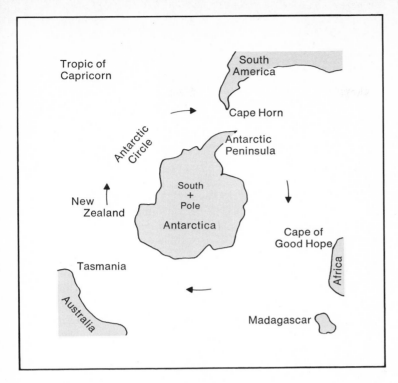

**Figure 16.12.** South polar currents.

8:50 P.M. and the high tide would follow shortly thereafter. On the third night you would observe the moon to pass overhead at 9:40 P.M. and again the high tide would follow soon thereafter. By this time you would clearly observe that the moon arrives on your meridan (overhead) 50 minutes later each night, and, likewise, the high tide occurs 50 minutes later each night. One would surely suspect that the moon causes the tides, but how can a single moon produce two high tides in a day? It would be easy to reason that the moon produces a bulge in the ocean due to gravitational attraction; but if it produced only one bulge, then we would rotate in relation to that bulge only once per day, hence only one high tide per day (see Figure 16.13).

This model simply does not agree with our observation of two high tides per day. So let's see what change would be necessary to bring the model into agree-

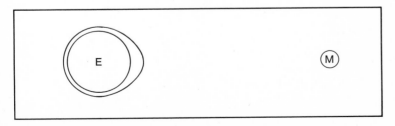

**Figure 16.13.** If the moon created only a single bulge in the earth's oceans, we would only experience one high tide per day.

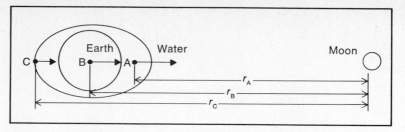

**Figure 16.14.** The moon produces a double bulge in the earth's oceans, hence we experience two high tides per day.

ment with our observations. If there were two bulges on opposite sides of the earth, then the approximate 6-hour interval between high and low tides and between low and high tides would be explained (see Figure 16.14).

To see if this is consistent with the way gravity works, consider a one gram mass at point $A$ (nearest the moon), and a one gram mass at point $B$ (in the center of the earth), and a one gram mass at point $C$ (farthest from the moon). Recalling that the force due to gravity falls off with the square of increasing distance, the force on the gram mass at $A$ is greater than the force on the gram mass at $B$ simply because the moon is closer to $A$ than to $B$. Likewise, the force on the gram mass at $B$ is greater than on the gram mass at $C$. These differing gravitational forces produce differing accelerations on each gram mass, as shown by the vectors (arrows) in Figure 16.14.

The net effect of these various accelerations can be seen by subtracting the acceleration on point $B$ from each of the three accelerations, as indicated in Figure 16.15. These accelerations produce a bulge not only on the side of the earth toward the moon but also on the opposite side of the earth, away from the moon.

The sun also exerts a gravitational force on the earth, but the sun is only about half as effective as the moon in raising tides. Under what conditions would the sun and moon work together to produce unusually high and low tides? [See Figure 16.16(a).]

You can see that when the moon is new or full it aligns most closely with the sun, and the two bodies would assist each other in raising extra high tides called *spring tides*. On the occasion of the first- or third-quarter phases, the moon produces tides in line with itself, the sun produces tides in line with itself, and the net result is smaller high tides called *neap tides*. Of course, the moon exerts the

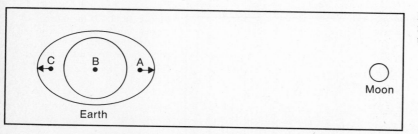

**Figure 16.15.** The net tidal forces that the moon exerts on the earth.

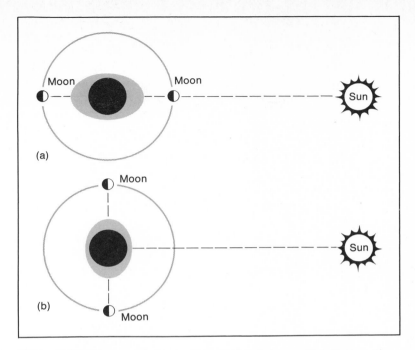

**Figure 16.16.** The sun and moon, each producing their own tidal effects on the earth, sometimes produce very high tides called spring tides (a) and at other times work against each other to produce lesser tides called neap tides (b).

stronger tidal force, and high tides are always produced in line with it. Because the moon's distance from the earth varies between 380,000 and 408,000 km and because the earth's distance from the sun varies between 147,600,000 and 152,400,000 km, what do you think might happen when the moon is closest to the earth and the earth is closest to the sun and they are lined up (new moon or full moon) in almost a straight line? All of these factors would tend to create extremely high tides—tides that have been known to flood some coastal cities and may even trigger earthquakes.

There are other factors that make tides deviate from the ideal model we have been considering, and one such factor is the shape and slope of the particular tidal basin being considered. There are places such as the Bay of Fundy (Nova Scotia) in which tidal changes may exceed 15 m, whereas other locations may experience a change of only 1 m or less.

## THE EFFECT OF TIDES

It would be quite accurate to say that the earth rotates in relation to its doubly bulged oceans and in so doing it experiences friction that tends to slow its rotation. As a result, the length of a day is increasing 0.002 sec per century. Although this seems to be an insignificant change, we need only look back in time 500 million years to find the day only 21 hours long. This effect has been confirmed by the rate of daily growth rings in South Seas coral.

Tides are effective even on planets and/or moons that do not possess oceans, for bulges are also raised on the surface of such a body. The earth has raised "land" tides on the moon from the time of the moon's formation. The effect of such tidal action has been to synchronize the rotation of the moon (its spinning) with its revolution around the earth, each in a period of approximately $27\frac{1}{3}$ days. If we suppose that the moon was once spinning faster, then the effect of tides raised there was to slow its rotation. If we suppose that the moon's spinning was slower than its revolution, then the effect of tides would have tended to increase its rate of rotation, for the bulges would have been carried around the moon (by the earth) at a rate faster than its rotation. In either case, the tidal effect is to make the rotation period and revolution period equal. Evidence that this has been accomplished is the fact that the moon always "points" the same face toward the earth.

**Figure 16.17.** The floor of the Atlantic Ocean. (Alcoa)

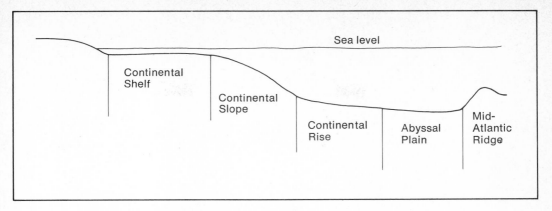

**Figure 16.18.** The profile of the ocean bottom.

## GEOLOGY OF THE OCEAN FLOOR

It is evident from Figure 16.17 that the bottom of the Atlantic Ocean is not featureless, for the Mid-Atlantic Ridge could be compared in height to significant mountain ranges above sea level. A cross section of the Atlantic Ocean floor would look something like that shown in Figure 16.18. Figure 16.17 also reveals the continental shelf as an extension of the continental land masses, and water over these areas is usually less than 30 m deep.

To see the full range of geological features in the ocean floor, we must also include the Pacific Ocean, for here we see trenches (resulting from subduction, which we studied in Chapter 15) extending downward to depths of 11 km below sea level. The highest mountains fall short of this dimension. Seamounts are another feature of the ocean floor. These are usually sharply pointed mountains rising from the ocean floor that do not reach the surface. They are thought to be of volcanic origin. Although they are not called seamounts, many of the island chains of the Pacific have very similar form (see Figure 16.19).

## WEATHER

The weather at a given time and place results from a rather complex interaction among many factors. Furthermore, weather analysis and prediction require very specialized training and experience; hence our effort will be directed toward an understanding of just a few basic principles. Let us enumerate several principles that we have already seen in this chapter.

1. Heating of air by solar radiation causes expansion and a reduction in density and pressure. Warm (less dense) air tends to be buoyed up by cooler (more dense) air.
2. Air tends to move from regions of higher pressure to those of lower pressure, thus producing the winds (convection currents).

**Figure 16.19.** The Hawaiian Islands: Sea mounts rising from the depths of the Pacific Ocean. (National Geographic)

3. Warm air will hold more moisture than cold air.
4. Moisture becomes available to air masses through evaporation of oceans, lakes, and plants as a result of solar energy.
5. As one moves up through the troposphere, there is a decrease in temperature.

In these few basic principles lies the explanation for clouds, rain, and snow. Typically, as warm moist air rises, it experiences cooling due to expansion and due to the lower temperature at higher elevations. As the air cools, it no longer can hold as much moisture (it becomes saturated), so droplets condense to make a cloud. As long as these droplets are still very small, they may remain as a cloud supported by the upward flow of air just described. It is interesting to note that experiments that should lead to cloud formation have been performed in clean (filtered) air and no clouds formed. Yet when minute smoke particles were introduced in the experimental chamber, clouds readily formed. Do such particles exist in the earth's atmosphere? Yes; they do exist in the form of ash from fires and volcanos, soil particles raised by the wind, and salt particles that are byproducts of the evaporation of oceans. These small particles, averaging only 0.2 micron (0.00002 cm) in diameter, serve as nuclei (centers) around which water vapor may condense.

Normal cloud droplets are so small (in the order of 20 microns) that a million or more such droplets would be needed to form one rain drop; hence all clouds do not produce rain. It appears that raindrops will form only if some larger object is present to act as a nucleus for that formation. One such object is an ice crystal that can form due to the supercooling effect produced in an expanding (rising) air mass. Droplets may stick to these ice crystals and freeze, producing snow flakes. When the snow flakes have reached such proportions that they can no longer be supported by the upward current of air, they descend and usually melt on the way down to fall in the form of rain. At higher elevations or in colder climates, of course, the moisture droplets may fall as snow. If the upward flow of air is able to support the frozen objects for longer periods, hailstones often form by the continued accretion (gathering in) of more and more water that freezes to form layer upon layer of ice.

## CLOUD SEEDING

You can see that if nuclei are so important, both to the formation of a cloud and ultimately to the formation of rain within the cloud, it may be possible to assist the process by artificially introducing such nuclei. Experiments have been conducted in which dry ice (frozen $CO_2$) has effectively been dropped into clouds, resulting in the formation of numerous ice crystals that then serve as nuclei for condensation. Silver iodide compounds may be burned, producing a smoke that contains these elements, and the smoke particles, if introduced into a cloud, will trigger precipitation. It is questionable whether cloud seeding can produce significantly more rain on a large scale, but it may have application through localizing rain where it will do

(a)

(b)

(c)

**Figure 16.20.** Cloud types: (a) cumulus, (b) cumulonimbus, (c) cumulonimbus (flattened top, rain producing). (National Oceanic and Atmospheric Administration)

the most good. However, you can see the potential for difficulty if one farmer robbed another of his rain by making the rain fall over the former's farm.

## CLOUD FORMS

The cloud form most likely to produce rain (as just described) is the cumulus → cumulonimbus sequence [see Figure 16.20(a,b,c)]. This sequence is associated with rising columns of air resulting from the heating of the earth. First cumulus clouds form in their typical flat-bottom shape. The rounded tops continue to blossom upward, as in Figure 16.20(b), until finally we see a leveling out of the tops and sudden release of moisture in the form of rain [see Figure 16.20(c)].

Other cloud forms are usually associated with different elevations as in Figure 16.21. We should also note that fog represents a cloud in which you are immersed.

(a)

(b)

(c)

**Figure 16.21.** Cloud types: (a) stratus (lower clouds); (b) altocumulus lenticularis (lens shaped); (c) cirrus (high clouds). (National Oceanic and Atmospheric Administration)

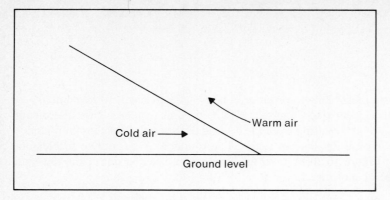

**Figure 16.22.** An advancing cold front.

## WEATHER FRONTS (SEASONAL VARIATION)

Next, let us picture weather on a large scale, for example, that of the United States or any other country of about the same latitude. You will see from Figure 16.10, page 280, that the warm airs from about 30° N latitude flow somewhat toward the north and the cool airs flow somewhat southward from the polar regions. These two movements meet head-on along a line we call a *front*. Sometimes the cold polar air advances southward; this is known as an *advancing polar front* (a cold front) (see Figure 16.22). Then the warm air is forced upward rather quickly; hence it cools rapidly, producing clouds, rain, and thunderstorms. On the other hand, a warm front sometimes advances, pushing the cold air northward and rising slowly to form clouds; but this phenomenon is not usually connected with rain and thunderstorms. These two kinds of fronts (cold and warm) may be correlated with the seasons. In the winter the warming trend is south of the equator, and as a result the cold front advances much farther south (see Figure 16.23).

Note that the cold front does not extend out over the oceans. This shows the great moderating effect of large bodies of water. If the air is colder than the water, the water will give up its heat to the air. If the air is warmer than the water, the

**Figure 16.23.** The cold front: (a) in winter; (b) in summer.

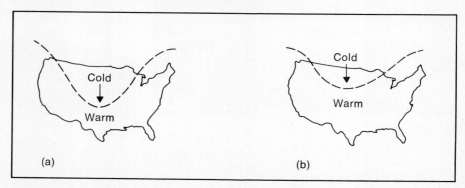

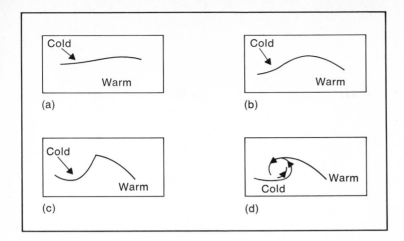

**Figure 16.24.** The evolution of a cyclone.

water will absorb heat, thus cooling the air. You will never see the extremes of temperature near large bodies of water as compared to the temperature inland over large continents.

The boundary between the cold and warmer air along a front does not always remain smooth but may develop wave patterns. First it may only appear as a ripple; but it if continues to develop, it may produce a wave that in a sense "breaks over" like the ocean wave to form a *cyclone*—a counterclockwise wind in the Northern Hemisphere (see Figure 16.24). Cyclonic winds usually center in low-pressure regions and produce a rising column of air that in turn produces clouds and rain. How do weather maps reveal the development of a cyclone? The answer will be given in the next section.

## WEATHER MAPS

A weather map shows many things, like wind direction, wind velocity, temperature, and current weather. It is basically a plot of points of equal pressure. The lines (curves) that connect points of equal pressure are called *isobars*, meaning "same pressure." In Figure 16.25 you will see isobars labeled 1008, 1012, 1016, 1020, etc., where 1000 millibars means normal sea-level pressure. Can you find a front and a developing cyclone in this plot of isobars? (See Figure 16.25.)

The front marks the boundary between the tendency toward lowering pressure from the south and a similar tendency from the north (see Figure 16.25). The place where these two fronts join shows a developing cyclone. A simultaneous photograph from a weather satellite confirms the existence of this developing cyclone. (see Figure 16.26).

Cyclones are also known as *hurricanes* when they occur in the Atlantic or *typhoons* when they occur in the Pacific. Mature hurricanes usually measure several 100 km across and are characterized by very high winds; however, in the center, the eye of the hurricane, it may be very calm and relatively clear.

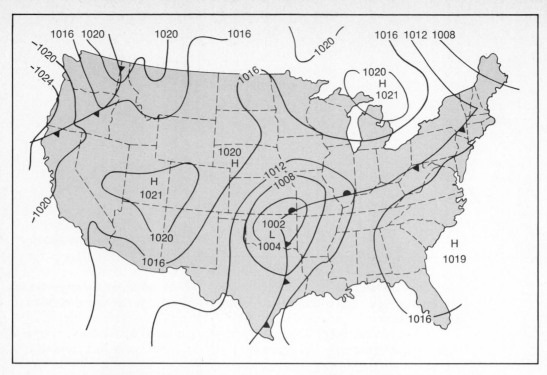

**Figure 16.25.** A weather map, showing a developing cyclone over northern Texas and Oklahoma. (National Oceanic and Atmospheric Administration)

**Figure 16.26.** A satellite photograph, showing a developing cyclone and correlated with the weather map of the same day (see Figure 16.25). (National Oceanic and Atmospheric Administration)

POLLUTION OF THE ATMOSPHERE

We have seen, in our discussion of the atmosphere, that it is quite natural for air to be in motion. It is this motion that normally disperses poisonous and noxious elements. This is accomplished in two ways: (1) by spreading harmful molecules horizontally and thus reducing their level of concentration and (2) by carrying them vertically to dissipate in the upper atmosphere. However, these natural processes do not always occur, particularly in certain geographic locations.

Early explorers naturally chose fertile valleys in which to settle, for there they could produce the necessary foodstuffs. Often such valleys were bounded by mountains on several sides, forming a natural basin of air. As urbanization began to take place in these basin valleys, two factors became evident. First, man, in his industrialization and increased use of fuels, introduced larger and larger quantities of toxic material into the atmosphere. Second, a basinlike valley surrounded by mountains often experiences atmospheric temperature inversion that prevents the normal dispersal of these toxic materials. To understand this inversion process, consider the following facts. Typically, air that is warmed by the land expands and rises into cooler regions, carrying the pollutants with it. However, a thermal inversion may occur overnight when the earth looses heat, leaving a cooler layer in contact with the earth. Because the cooler air is more dense, it tends to stay where it is, becoming increasingly polluted. The mountains prevent the horizontal winds from relieving this situation. Therefore, because we cannot manage the weather very effectively, it seems that our best hope for solving the pollution problem is to attack its sources.

If you live in almost any large city and have stood near a busy intersection, no one needs to tell you that the automobile is a primary contributor to pollutants that irritate the eyes and bronchial tract. Newspapers often carry the account of someone who has ended his life by breathing the exhaust of an automobile, which illustrates just how deadly automobile emissions can be when sufficiently concentrated.

## CARBON MONOXIDE

Gasoline is a rather complex molecule composed of carbon and hydrogen (with additives such as lead to improve its performance in the internal combustion engine). Whenever a carbon compound is burned in the presence of oxygen, carbon monoxide (CO) is formed. The carbon monoxide reacts with the hemoglobin of the blood ($HbFeO_2$) to remove oxygen, thus robbing the blood of its vital function:

$$CO + HbFeO_2 \longrightarrow HbFeCO + O_2$$

## HYDROCARBONS

As a result of incomplete burning (combustion) in the engine, many different hydrocarbon compounds or derivatives are formed, and we will see their role in the formation of photochemical smog shortly. Other sources of hydrocarbons include the transportation, storage, and processing of petroleum products. Certain pesticides contain hydrocarbons that have a long-lasting and injurious effect on animals. And sanitary landfills (dumps) produce large quantities of the hydrocarbon compound methane ($CH_4$), often called *marsh gas*. This results from the decay of organic material buried in such fills; usually the methane escapes into the air to contribute to the pollution of the atmosphere.

## NITRIC OXIDES

Under normal conditions the nitrogen of the earth's atmosphere does not react (bond) readily but acts almost as an inert gas. In the cylinder of an automobile engine, however, the high pressure and temperature that prevail cause a nitrogen atom to bond with either one or two oxygen atoms to form nitric oxide (NO), or nitric dioxide ($NO_2$), or in general $NO_x$. As these molecules are emitted into the atmosphere, several other very reactive compounds may be formed, as shown in the following section.

## PHOTOCHEMICAL SMOG

When a nitric oxide molecule absorbs the energy of the sun's ultraviolet radiation, it is decomposed as follows:

$$NO_2 + \text{ultraviolet light} \longrightarrow NO\cdot + O\cdot$$

The existence of a single oxygen molecule (O) is unusual since oxygen more naturally occurs as $O_2$. The $O\cdot$ (indicating high reactivity) bonds with an $O_2$ molecule to form $O_3$, ozone. Although ozone performs a beneficial function in the stratosphere, it is highly destructive when formed near the surface of the earth—most obvious examples are the breakdown of rubber products like the stripping around automobile windows and the sidewalls of tires. To complete a cycle of reactions, the ozone atom may again react with nitric oxide to form nitric dioxide and oxygen-2.

$$O_3 + NO \longrightarrow NO_2 + O_2$$

In this cycle, it appears that the ozone would be destroyed as fast as it is formed; but it is at this point that the hydrocarbons mentioned earlier may explain the build up of ozone that we experience. Consider these reactions, where "Hc" means *hydrocarbon:*

$$NO_2 + (UV) \longrightarrow NO\cdot + O\cdot$$
$$O\cdot + O_2 \longrightarrow O_3$$

$$O_3 + NO\cdot \longrightarrow NO_2 + O_2$$
$$O\cdot + Hc \longrightarrow HcO\cdot$$
$$HcO\cdot + O_2 \longrightarrow HcO_3\cdot$$
$$HcO_3\cdot + O_2 \longrightarrow O_3 + HcO_2\cdot$$

## PAN

One of the most eye-irritating compounds in smog is called *PAN*, short for *peroxyacylnitrates*. See how several products of the automobile exhaust combine to produce PAN:

$$HC + NO_2 \longrightarrow \text{radical}-\overset{\overset{\displaystyle O}{\|}}{C}-O-O-NO_2$$
$$\text{(PAN)}$$

## LONDON-TYPE SMOG

In 1911, over eleven hundred Londoners died in a severe smog episode that was triggered by a thermal inversion that allowed sulfur compounds to build up in the foggy atmosphere. These sulfur pollutants derive from the burning of sulfur-bearing coal and, therefore, are typical of highly industrialized regions. The typical reactions that take place during burning followed by a reaction with air are shown as follows:

$$S + O_2 \longrightarrow SO_2$$
$$SO_2 + \text{air} + \text{sunlight} \longrightarrow SO_3$$
$$SO_3 + H_2O \longrightarrow H_2SO_4$$
$$\text{(sulfuric acid)}$$

Sulfuric acid readily dissolves in water, and much of this pollutant finds its way into streams and rivers only to increase the acidity to the point where fish can no longer survive. Furthermore, sulfuric acid in moist air causes great deterioration of metal surfaces.

## PARTICULATE MATTER

Particulate matter includes those particles that are large enough to be seen—in the order of 1 to 10 microns (1 micron = $10^{-6}$ meters). Sources of such particles include jet aircraft, forest fires, and dust-producing processes such as milling, cement processing, and fertilizer plants. Driven by the wind, these particles become abrasive to all surfaces. Furthermore, these particles form centers on which other types of pollutants can form. Hence, they complement the detrimental effect of those pollutants.

## SOLUTION TO POLLUTION

Two basic approaches to solving the pollution problem would include: (1) the search for pollution-free sources of energy (solar energy, wind energy, wave and tidal energy, natural waterfall, etc.) and (2) the removal of harmful elements in fuel and the more complete combustion of fuels used. Our lives may very well depend on finding a solution to the problem of pollution.

## QUESTIONS

1. The various layers of the earth's atmosphere are defined somewhat by the changes in temperature. Tell what trend (increasing or decreasing) accompanies an increase in elevation through each of these layers: (a) troposphere, (b) stratosphere, (c) mesosphere, and (d) thermosphere.
2. Tell why almost 90% of the earth's atmosphere is packed into the first 16 km—into the troposphere.
3. Describe the instrument whereby the atmospheric pressure may be measured—namely, the mercury barometer.
4. Why is it possible for a pilot to use an air barometer to indicate his elevation?
5. If ocean water has a density of 1.1 $g/cm^3$, find the pressure exerted on each square centimeter of a surface submerged to a depth of 10 m. Compute only water pressure, but remember that atmospheric pressure must be added in order to obtain the total pressure exerted on the surface.
6. Ninety-nine percent of the earth's atmosphere (by volume) consists of only two elements. Name these elements in the order of abundance.
7. If the earth's atmosphere contained no carbon dioxide, describe the factors that would be substantially different from the present.
8. What factors determine the amount of moisture that can be held by air?
9. What part of the earth is most effectively heated by the sun's rays on June 21? Why?
10. What part of the earth experiences the most effective heating from the sun's rays on the average, throughout the year? Why?
11. Describe the effect of the earth's rotation on the air currents within the troposphere and on the ocean currents.
12. Why are ocean tides correlated with the position of the moon rather than that of the sun, when the sun contains approximately 24,000,000 times as much mass as the moon?

13. Twice each month the earth experiences unusually high tides. Describe these times in relation to the phases of the moon.
14. By creating tides of the moon, the earth has brought the moon into very special motion. Describe the moon's rotation and revolution from the point of view of the stars.
15. List as many different pieces of evidence the scientist utilizes to conclude that continents are continuing to "drift" about, some continental plates colliding and others separating.
16. What is the single most effective factor in causing air masses to move from one place to another?
17. Describe two or more factors that cause clouds to form.
18. How does "cloud seeding" sometimes induce rain?
19. List several kinds of weather disturbances that result from a cold air mass "running into" a warmer air mass.
20. What conditions tend to produce thermal inversion and how does such an inversion contribute to increased air pollution?
21. Name elements (and compounds) that contribute to smog in a large city.

# earth–moon, a binary system

Our understanding of the position and separate motions of the earth-moon system seems a far cry from Aristotle's conclusion that the earth was the center of the universe and did not move. How can we be confident that the earth spins on its axis in a day, that it revolves about the sun in a year? What are the consequences of such motion?

## ROTATION OF THE EARTH

Not until the midnineteenth century did scientists have concrete evidence that the earth does rotate. In 1852, Jean Foucault, a French physicist, suspended from the dome of the Pantheon in Paris a weighted pendulum that he displaced to one side of the building and released. As the pendulum continued to swing for some time, he noticed a slight change in its orientation. Although he had started it swinging toward the main entrance, it later was swinging toward another entrance. The change in direction occurred at a given rate, like the hands of a clock. Since the only force that acted on the pendulum was the force of gravity and such a force could not influence the direction of swinging, it was concluded that the earth must be rotating, thereby creating this illusion (see Figure 17.1).

As a consequence of the earth's rotation, star trails near the north polar region will appear circular in form when a time exposure is made on a dark night (see Figure 17.2). However, these circular trails could not be considered a proof of the earth's rotation, for they could be explained equally well by the turning of the celestial sphere. On the other hand, the Foucault pendulum represented a proof because no alternative explanation was possible, short of the actual rotation of the earth.

## REVOLUTION OF THE EARTH

If the earth revolves about the sun and rotates on a tilted axis, as shown in Figure 17.3, the seasonal changes can be explained quite easily. Note how the sun's rays strike the earth north of the equator on June 21, producing summer in the Northern Hemisphere, and how they strike the earth south of the equator on

**Figure 17.1.** The Foucault pendulum in the Pantheon, Paris. (Science Museum, London)

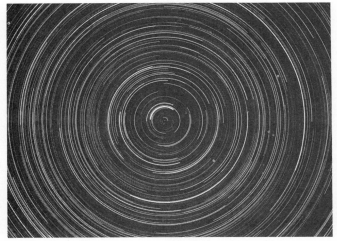

**Figure 17.2.** Circular star trails, caused by the rotation of the earth. (Lick Observatory)

December 21, producing winter in the Northern Hemisphere. On March 21 (vernal equinox) and on September 21 (autumn equinox) the sun's rays are directed toward the equator. Thus we have explained seasons by assuming that the earth revolves about the sun. However seasons cannot be called a proof of the earth's revolution, for it could be possible for the sun to orbit the earth in such a way as to produce a similar effect.

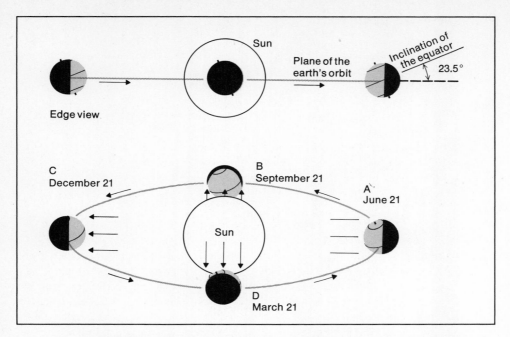

**Figure 17.3.** The seasons, caused by the tilt of the earth's axis and its revolution around the sun.

What then can be called a proof of the earth's revolution? If the earth revolves around the sun, then the alignment between the earth, a nearby star, and a distant star may be seen to change (see Figure 17.4). Such a shift is visible but only with a very special measuring device, for the angle ($\alpha$) is very small—in the order of 1/4000th of a degree even for the closest star, Alpha Centauri. To appreciate this fact we must realize that the distance to the nearest star is approximately 270,000 times as far from the sun as is the earth. The apparent shift in the position of a nearby star against the background of distant stars is referred to as *stellar parallax* and constitutes a proof of the earth's revolution. If the earth were stationary at the center of the system, no shifting would occur.

**Figure 17.4.** Stellar parallax, due to the earth's revolution.

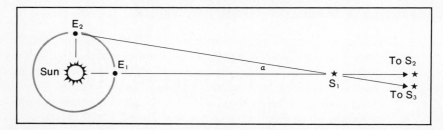

One of the consequences of our point of view as we ride on a moving earth that revolves around the sun is that the sun seems to continually change its alignment among the stars. The 12 constellations through which the sun seems to move in a year are designated as the signs of the zodiac (see Figure 17.5). The apparent path of the sun against the background of stars is called the *ecliptic* because it is also the path along which eclipses occur.

The sun appears to move completely around the sky (360°) in a period of one year (365¼ days); therefore, its average motion amounts to approximately 1°/day:

$$\frac{360°}{365\frac{1}{4} \text{ days}} \approx 1°/\text{day}$$

In relation to the earth's rotation, 4 minutes are required to turn through 1°. Because we make our clocks run in time with the sun, any given star appears to rise (and set) 4 minutes earlier each night. This does not seem to be very significant; however, 4 min/day amounts to 120 min/month or 2 hr/mo and 24 hr/year. So we see that the portion of the sky that is overhead in the early evening changes continually and rather swiftly. For this reason it makes sense to designate each quarter of the sky by the season in which it appears in the early evening (see Figure 17.6). Note that three of the 12 signs of the zodiac are shown on each map.

**Figure 17.5.** As the earth revolves about the sun, the sun appears to align itself with each of the signs of the Zodiac in succession.

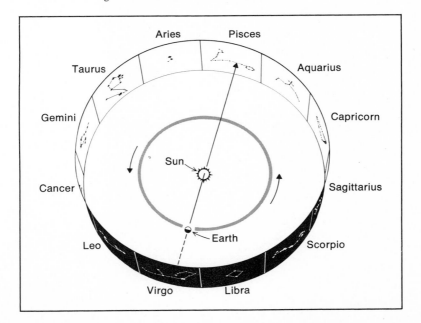

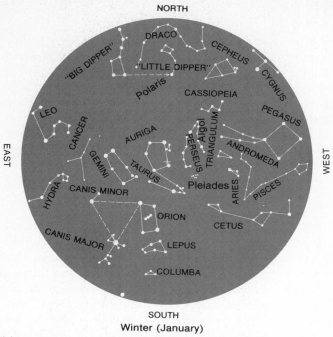

(a)

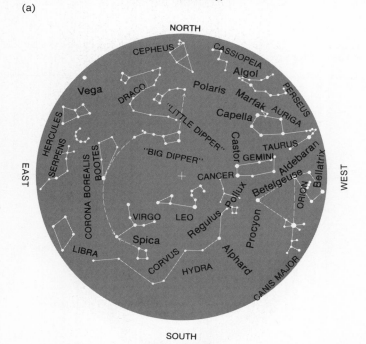

(b)

**Figure 17.6.** Seasonal sky maps. (Griffith Observatory)

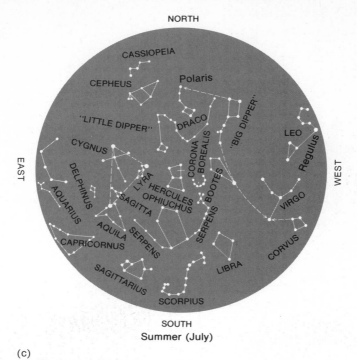

(c) Summer (July)

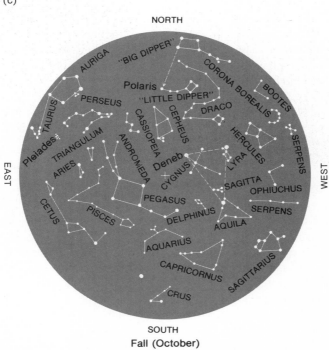

(d) Fall (October)

307

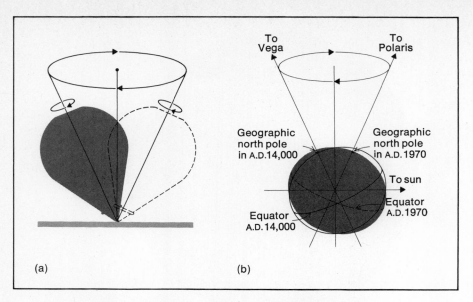

**Figure 17.7.** Precession: (a) of a toy top; (b) of the earth.

## PRECESSION OF THE EQUINOXES

We have discussed the earth's rotation in one day, its revolution around the sun in a year, and now we note another basic motion of the earth called *precession*—a motion that requires 26,000 years to complete. Because the earth is spinning on a tilted axis and is bulged around its equator because of that spinning, it experiences a gyration effect like a toy top (see Figure 17.7). One of the consequences of this motion is that observers must continually search out new north stars throughout a period of 26,000 years, since the axis of the earth constantly points to new places in the sky. However, during the life span of a single human, the change could hardly be detected without a telescope.

## AGE OF AQUARIUS

If the earth's equator is extended outward to intersect the celestial sphere (the imaginary sphere of stars), it will determine a celestial equator. Likewise, if the plane of the earth's orbit is extended to the celestial sphere, it will determine the ecliptic as shown in Figure 17.8. Because the earth rotates on an axis that is tilted $23\frac{1}{2}°$, the two circles (the celestial equator and the ecliptic) are distinct and they cross each other at two distinct points. These points are called the *vernal equinox* (the apparent position of the sun on March 22) and the *autumn equinox* (the apparent position of the sun on September 22). Historically, the location of the vernal equinox in the sky has been taken as determining the age in which we are living. Because of the precession of the earth, this point seems to "slip" westward

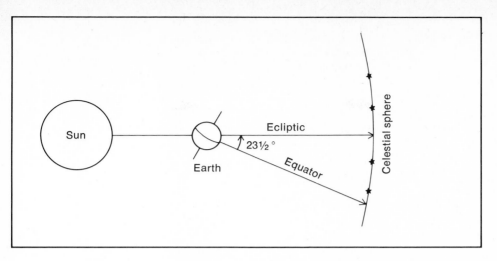

**Figure 17.8.** The celestial equator and ecliptic in the sky are simply an extension of the earth's equator and its plane of orbit, respectively, onto the celestial sphere.

among the signs of the zodiac. At the present time the vernal equinox is located in Pisces as shown in Figure 17.9. Hence, astronomically we live in the "Age of Pisces."

Several thousand years ago when astrology had come to a certain state of refinement, the vernal equinox was in Aries. Hence much of the literature of astrology refers back to this time by referring to the vernal equinox as "the first

**Figure 17.9.** The vernal equinox is presently located in Pisces, but it is slipping westward into Aquarius because of the earth's precession.

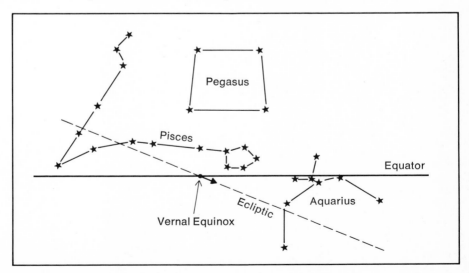

point of Aries." The "Age of Aquarius" can only begin, from an astronomical point of view, when the vernal equinox slips into Aquarius.

We have looked at three basic motions of the earth, but let's not forget that the sun participates in the rotation of the Milky Way galaxy and moves at the rate of almost 250 km/sec. Whatever motion the sun has, of course, the earth goes along with it. Even the Milky Way galaxy has a motion relative to other galaxies, and both the sun and earth participate in that motion. Now let's return to a closer look at the earth's motion as it is influenced by the presence of the moon.

## EARTH–MOON, A BINARY SYSTEM

To speak of the motion of the earth or of the moon individually would be to miss what is actually happening, for the earth and moon influence each other's motion because of their mutual gravitational force. Think of two masses attached to the

**Figure 17.10.** The balancing of various masses—a model for the earth–moon system.

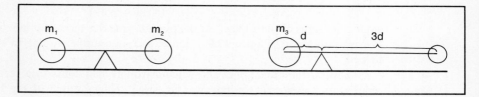

**Figure 17.11.** The earth–moon, a binary system.

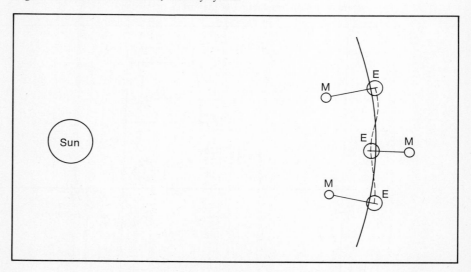

earth–moon,
a binary
system

ends of a rod. If the two masses ($m_1$ and $m_2$) are equal, then they will balance at the center of the rod. But if one mass ($m_3$) is three times the other mass ($m_4$), then they will balance at a point that is closer to the larger mass by a factor of three (see Figure 17.10). The balance point is called the *barycenter*, and it is this point around which the objects would tend to revolve if set into orbital motion.

When observers study the earth-moon motion, they realize that the earth deviates from a regular elliptical path in the fashion shown in Figure 17.11. The deviation amounts to 3000 miles on either side of the elliptical path; this is because of the presence of the moon. The earth and moon can be said to move about a barycenter located 3000 miles from the center of the earth and approximately 240,000 miles from the moon. Furthermore, the location of the barycenter indicates that the mass of the moon is only 1/80th that of the earth (240,000/3000 = 1/80).

## PHASES OF THE MOON

From the point of view of an earthbound observer, the moon appears to revolve around the earth, and one of the consequences of this motion is that the moon appears to go through phases. Actually, half the moon is lighted by the sun at any given time, but the phase we see depends on how much of that lighted half we can see. Sometimes we see only a thin crescent; but if we observe that same moon night after night, we see the phase wax (grow) to full and then wane (decrease) back to crescent and then to new again. Note that the phase labeled "new moon" (see Figure 17.12) is not visible because of the close alignment to the sun. What most people call a "new moon" is labeled "waxing crescent."

Study Figure 17.12 very carefully to see if you can discover where and when a given phase might appear in the sky. Remember that the earth rotates counterclockwise in this view; hence if the moon is in any of the waxing phases, we will see it follow the sun across the sky and it should be more easily visible in the evening hours after sunset. In the waning phases the moon could be said to lead the sun across the sky; hence we should look for these phases in the late night or early morning hours before sunrise. Of course, the moon is often visible in the daytime as well, and this is not unusual when it is relatively closely aligned with the sun.

We have all observed the moon's apparent motion during one day (or night)—it is toward the west, along with all other objects in the sky. This, of course, is due to the earth's rotation, but what of the moon's motion relative to the stars? Try making a drawing of the moon and nearby stars or planets on a given night. Then look again the next night. The moon will have moved toward the east in relation to those stars. If you continue this procedure for $27\frac{1}{3}$ days, you will find that the moon has "traveled" entirely around the sky and has returned to the beginning stars. But a curious thing will be apparent: the moon will not have returned to the same phase as it had when you started. In fact, $2\frac{1}{6}$ days additional time, or $29\frac{1}{2}$ days altogether, are required for each cycle. Figure 17.13 shows the reason for this difference of $2\frac{1}{6}$ days.

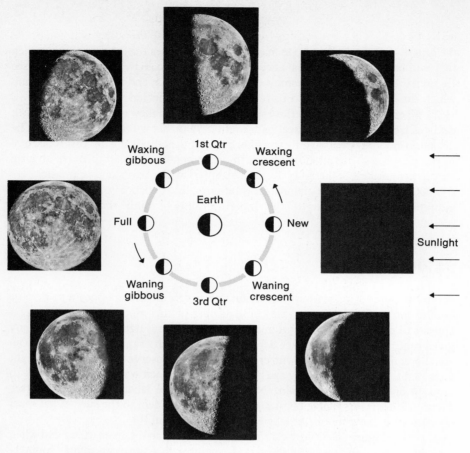

**Figure 17.12.** Phases of the moon. (Lick Observatory)

**Figure 17.13.** The moon's apparent revolution around the earth—the determination of the month.

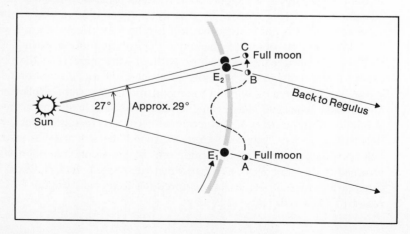

312

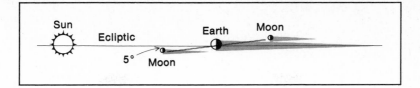

**Figure 17.14.** The moon's orbital plane is inclined 5° to that of the earth.

As you can see, we start with a full moon aligned with the star Regulus. While the moon orbits the earth, the earth is also orbiting the sun (approximately 30°). When the moon returns to its alignment with Regulus, it is not full, for it is not directly opposite the sun. So the moon must orbit an additional 30° (requiring $2\frac{1}{6}$ days) to again be full. Because the moon revolves through 360° in $27\frac{1}{3}$ days, it appears to move among the stars about 13°/day; therefore, it rises 50 minutes later each night.

As you watch the moon move to new star alignments each night, note that it usually falls within one of the signs of the zodiac. We saw (earlier in this chapter) that the signs of the zodiac are constellations that fall in the plane of the earth's orbit—constellations through which the sun appears to move. Because the plane of the moon's orbit is inclined only 5° with respect to that of the earth, the moon moves always within 5° of this same path (see Figure 17.14). Therefore it is perfectly logical to say, "The moon is in Taurus," or "The moon is in Gemini," on a given night.

## ECLIPSES

Eclipses occur when the shadow of one object falls on another. When the moon is new (see Figure 17.15), it appears that the moon's shadow should fall on the earth, creating an eclipse of the sun. However, when the 5° inclination of the moon's orbit is considered, we see that the shadow misses the earth. Likewise, the shadow of the earth misses the moon.

Note that there are two periods in a year when the alignment does permit eclipses. These positions are identified as eclipse seasons in Figure 17.15. One may then expect an eclipse season every six months; in any given season usually at least one solar and one lunar eclipse occur. Figure 17.16 shows that when the moon's shadow strikes the earth, it does so in only a limited region covering only a few hundred square miles at any moment; however, primarily due to the rotation of the earth, this shadow spot moves over the surface of the earth at several hundred miles per hour. If you are standing along the "path of totality," you will see a total solar eclipse as the spot passes over your position. The paths of totality are predictable from our knowledge of the regular motions of the sun, earth, and moon; some are shown in Figure 17.17.

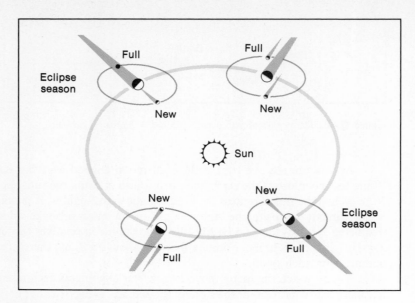

**Figure 17.15.** Eclipse seasons.

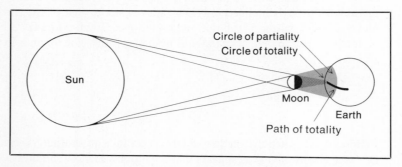

**Figure 17.16.** A solar eclipse. Only the observers within the path of totality will witness a total eclipse of the sun.

As the shadow spot approaches an observer, he would see only partial phases of the eclipse. When the spot reaches the observer and the bright disk of the sun appears totally covered by the moon, the corona becomes visible (see Figure 17.18). This represents the outer atmosphere of the sun that is usually invisible due to the brightness of the disk. Strange things seem to happen during a total eclipse of the sun: unusual winds blow, dogs howl, chickens go to roost. Observers on either side of the path of totality would see only partial phases.

It is essential when looking at any solar eclipse to have your eyes protected properly. Without protection the sun is capable of burning your eyes (with no indication of pain), and blindness can result. Two safe ways of viewing are shown in Figure 17.19.

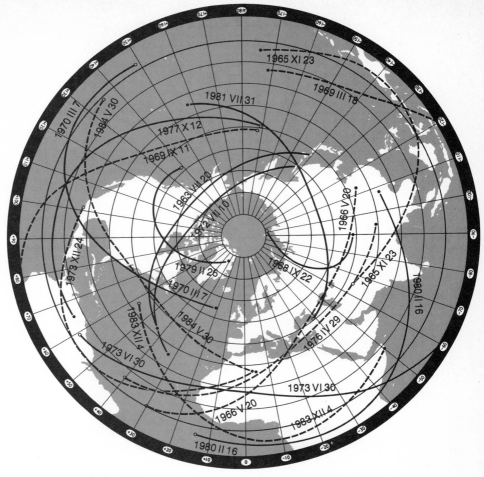

**Figure 17.17.** Paths of totality predicted through 1984. (J. Meeus, K. Grosjean and W. Vanerleen, *Canon of Solar Eclipses.* Elmsford, N.Y.: Pergamon Press, 1966)

**Figure 17.18.** The corona of the sun—its extended atmosphere seen only during times of total eclipse. (Yerkes Observatory)

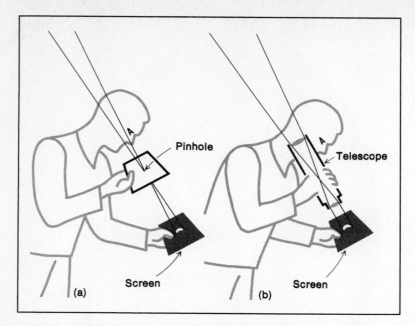

**Figure 17.19.** Observing a solar eclipse safely: (a) using a pinhole and (b) using a small telescope.

## LUNAR ECLIPSE

The moon will be eclipsed whenever it enters the central shadow of the earth. Because that shadow is so large, the moon may stay in the shadow for about one hour when good alignment occurs (see Figure 17.20).

Even when totally eclipsed, the moon does not darken completely but takes on a reddish hue. This is because the earth's atmosphere refracts (bends) the sunlight and scatters blue light, leaving the red hues to pass on to the moon (see Figure 17.21).

**Figure 17.20.** The lunar eclipse.

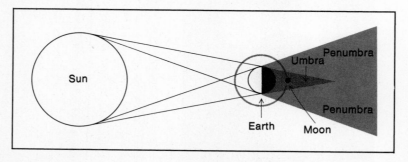

316

**Figure 17.21.** The moon is seen moving into the dark portion of the earth's shadow. (R. T. Dixon)

A TRIP TO THE MOON

In the last decade, the moon has yielded many of its secrets to our insatiable curiosity. If you could travel to the moon, you could find a surface terrain equally as rugged as the earth's. There are flat lowland regions called *maria,* where water would collect if it existed on the moon. The moon soil has the consistency of damp sand; however, it is not damp, for no surface moisture is retained on the moon in the absence of an atmosphere. The highlands of the moon range from dune forms to full-fledged mountains. The moon's surface is dotted with craters, probably of impact origin, yet we would not exclude volcanic activity as a possible cause. Smaller features include the rilles, which appear to be fault lines, for they "run" right up over the rims of craters and into the crater bottoms (see Figure 17.22). Only in a very limited way do we see anything that resembles a riverbed, because, as noted above, there is no liquid water on the moon.

**Figure 17.22.** A lunar crater showing rilles, which cross its rim. (NASA)

AGE AND HISTORY OF THE MOON

The Apollo manned landings have made possible the return to earth of large amounts of moon rock and moon soil. With such material in hand, scientists on earth have dated the various samples by comparing the present percentages of radioactive elements to those of their daughter products (i.e., uranium 238 to lead 206) as explained in Chapter 15.

In general, materials from the maria (lowlands) dated back approximately $3\frac{1}{2}$ billion years; whereas the oldest rocks, from the highlands, dated back nearly $4\frac{1}{2}$ billion years. This later date coincides with other estimates of the age of the entire solar system. But what can be said for the fact that virtually nothing has been found to be much younger than $3\frac{1}{2}$ billion years old? Many observers believe that the major features of the moon were formed during its first billion years of existence and that it has been "frozen" since that time except for impacts of smaller objects that provide almost the only form of erosion on the moon. Gravity also tends to make loosened particles run downhill, but on the moon the force due to gravity is only one-sixth that on earth. Other erosion factors such as wind, rain, rivers, and so forth, simply do not occur on the moon. The lunar geologist values the fact that we have such a close neighbor that still shows us what conditions may have been like during the first billion years of the solar system.

If you could visit the moon, you certainly would be concerned about temperature changes. In bright sunshine, temperatures range upward to 120°C (250°F); but if the sun goes down on your location, the temperature drops to $-155°C(-250°F)$. Furthermore, if you lived on the moon, you could expect two

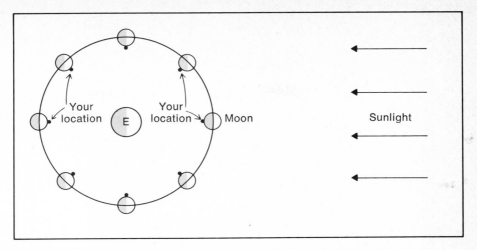

**Figure 17.23.** If you lived on the moon, in the position shown, you would always see the earth in the sky, but it would seem to go through phases, even as the moon appears to go through phases as we see it from the earth.

weeks of daylight and then two weeks of darkness. This fact is consistent with another: the same side of the moon always faces the earth. Does this mean that the moon does not rotate? The answer to this question will be found in the following paragraph.

Consider your location to be marked by the dot on the moon in Figure 17.23, always facing toward the earth. If viewed from a very great distance, say from a star, the moon does rotate—it simply rotates with a period equal to its period of revolution. This is called *synchronous rotation and revolution.* This synchronized motion is not a coincidence but rather a direct effect of friction that is created by the tides, as was discussed in Chapter 16, page 285. Over a long period of time, perhaps a billion years, tides on the moon created by the earth changed the moon's rotation until it became synchronized. In a similar manner, the earth tides created by the moon tend to slow the earth's rotation.

As we look at the solar system in the next chapter, we will find that there are 35 moons in all, and probably many of these moons also have synchronized rotation and revolution.

## QUESTIONS

1. List several consequences of rotation (the earth's spinning) not mentioned in the text.
2. If the earth stood still in the center of the universe, could you explain how seasons might still occur?
3. Why are the 12 signs of the zodiac significant to astronomers as well as to astrologers?

4. Would you expect to see the same constellations at 9 P.M. in the summer as you do at 9 P.M. in the winter?
5. True or false: Polaris, our present north star, will always be our north star, even 10,000 years from now.
6. We call the earth and moon a binary system. What does this mean in terms of the motion of the earth around the sun?
7. True or false: One-half of the moon is lighted at any given time.
8. The moon appears full when it is directly opposite the sun. At what time of night would the moon appear overhead?
9. A total solar eclipse can occur only when the moon is in what phase?
10. True or false: Usually more people get to see a total eclipse of the moon than they do a total eclipse of the sun.
11. What facts must be known in order to predict an eclipse of the sun?
12. What is the indicated age of the moon?
13. How can you account for the fact that the moon always points the same side toward the earth? Does this mean that the moon does not rotate with respect to the stars?
14. List the major types of features on the moon's surface.

# 18

## the solar system

Could you recognize a planet if you saw one in the night sky? Planets do appear very much like the stars when viewed with the naked eye, but there are several telltale signs that may assist your recognition of planets. First, stars appear to twinkle because of turbulence in the earth's atmosphere—motions of the air molecules refracting light first one way and then another. On the other hand, planets do not appear to twinkle as much as stars but seem to shine with a steady light. This is because the planets are vastly closer to us than stars and, therefore, present an extended source of light, a source with an apparent diameter. This becomes obvious when you use a telescope to check out an object you believe to be a planet. If the object is truly a planet, it will appear as a disk in the telescope. We can, therefore, explain the steady light of a planet by thinking of its disk as having many point sources of light, each one twinkling, yet with the combined effect of canceling out the individual twinkles and producing a steady light.

## WANDERERS IN THE SKY

The ultimate proof that you are observing a planet comes as you see it move in relation to the starry background. Early observers called the planets *wanderers* for this reason. This does not mean that their movement in relation to stars is easily detected in one evening. However, if the location of a planet is carefully noted in relation to surrounding stars, its motion may be detected in the course of several nights or weeks. If the motion of planets were traced over a period of months or years, it would become apparent that they do not move in many different directions or orientations in the sky; rather they generally move eastward in relation to the stars, sweeping out a band in the sky we call the *signs of the zodiac*. These are the 12 constellations through which the sun appears to move in one year and through which the moon appears to move in one month. Which of the following systems does this apparent motion suggest? (See Figure 18.1.)

If the system looked like Figure 18.1(a), then we would see planets moving in many different directions in the sky—some northward, some southward, some eastward, and some westward. The fact that the usual motion is eastward along

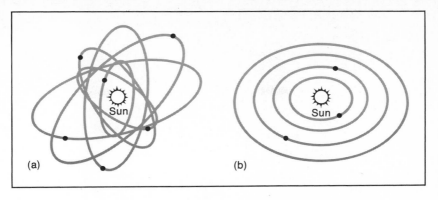

**Figure 18.1.** (a) The orientation of planetary orbits that might be expected if they had been captured at random. (b) The actual disklike nature of planetary orbits.

approximately the same path suggests a system more like Figure 18.1(b). This sun-centered model is consistent with the model suggested by Copernicus in 1542, as discussed in Chapter 2, and it is consistent with modern observations of stellar parallax, as discussed in Chapter 17, but one aspect of the Copernican model is known to be incorrect. The planets do not revolve around the sun in circular orbits. This fact was revealed through the combined talents of Tycho Brahe and his assistant Johannes Kepler.

## SIXTEENTH-CENTURY OBSERVATIONS

Tycho Brahe (1546–1601), a Danish nobleman under the support of the king, established an island observatory off the Danish coast. In spite of the fact that the telescope had not been invented yet, Brahe made thousands of measurements of planet positions among the stars, thus recording their motion. What he lacked in sophisticated equipment (see Figure 18.2), he made up for in technique. He made repeated naked-eye observations of star and planet positions; then he averaged his results in an effort to minimize observational errors.

## SEVENTEENTH-CENTURY OBSERVATIONS

In the late seventeenth century, the German astronomer and mathematician Johannes Kepler (1571–1630) worked as Brahe's assistant for the purpose of interpreting the record of planetary motions. It was soon apparent to Kepler that the planets do not travel about the sun in uniform circular motion as Copernicus had assumed. Brahe's observations had already shown that a given planet travels

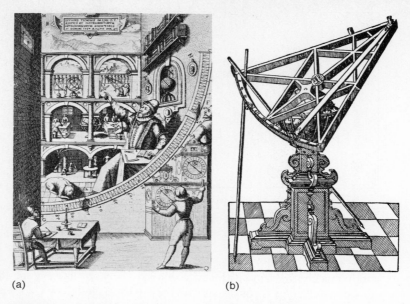

**Figure 18.2.** (a) Tycho Brahe observing with his mural quadrant; (b) sextant used to measure angles between celestial objects. (Rare Book Division, the New York Public Library; Astor, Lenox and Tilden Foundations)

more slowly at times, hence was not uniform in its motion. This suggested an orbit other than that of a circle. In an attempt to "fit" an orbit to the planetary records of Brahe, Kepler hit on the idea of an ellipse. An ellipse may be formed by cutting a cone as in Figure 18.3(a), or by attaching a string to two thumbtacks ($F_1$ and $F_2$) and holding the string taut, and then drawing the curve as in Figure 18.3(b). Kepler explained, by his first law, that:

> *A planet revolves around the sun in an elliptical-shaped orbit with the sun at one focal point*

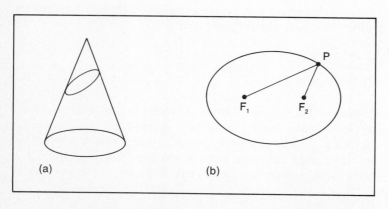

**Figure 18.3.** The ellipse: (a) as a cross-section of a cone; (b) as drawn by means of two thumbtacks and a piece of string.

See focal point $F_1$ in Figure 18.3(b). Kepler accounted for two observed facts: (1) the planet's distance from the sun varies with time and (2) so does its velocity. He correlated these two facts in his second law, which states:

*An imaginary line joining the sun and any given planet sweeps out equal area in equal time.*

In Figure 18.4 each sector represents the area swept out in two months by a line joining the sun and Mars. As you can see, a planet moves fastest when closest to the sun and slowest when farthest from the sun. This second law of Kepler applies to only one planet at a time; but he went on to recognize a mathematical relationship among all planets, which will be shown as his third law below. Based on Brahe's observations and some of his own deductions, Kepler compiled the facts shown in Table 18.1. It is easy to see from this table that the farther a planet is from the sun, the longer it takes to revolve around the sun.

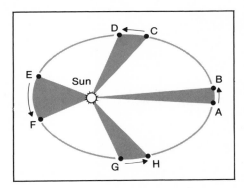

**Figure 18.4.** Illustration of Kepler's law of equal areas in equal time.

**Table 18.1**

| PLANET | $r$ (AVERAGE DISTANCE FROM SUN IN A.U.*) | $r^3$ | $p$ (PERIOD OF REVOLUTION IN YEARS) | $p^2$ |
|---|---|---|---|---|
| Mercury | 0.39 | 0.058 | 0.24 | 0.058 |
| Venus | 0.72 | 0.378 | 0.62 | 0.378 |
| Earth | 1.00 | 1.000 | 1.00 | 1.000 |
| Mars | 1.52 | 3.54 | 1.88 | 3.54 |
| Jupiter | 5.20 | 140.7 | 11.86 | 140.8 |
| Saturn | 9.54 | 867.7 | 29.46 | 867.9 |

*Astronomical unit (A.U.), is defined as the average distance from the sun to the earth.

Because he was a mathematician, Kepler looked for a very specific mathematical relationship. He probably tried many factors before realizing that if he cubed the average distance numbers (in A.U.) and squared the period numbers (in years), he very nearly obtained the same number for any given planet (compare the $r^3$ and $p^2$ columns of Table 18.1). Therefore, his third law is stated $p^2 = r^3$, or more generally:

$$\frac{p_1^2}{p_2^2} = \frac{r_1^3}{r_2^3}$$

where

$p_1$ = period of planet 1

$p_2$ = period of planet 2

$r_1$ = average distance of planet 1

$r_2$ = average distance of planet 2

Kepler thus discovered the effects of mutual gravity between the sun and a given planet before anyone had even recognized the existence of gravity.

## PHYSICAL PROPERTIES OF THE PLANETS

As we look at the solar system in general, we see a natural grouping of planets by size and by other characteristics. Counting from the sun, the first four planets are relatively small, then comes the asteroid belt, then four large planets, and finally Pluto, which is small and moonlike (see Figure 18.5).

The first four planets (Mercury, Venus, Earth, and Mars) are in many ways like Earth; hence they are called the *terrestrial planets*. The four planets that lie

**Figure 18.5.** The solar system, showing relative sizes of the planets.

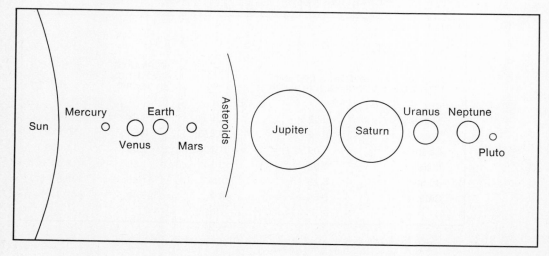

beyond the asteroid belt (Jupiter, Saturn, Uranus, and Neptune) are similar to Jupiter in many respects, so they are called the *Jovian planets*. In Table 18.2 we have compared some of the physical properties that differentiate the two types of planets. Check these characteristics against the tables of planetary data in Appendixes 1 and 2. Let us see how these distinctly different characteristics can be explained?

We have already discovered that the temperature of a gas is a measure of the average velocity of molecules in that gas. We would naturally expect the planets nearer the sun to possess higher temperatures; therefore, the molecules that compose their atmospheres would be accelerated to higher velocities. Whether a given molecule will be able to escape from a planet depends on the effect of the mutual gravity between the molecule and the planet. The minimum velocity needed to escape the gravitation influence of a given planet is given in Appendix 2, page 391, under the heading Escape Velocity. Note that the terrestrial planets have very low escape velocities; hence it would be reasonable to expect that the energy of the sun would be sufficient to accelerate at least the lighter molecules like hydrogen and helium to the escape velocity, and those molecules would then be lost. Surface temperatures for the terrestrials have highs of from $20°C$ to $500°C$. On the other hand, the temperatures at the visible surface (cloud tops) of the Jovians range from $-130°C$ downward to $-230°C$, and because of their higher masses escape velocities for these planets are high. Hence, we would expect that the Jovian planets have been able to "hang onto" most of the atmosphere that they acquired at the time of origin.

By the above line of reasoning, the terrestrials probably lost most of their original atmospheres. Then how have they acquired their present atmospheres? Venus has a very significant amount of atmosphere, primarily composed of carbon dioxide gas. Mars has a very thin carbon dioxide atmosphere, and the earth has a moderate amount of atmosphere composed of nitrogen and oxygen. It is believed that this secondary atmosphere resulted from outgassing of the planet, including volcanic activity. We have direct evidence of volcanos on both the earth and Mars

Table 18.2

| TERRESTRIAL PLANETS | JOVIAN PLANETS |
|---|---|
| Smaller | Larger |
| Hard crusty surface | More like a ball of gas |
| Thin or no atmosphere | Very thick and dense atmosphere |
| Hotter | Colder |
| More dense | Less dense |
| Slow rotators | Fast rotators |
| Few moons | Many moons |

(see Figure 18.12 on page 333), and we know that typically volcanos pour out carbon dioxide and water vapor when they erupt. This explanation does not account for the nitrogen-oxygen atmosphere of the earth, however, for why should the earth's atmosphere be almost free of carbon dioxide? Had the earth's atmosphere retained its carbon dioxide, the earth would have been far too hot for man's existence. Carbon dioxide in any atmosphere creates a greenhouse effect in that it tends to absorb heat energy that the planet radiates. For example, as a result of Venus's greenhouse effect, its surface temperature ranges upward to 500°C, even hotter than the surface of Mercury. Some observers believe that the rocks of the earth may have absorbed much of the carbon dioxide of its atmosphere, reducing the greenhouse effect and allowing cooling to occur. Water vapor then could condense and may have supplied the liquid water for oceans and lakes. Today the plants and certain animals of the sea continue to absorb $CO_2$ as a part of their life process. The burning of fossil fuels (like gasoline) introduces additional CO into the atmosphere; however, the processes mentioned above tend to maintain suitable levels. Thus, only the earth possesses an atmosphere conducive to life processes as we know them.

Now let's look at some of the characteristics that make each planet distinctive.

## MERCURY

Because of its low mass ($\frac{1}{20}$ that of the earth), Mercury cannot retain an atmosphere under the tremendous solar energy it receives. As a result, it experiences a great difference in surface temperature on its sunlit side (427°C) as opposed to its shady side ($-183$°C). No other planet shows that much variation in temperature. Until a few years ago most observers thought that Mercury always pointed the same side toward the sun, rotating in 88 days, the same period in which it revolves. However, in 1965 astronomers succeeded in bouncing radar signals from the planet. Sending a signal of only one specific wavelength, the observers noticed a variety of wavelengths returning. They attributed this observation to the rotation of Mercury. This is to say that due to rotation, some parts of the planet are moving toward the

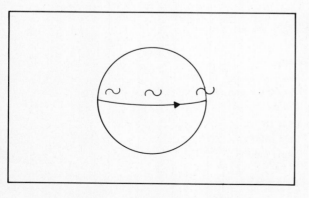

**Figure 18.6.** The rotation of Mercury, as discovered by the Doppler Effect on radar waves reflected by the planet.

earth and would shorten the waves due to the Doppler Effect; and some parts would be moving away and would lengthen the waves (see Figure 18.6). From the amount by which the wavelength was changed, the rotation rate was computed and equated to one rotation every 58.64 days.

In 1974 the space probe, Mariner 10, flew within 700 km of Mercury and for the first time revealed its densely cratered surface (see Figure 18.7(a)). At first glance Mercury resembles the moon in that it has large, relatively smooth areas somewhat like the maria of the moon. It has a variety of crater forms ranging from those that appear to be very old with rounded rims to those that appear to be very young with sharp rims. Some craters have central peaks and others show the effects of repeated impacts.

Mariner 10 measured surface temperatures as high as 400°C and as low as −200°C, a range of 600°C. This wide range of temperature confirms the fact that Mercury has virtually no atmosphere. What little atmosphere was found contains helium, argon, neon, and xenon.

After passing Mercury, Mariner 10 was placed in a solar orbit with a period

**Figure 18.7.** Mercury, as photographed by Mariner 10 in March 1974: (a) A dark, smooth, relatively uncratered area with some suggestion of lava flows; (b) fractured and ridged plains of the Caloris Basin, as seen on Mariner's third encounter with Mercury; (c) heavily cratered area also shows a valley approximately 100 km long; (d) a photomosaic, constructed from 18 photos taken 6 hours after Mariner 10 fly by Mercury. (NASA-JPL)

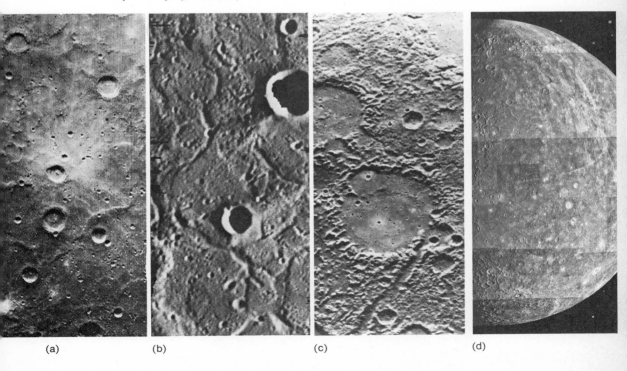

(a)  (b)  (c)  (d)

of 176 days. This period, just twice the period of Mercury's revolution, brought the spacecraft near Mercury for two additional observation passes.

The third pass carried the probe within 200 miles of the surface of the planet, yielding the best photographs, which show details down to 50 m in diameter [see Figure 18.7(b)]. This third fly-by also gave an opportunity to refine scientists' understanding of the planet's magnetic field, a field that is somewhat stronger than had been expected on the first pass.

## VENUS

Very much like the earth in properties such as diameter, volume, mass, and density, Venus is quite different in its very dense cloudy atmosphere, through which no earth-based observer has seen. In 1962, the clouds of Venus were penetrated by radar waves, revealing a retrograde rotation rate of 243.16 days by the Doppler Effect. More recent radar observations have scanned the planet to reveal rough terrain and craterlike structures (see Figure 18.8).

The effort to probe this cloud-covered planet by means of space probes also began in 1962 when the United States sent its first fly-by probe, Mariner 2, followed by Mariner 10 in 1974 (the same probe that went on to view Mercury). In contrast, the U.S.S.R. has concentrated on a landing approach to probe the planet. It is believed that the first Soviet lander was crushed by the tremendous atmospheric pressures of the planet while still 24 km above the surface, communications ceasing at that point. This problem, however, was overcome in later landers, and scientific data together with the first visual transmission of Venus have been received. The findings of the United States and U.S.S.R. efforts are summarized below.

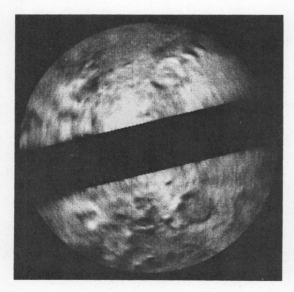

**Figure 18.8** Radar signals that penetrated the clouds of Venus and were reflected by its crusty surface reveal the multitude of craters and possible mountain ranges that characterize its surface. The largest craters are estimated to be 160 km across and 500 m deep. (NASA-JPL)

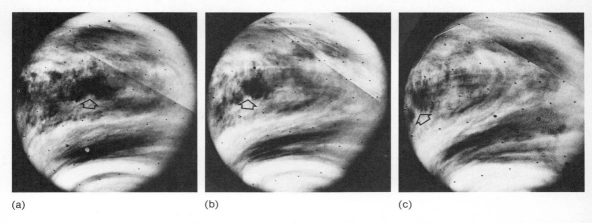

(a)                             (b)                           (c)

**Figure 18.9.** A series of ultraviolet photographs of Venus, taken by Mariner 10, during a 14-hr period, showing the rapid rotation of the upper cloud deck. (NASA-JPL)

The atmosphere is 95% carbon dioxide with small percentages of water vapor, hydrogen, helium, carbon, oxygen, ammonia, and several acids (HCl, HF, and $H_2SO_4$). The atmospheric pressure at the surface is 90 times that on the surface of the earth, correlating to the pressure that would be experienced by a diver at a depth of 800 m beneath the surface of the ocean. Temperatures on the surface range upward to 500°C, whereas at the cloud tops temperatures are slightly below the freezing point of water ($-23°C$). The Mariner 10 view of the cloud tops, in ultraviolet wavelengths, revealed high winds with the cloud tops rotating in 4 earth days (as compared to a surface rotation period of 243.16 earth days) (see Figure 18.9).

To the layman, Venus is still the bright object that shines in the evening sky, called the *evening star*. An amateur telescope will reveal the fact that Venus goes through phases, somewhat like the moon.

## MARS

Second only to the earth, Mars has been scrutinized more closely than any other planet. The history of man's closer and closer looks began with his naked-eye observations of a reddish dot in the sky that sometimes looked very bright and at other times appeared to dim and disappear into the sun's glare. The observer was seeing the result of an orbital period that takes almost twice the time of the earth's yearly trip around the sun. This orbit brings Mars near the earth every 26 months. Figure 18.10 shows intervals of successive oppositions—times when Mars is in the opposite direction to the sun, from our point of view. These became the best times for pointing a telescope toward Mars, revealing the polar caps of the planet and the general markings that appear to change with the Martian seasons. In 1877 Giovanni Schiaparelli of Milan made a drawing of what he saw over long periods of time. His patient observations were rewarded by momentary times of clearing in

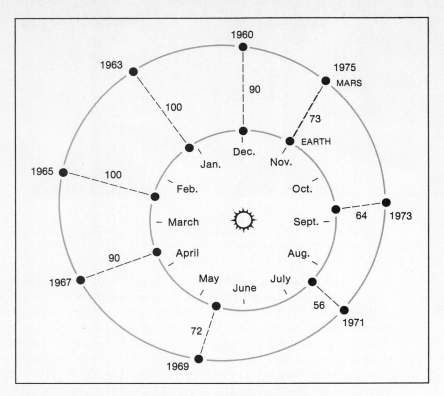

**Figure 18.10.** Successive oppositions of Mars with the earth, occurring every 26 months. The distances are shown in millions of kilometers.

both the earth's atmosphere and that of Mars itself. Schiaparelli asserted that he saw a network of fine straight lines that seemed to interconnect larger features on the surface of the planet. (See a modern version of visual observations in Figure 18.11.)

Percival Lowell (1855-1916), an American astronomer, suggested that these lines represented canals (waterways) built by little Martians to irrigate their crops. Although this remark may have been made with "tongue-in-cheek," it certainly sparked people's imagination and their interest in the possibility of life on Mars. A new era began in 1965 when Mariner 4 flew by the planet to reveal a rather drab, somewhat cratered Mars. This exploit was followed by Mariners 6 and 7 in 1969 and by Mariner 9 in 1971. Note how Figure 18.10 shows these dates to represent successive times of opposition, with 1971 representing an exceptionally close approach of 56 million km—an ideal time to send a probe to this planet. The photographic record of Mariner 9 far outstripped that of earlier missions, revealing a planet with several very distinctive surface features—including a shield volcano cone 600 km broad and 25 km high (larger than any feature on the earth) (see Figure 18.12), some 20 lesser volcanic cones, and a canyon so long as to stretch across the continental United States, up to 100 km wide and 6 km deep in places

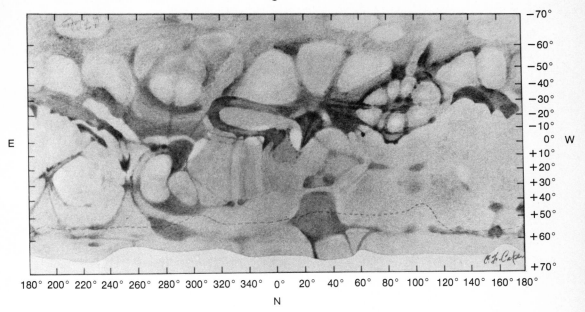

**Figure 18.11.** A high resolution photovisual map of Mars showing the seasonal aspects of late Martian summer. This map was produced from measurements of 20 photographs, 13 visual drawings and 15 telescopic micrometer observations. South is at top, as seen in a telescope. (Charles Capen, Braeside Observatory)

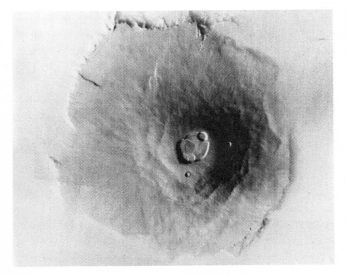

**Figure 18.12.** Mariner 9 photo of Nix Olympia, the Martian volcanic mountain that is larger than any feature on earth. (NASA-JPL)

(see Figure 18.13). The planet may further be subdivided into regions of volcanic plains, cratered plains, chaotic terrain, mountainous terrain, and channeled terrain (see Figure 18.14).

Of particular interest are regions that show evidence of the action of water or some similar low-viscosity fluid. The braided channel, shown in Figure 18.15, is very similar to a riverbed formed by sudden flood waters. We know that liquid water cannot remain on the surface of Mars for any length of time because of the planet's very thin $CO_2$ atmosphere, resulting in a pressure about 1/200th that of the earth. Even within the temperature range of $+20°C$ to $-130°C$, which characterizes Mars, liquid water would vaporize very readily. Yet vast amounts of frozen water are stored within the polar caps of Mars and perhaps beneath the surface generally. A huge impact might release such stores of water, or perhaps atmospheric and climatic conditions were vastly different in the past, permitting water to exist on the surface.

As 1976 brought the prospect of actually landing on Mars, sites were chosen from close-up orbiter photos, with water in mind. Viking 1, which landed on July 20 in a region called Chryse (see Figure 18.14), found water in the form of water-ice vapor (clouds), clouds that revealed themselves as bright spots on a daily cycle—indicating a daily exchange of water vapor from the surface to the atmosphere. Samples of Martian soil also revealed copious amounts of water released upon heating in one of the life-search experiments. Viking 2 landed in September 1976 in a site that exhibited evidence of braided flow channels—in an area called Utopia, much closer to the northern polar cap (see Figure 18.14).

The first reports from the life-search experiments fluctuated from positive to negative almost daily. The striking release of oxygen when nutrients were first supplied to the Martian soil triggered a positive response; however, scientists immediately conceived of nonbiological causes for oxygen release and began experiments in earthly laboratories to see if these results could be duplicated. In a second experiment in which nutrients were tagged with a radioactive tracer, the initial response was dramatic but tapered off in a way not typical of living, multiplying organisms. Scientists have grown more pessimistic in their search because a third experiment designed to recognize complex organic molecules has found none; however, their work continues undaunted by these seemingly negative responses. For example, on command the scooper arm of a probe has moved a rock that may have been in place for 100 million years and has retrieved a sample of soil from beneath that rock (see Figure 18.16). Here living organisms may have been protected from the sun's direct ultraviolet radiation. Digs that are deeper yet are planned.

Vikings 1 and 2 have already provided data that will take years to fully decipher—data recording two significant Marsquakes, data regarding the exact size and shape of the planet, data that tend to reveal the interior core and mantle under a thin crust of iron-rich soil. Viking 2, because of its proximity to the north polar region, has revealed a very extensive sand dune belt surrounding the polar cap. This, together with the sandwiched and terraced appearance of the cap, may reveal much of the history of Mars. The successive layers of ice and dust reveal not only

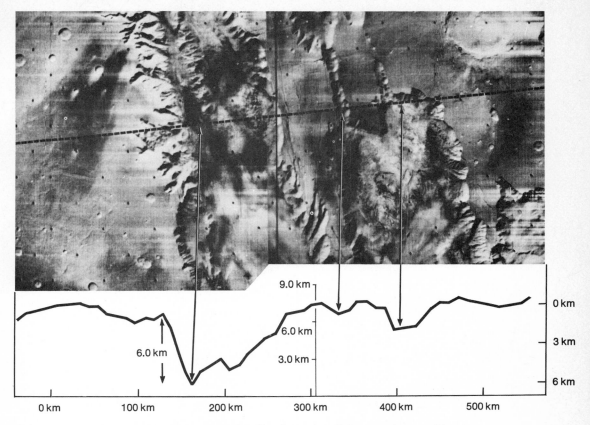

**Figure 18.13.** A close-up and profile of a portion of the great chasm on Mars. The length of the entire canyon is approximately 4000 km. (NASA-JPL)

seasonal changes but hint at long-term cycles in which the tilt of the planet's axis may have varied as much as 13° from its present 25° tilt. Such changes could have caused significant thawing of what are now permanently frozen-water icecaps. The seasonal changes that early observers saw in the polar caps represent only a $CO_2$ frost covering that develops in the winter and sublimates into the atmosphere as summer approaches. The spectacular picture of the Martian landscape, relayed to the earth from Viking 1 (see Figure 18.17), clearly indicates the effects of winds up to 200 km/hr, which are typical on this planet.

Mars has two small moons (see Figure 18.18) that circle the planet in almost the same plane as its equator and in almost circular orbits, strongly suggesting that they had an origin in common with the planet itself. Recent evidence from the Viking 1 orbiter, however, shows a composition characteristic of objects formed farther out in the solar system.

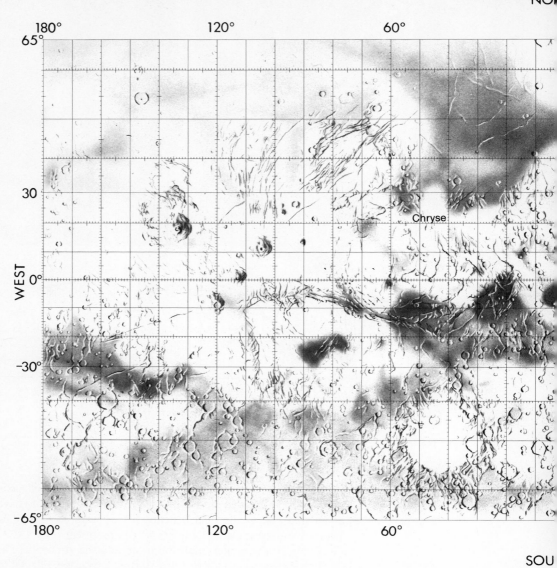

**Figure 18.14.** A map of Mars, based on the photographic survey of Mariner 10. (NASA-JPL)

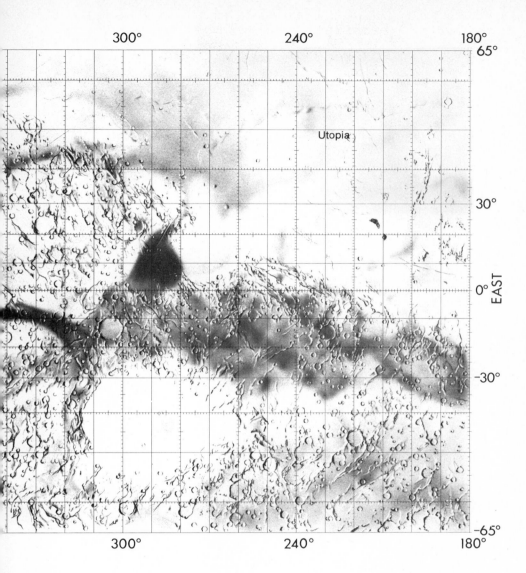

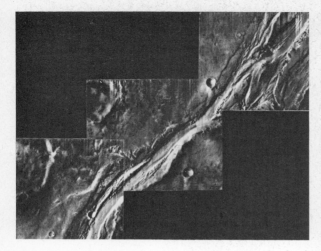

**Figure 18.15.** (above) The Amazonis channel, thought to have been formed by running water once present on Mars. The section shown is 100 km long. (NASA-JPL)

**Figure 18.16.** (right) The scooper arm of Viking 1, used to retrieve Martian soil samples and to move a rock in at least one case. (NASA-JPL)

**Figure 18.17.** A spectacular view of the Martian landscape, as seen by Viking 1 Lander, shows sharp dune crests, which indicate recent wind storms. The small deposits downwind of rocks also indicate wind direction. The large boulder at left measures about 1 m by 3 m. The sun rose two hours before this photograph was taken. (NASA-JPL)

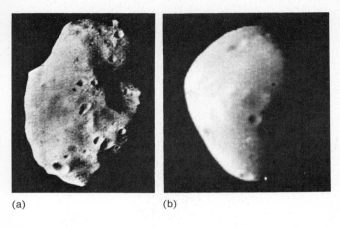

**Figure 18.18.** The moons of Mars: (a) Phobos and (b) Deimos. (NASA-JPL)

## JUPITER

The Jovian planets are of particular interest to the astronomer because we have reason to believe that they resemble the primordial clouds from which they were formed. At their great distance from the sun, temperatures are so low that virtually no atmosphere has been lost due to solar heating. The atmospheric constituents of Jupiter were measured by the Pioneer 10 and 11 spacecrafts and found to be 82% (by volume) hydrogen and almost 17% helium, with traces of methane ($CH_4$) and ammonia ($NH_3$). The relatively high percentage of helium seems to correlate rather well with the percentage of helium generated in early stages of the universe. This is to suggest that Jupiter may resemble a star in its composition. Jupiter also resembles a star in that the planet radiates about $2\frac{1}{2}$ times as much thermal energy as it receives from the sun; hence it must have a source of energy of its own. The mass of the planet is 318 times that of the earth but only $\frac{1}{1000}$th that of the sun, so we would not expect any starlike processes to occur. However, Jupiter may still be contracting and, thereby, giving up gravitational (potential) energy to produce heat.

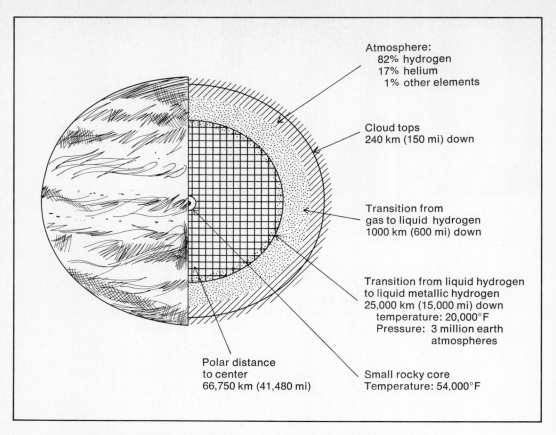

**Figure 18.19.** A model of Jupiter's interior. The planet is mainly liquid hydrogen. (NASA)

Many models for the interior of Jupiter have been proposed, one of which is shown in Figure 18.19. Note that the planet is gaseous in nature to a depth of 1000 km. Then, primarily due to pressure that increases with depth, the state of matter changes to that of a liquid and finally to a solid in the central core. This model does not derive from direct observation of the interior of the planet but from measurements of diameter, mass, and surface temperature together with a knowledge of gas laws and pressure increases that result from the gravitational forces that act in any given body.

The Great Red Spot on Jupiter, seen in Figure 18.20, remains somewhat of a mystery, varying in size and brilliance from time to time. In spite of the rather tenuous nature of cloud-top features, this spot has remained visible for hundreds of years. Pioneer 10 and 11 measurements show the temperature of the spot to be at least 2°C cooler than surrounding clouds, and this fact has been interpreted as meaning that it lies slightly higher than its surroundings. Perhaps it is a whirlpool of gases, somewhat like a cyclone in the earth's atmosphere. Temperatures are

**Figure 18.20.** Jupiter, its Great Red Spot at left, and the shadow of its moon Io, as seen by Pioneer 10. (NASA-Ames)

thought to increase from about $-173°C$ at the cloud tops to $-23°C$ at a depth of several hundred kilometers.

Jupiter rotates in just under 10 hours, yet is large enough to contain over 1300 earths in its volume. When we consider these two facts, it is easy to see that a point on Jupiter's equator has a lineal speed of 40,000 km/hr due to rotation, compared to 1500 km/hr on the earth.

Associated with its rapid turning is its very strong magnetic field, of which the total magnetic energy is approximately 250,000 times that of the earth. Entrapped in this field are high-energy charged particles, forming Van Allen-type belts with radiation strengths 100 times that which would be lethal to man.

Jupiter "rules" the largest number of moons—14 in all. Four of these moons—Io, Europa, Ganymede, and Callisto—can be viewed easily in a small telescope. These moons are often called the *Galilean satellites* after Galileo, who first observed them in his early telescope. These are large moons, with two of them exceeding the diameter of the planet Mercury. This fact should help portray the idea that Jupiter and its moons constitute a very large system. The relatively large mass of the Galilean satellites together with the low temperature of their surfaces (about $-150°C$) has evidently allowed them to "hang onto" a thin atmosphere, which Pioneer 10 showed to include ammonia and sodium.

As Pioneer 11 passed Jupiter in December of 1974, it received a gravitational boost (a slingshot effect) by that planet and is now on its way to Saturn with an estimated time of arrival in 1979.

## SATURN

Second in size, Saturn is much like the larger planet of Jupiter. Primarily composed of hydrogen and helium gas, Saturn rotates in just over 10 hours. On the other hand, it is unlike any other planet in terms of its extensive ring structure. In 1972 American astronomer Richard M. Goldstein, of the California Institute of Technology, succeeded in bouncing radar signals off the rings themselves, then interpreted the very efficient return of signal to mean that the rings are composed of chunks of material at least 1 m across. These chunks are probably composed of rocklike materials, bound together by frozen gases. You can see natural divisions in the rings in Figure 18.21. The names given to each ring along with its period of rotation are shown in Figure 18.22. The motions of each ring particle can be described by Kepler's third law ($p^2 = r^3$), with the innermost particles taking the least time to orbit. The gaseous nature of the planet itself is shown by the different rotation times for various latitudes.

Saturn has ten moons, the tenth having been discovered in 1966 when the rings were seen "edge on." Its name is Janus, and before 1966 it was not noticed because of its proximity to the outermost edge of the ring. Due to the tilt of the axis of the planet and its rings, we see those rings from an edge view every 15 years.

In 1979 Pioneer 11 will pass Saturn, revealing more data concerning this planet than has been collected since the advent of the telescope. By late 1977, two additional probes to the outer planets had been launched—Voyagers 1 and 2.

**Figure 18.21.** Saturn, as photographed with the 2.6-m Mount Wilson telescope. (Hale Observatories)

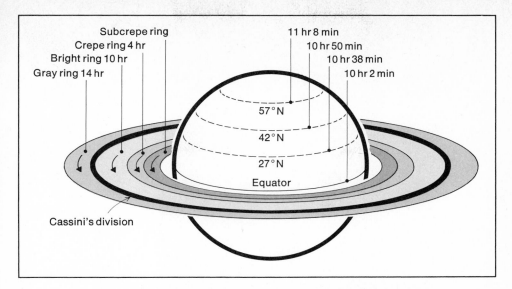

**Figure 18.22.** The rotation periods of the clouds of Saturn at various latitudes and of selected particles in the rings.

These probes will concentrate on Jupiter and five of its moons early in 1979 and then fly on to scrutinize Saturn and six of its satellites in 1980 and 1981, respectively. One further option is possible for Voyager 2. It may be directed to Uranus, arriving there in early 1986 and to Neptune, arriving there in 1989.

## URANUS

Uranus represents the first planet of our study that was not known to the ancients. It is not usually considered to be visible to the naked eye; however, under ideal conditions it may be seen quite easily using only binoculars. Uranus was discovered almost by accident. Although it had been seen and its position charted as early as 1690, it was only in 1781 that the English astronomer Sir William Herschel (1738-1822) recognized it as a planet; he also discovered two of its larger moons in 1787. Herschel, who was appointed court astronomer to King George III in 1782, had one of the best telescopes of his time, thus making his accomplishment possible (see Figure 18.23). Other observers had termed the object a star, but Herschel was able to recognize its disklike appearance (see Figure 18.24).

Uranus orbits the sun at an average distance just under 20 A.U. and in a period of 84 earth years. If a man could spend his entire lifetime on Uranus, he would die at the age of only one Uranus year. One of the most distinctive characteristics of this planet is the tilt of its axis, which is 98°. If we imagine a planet whose axis is not tilted at all, we can say that its equator is in the same plane as its orbit. Now imagine that this same planet rotates in a counterclockwise direction, a point on its equator traveling in an eastward direction. Tilt the planet's

**Figure 18.23.** Sir William Herschel's 1.25-m reflecting telescope. (Yerkes Observatory)

**Figure 18.24.** Uranus, with three moons. (Lick Observatory)

axis of rotation by 90°; the axis now lies in the plane of its orbit. Continue to tilt the axis of the planet 8° more, making a total of 98°. In what direction will it now be rotating? To answer this question, use your finger to represent the planet rotating in a counterclockwise direction with no tilt at all; then gradually tilt your finger until you have reached the 98° position. You will see that your finger is now rotating in a clockwise direction; similarly, the rotation of Uranus is clockwise (retrograde).

Uranus has five moons, all of which orbit in the same plane as the planet's equator and in the same direction as the planet's rotation. Because of the planet's extreme tilt, its moons are sometimes seen circling it in a plane perpendicular to our line of sight; at other times the plane of the moons' orbits seems to align itself with our line of sight, so we see their motion edge on.

In 1977, a rather startling discovery was made—Uranus has rings. The discovery was made as follows. It is a rather special occasion, from the astronomer's point of view, when a planet passes directly in front of a star. We say that the planet *occults* the star when this happens. As Uranus was approaching the given star, the star appeared to blink off and on again five times. Then following the passage of Uranus in front of the star, the star again blinked off and on five times. The interpretation given this observation suggests that five rather narrow rings surround the planet. As each of the five rings pass in front of the star, the star appears to blink off and on. This discovery prompted observation of the planet in the infrared portion of the spectrum and not only have the five rings been confirmed but eight or ten or even more additional rings are indicated. If the option to send Voyager 2 on to Uranus is exercised, then we may be able to view these rings directly. Certainly this recent discovery illustrates the fact that we have much to learn about the outer planets.

## NEPTUNE

Following the discovery of Uranus, it became evident that its orbit was being perturbed by still another unknown object. Two astronomer-mathematicians, John Adams (1819-1892) of England and Joseph Leverrier (1811-1877) of France, working independently, took a mathematical approach to the problem of finding this unknown object. Each man asked himself where an object must be in order to cause the perturbations that had been observed in the orbit of Uranus. When their calculations had been completed, using Newton's laws of gravitation, a search was made of the area in which they had predicted the presence of another planet, and in a very short time the planet was found (see Figure 18.25). Here again it was a matter of recognizing an object, previously called a star, as being truly a planet. The date of this discovery was July 1846, and it represented a real triumph for a theoretical approach that brought together astronomical observations, the known laws of motion and gravitation, and the mathematics necessary to solve the problem.

Using Appendix 2, on page 391, compare the size, mass, density, rotation, and albedo (reflectivity) of Uranus and Neptune. Note the almost twinlike nature

**Figure 18.25.** Neptune and its largest moon, Triton. (Lick Observatory)

of these two planets. Recent observations of Neptune show a cloud-top temperature of about $-228\,°C$.

Neptune has two moons. The larger, named Triton, has a diameter of 4000 km, greater than that of the earth's moon; it orbits the planet in a retrograde direction. Triton is the only moon of significant size that has a retrograde motion in relation to the rotation of its parent planet. All the moons of Uranus orbit in a clockwise direction, but that would be expected because the planet itself rotates in a retrograde (clockwise) direction.

## PLUTO

The orbital motions of both Uranus and Neptune were followed with much interest by numerous observers. In the early 1900s the American astronomer Percival Lowell (1855-1916), founder of Lowell Observatory in Flagstaff, Arizona, concluded that Neptune could not account for all the perturbations in the orbit of Uranus. Again, by mathematical calculations, he predicted that still another planet might be found in one of two possible locations. In 1930 a planet was found by the American astronomer Clyde Tombaugh, (1906-    ), as assistant astronomer at the Lowell Observatory, who had photographed these regions and later analyzed his plates on a device called the *blink microscope*. Two plates, taken at different times (see Figure 18.26), are placed in the instrument and then alternately illuminated, first the left-hand plate, then the right, and so on back and forth. The illumination is alternated quite rapidly so that any object that has moved against the background of distant stars will appear to jump back and forth in the blink microscope. Pluto was found within 6° of one of the positions predicted by Lowell. Some observers would discredit his calculations, saying that he had too little information with which to work and that therefore the almost perfect coincidence of prediction and discovery was only accidental. This is a question that may never be completely resolved.

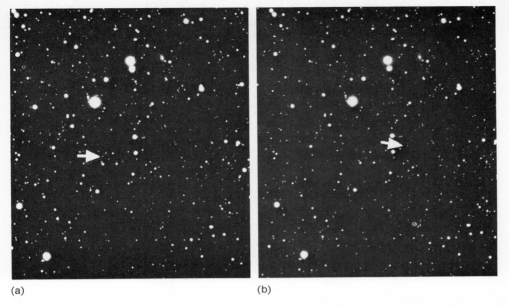

(a)  (b)

**Figure 18.26.** The discovery plates of Pluto, taken 6 days apart, show the motion of a new planet among the stars (see arrows): (a) January 23, 1930; (b) January 29, 1930. (Lowell Observatory)

Pluto has the most unusual orbit of all the planets, one that, besides being the most eccentric, its orbital plane is inclined 17° to the plane of the earth's orbit (ecliptic). Pluto orbits the sun at an average distance of 40 A.U., or 5.5 billion kilometers, but travels within less than 30 A.U. at perihelion (its closest approach to the sun) and out to almost 50 A.U. at aphelion (its most distant point away from the sun). The planet appeared at aphelion in 1865 and will again appear at this point in 2113. It will appear at perihelion in 1989. The orbit of Pluto is so eccentric as to bring it within the orbit of Neptune during the interval from 1979 to 1998 (see Figure 18.27). During this period, Neptune will be the farthest known planet from the sun. A collision between these two planets is impossible however, for while their paths appear to intersect, the planes of orbit do not coincide. Pluto requires almost 248 earth years for one revolution; hence it has not moved very far against the background of stars since its discovery.

Pluto appears only as a tiny speck in even the largest telescopes, and therefore its diameter may only be estimated at about one-half the diameter of the earth; its probable mass is roughly one twenty-fifth that of the earth. Recent observations of its rotation period, based on variations in brightness, produced what are considered to be quite reliable results: 6 days 9 hr 17 min. The temperature of the planet is unknown because of its distance but has been approximated at $-238°C$.

Both the eccentricity and inclination of Pluto's orbit suggest that it may represent an object that was not a member of the original solar system but was captured by the sun at a later time. Adopting this supposition, we will not try to

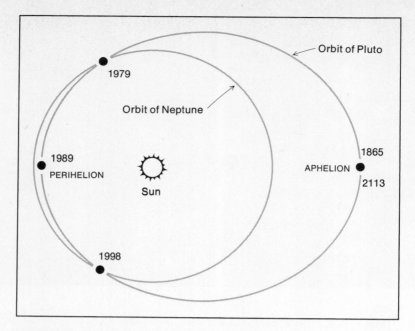

**Figure 18.27.** The orbits of Pluto and Neptune.

account for the planet's orgin in the section that follows. Some have suggested that Pluto may be an errant moon of Neptune, or at least that it may have interacted gravitationally with Triton, Neptune's largest moon. In fact, the physical characteristics of both Triton and Pluto are similar.

One of the most unexpected discoveries of 1978 was that Pluto has a moon. This fact is shown by a very dim elongated image on photographic plates. Named Charon, for the mythological oarsman who ferried souls of the dead across the river Styx, the moon orbits Pluto in just over six days. By its motion, astronomers have reassessed the mass of Pluto, making it the least massive planet of the solar system.

## ASTEROIDS

In 1772 the German astronomer Johann E. Bode (pronounced Boda) (1747–1826) published a mathematical relationship for the spacing of the planets. They developed this relationship by first writing a slightly modified geometric progression generated by doubling each number after 3 (see Table 18.3). The number 4 is then added to each term and the result divided by 10. Comparing the outcome with actual distances, we see very close agreement; however no object was known to exist at 2.8 A.U. Because of the close agreement of all other known planets, observers began to search for a possible planet orbiting at the average distance of 2.8 A.U. from the sun. In 1801 an object was found at approximately that

**Table 18.3** The Bode relationship

| PLANET | FIRST COLUMN | SECOND COLUMN | BODE NUMBERS | ACTUAL DISTANCES (A.U.) |
|---|---|---|---|---|
| Mercury | 0 | 4 | 0.4 | 0.39 |
| Venus | 3 | 7 | 0.7 | 0.72 |
| Earth | 6 | 10 | 1.0 | 1.0 |
| Mars | 12 | 16 | 1.6 | 1.5 |
|  | 24 | 28 | 2.8 |  |
| Jupiter | 48 | 52 | 5.2 | 5.2 |
| Saturn | 96 | 100 | 10.0 | 9.5 |
| Uranus | 192 | 196 | 19.6 | 19.2 |
| Neptune | 384 | 388 | 38.8 | 30.0 |
| Pluto | 768 | 772 | 77.2 | 40.0 |

distance. It was named Ceres and at first was considered to be the missing planet; however a year later another object (Pallas) was found, leading observers to suspect that there were more. In 1804, Juno, and in 1807, Vesta, were added to the list; today the estimates run to 25,000 such objects. These bodies range in diameter from about 800 km down to $\frac{1}{2}$ km and are referred to as *asteroids* (minor planets). Although most such objects orbit between Mars and Jupiter, in a band centered on 2.8 A.U., some asteroids orbit in rather eccentric elliptical orbits that carry them close to the sun at times and then almost out to Jupiter's orbit before they return (see Figure 18.28).

These asteroids are of particular interest because periodically some come within $\frac{1}{2}$ million km of the earth. As a given asteroid is viewed, it varies in apparent brightness in a periodic fashion. The interpretation of this observation is that asteroids must be nonspherical, probably more like a boulder in shape, tumbling in orbit around the sun. The question that arises from such evidence is in regard to the origin of the asteroids. Do they represent fragments of a planet that exploded or collided with another object, or are they simply irregular objects that remain today much as they were formed? Normally gravity tends to form spheres. Although we cannot answer the question with certainty, it is possible to estimate that the total mass of all the asteroids combined is less than that of the moon.

## COMETS

Like the planets and asteroids, comets move in elliptical-shaped orbits around the sun; hence comets are also true members of the solar system. Their orbits are

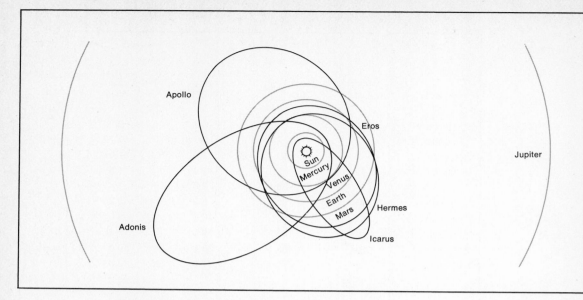

**Figure 18.28.** The orbits of selected asteroids.

highly elongated so that typically they pass within a few million miles of the sun, then recede to distances measured in billions or trillions of miles. Halley's Comet returns to the vicinity of the sun every 76 years. However, some comets return only after several thousand years. Kepler's laws apply to comets as they do to the planets and asteroids—elliptical-shaped orbits sweeping out equal area in equal time and $p^2 = r^3$ (see Figure 18.29).

Applying the law of equal area in equal time to such an eccentric orbit reveals that a comet may pass near the sun at 100 km/sec (360,000 km/hr) while traveling only 1 km/sec when at its most distant point. Since a comet is only visible when relatively near the sun, you can see that it spends most of its time in an invisible state.

**Figure 18.29.** Kepler's second law of equal areas in equal times, as applied to comets, reveals that they move very slowly when far from the sun and very rapidly when they are close to the sun.

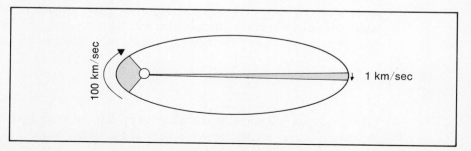

When far from the sun, a comet consists only of frozen gases that bind together bits of rocklike material, such as dust, sand, pebbles, rock, and perhaps a few boulders. This nucleus is like a swarm of dirty snowballs, typically a few kilometers across. As this nucleus moves to within 5 or 10 A.U. of the sun, the sun's radiant energy begins to vaporize some of the frozen gases and thus the comet grows a fuzzy head (of hair) called the *coma*, which may easily be as large as the earth. As the comet continues its approach to the sun, the outflow of atomic particles from the sun (the solar wind) drives off some of the coma gases to develop a tail. This tail may grow to a length of 100 million to 150 million km, always pointing away from the sun. After rounding the sun, the tail of a comet precedes the coma.

In any given year many comets become visible to the professional astronomer, probably as many as 20; however, only a few become bright enough to be obvious to the layman. On the other hand, it is possible for an amateur who is patient and persistent to be the first to discover a comet. If it should prove to be a new-found comet rather than a known (returning) comet, the amateur's name will be given to the comet.

## METEORS

The streaks of light (meteors) that are often seen in the night sky appear to be related to comets, because each time the earth in its orbit passes through the orbit of a comet, bits of material left behind by the comet enter the earth's atmosphere and are heated due to friction. The hot particles ionize the atoms that are encoun-

Table 18.4  Meteor showers

| APPROXIMATE DATE | NAME OF SHOWER | ASSOCIATED COMET | CONSTELLATION OF APPARENT ORIGIN |
|---|---|---|---|
| January 1–4 | Quadrantids | | Bootes |
| April 19–23 | Lyrids | 1861 I | Lyra |
| May 1–6 | May Aquarids | Halley | Aquarius |
| July 26–31 | Delta Aquarids | | Aquarius |
| August 10–14 | Perseids | 1862 II | Perseus |
| October 9–11 | Draconids | Giacobinizinner | Draco |
| October 18–23 | Orionids | Halley | Orion |
| November 1–15 | Taurids | Encke | Taurus |
| November 14–18 | Leonids | 1866 I | Leo |
| December 10–16 | Geminids | | Gemini |
| December 21–23 | Ursids | Tuttle | Ursa Minor |

tered; then the ionized atoms recombine with free electrons and emit light by downward transitions. Because the earth passes through the orbit of a comet on approximately the same date each year, meteor showers are predictable, as shown in Table 18.4. The comet with which a given shower is related is also shown.

Meteors are more readily visible when viewed in dark skies, away from city lights, and usually the streaks of light appear brighter in the early morning hours before dawn, for at this time the rotation and revolution of the earth carry it more directly into their path. Most *meteoroids* (objects that make the streaks of light) disintegrate in flight, before hitting the ground; however, if the object is larger than average, it may survive its flight and fall to the ground. If this happens, we then call the object a *meteorite*. Most meteorites have an appearance so similar to an ordinary stone that they may not be recognized as being a meteorite. More easily recognized are stony-iron and iron meteorites as illustrated in Figure 18.30 (b) and (c), respectively.

Typically, when meteorites are dated by means of radioactive decay, they show ages in the order of 4.5 billion years. They are particularly valuable for scientific analysis, for they may represent the primal material from which the entire solar system was formed.

**Figure 18.30.** Meteorite types: (a) Stony, with dark fusion crusts; (b) stony-iron and (c) the polished section of an iron meteorite, showing Widmannstatten lines. (Ron Oriti, Griffith Observatory)

# ORIGIN OF THE SOLAR SYSTEM

Perhaps you have noticed a number of facts concerning the solar system that seem to suggest a common, unified origin—the fact that all planets revolve about the sun in the same direction and on almost the same plane, the fact that moon rocks and earth rocks seem to share a dating between 3.5 billion and 4.5 billion years, the fact that the composition of the sun and of the Jovian planets is so similar, etc.

Most astronomers agree that stars are formed from gases that condense under the influence of gravity. The gas cloud (nebula) from which the sun formed had a component of rotation, and as it condensed, the nebula flattened due to its rotation. From this flattened nebula, the sun formed at the center and the planets formed by lesser condensations in the flattened disk. The moons formed in a similar way. Note that most of the 35 known moons of the solar system revolve around their parent planet in the same direction as the planet rotates. There are a few exceptions, i.e., Jupiter has five moons that orbit in retrograde direction and at various inclinations. These moons are thought to have been captured by Jupiter after its origin.

How may the origin of comets be accounted for in this model? Perhaps the comets formed before the planets and moons while the nebula was still very large, e.g., thousands of A.U. across. This would suggest that the solar system has a large supply of comets, orbiting the sun at very great distances, and only when they may be perturbed by another body do they change their orbit and move in toward the sun.

We have looked at our own solar system in some detail, but let us recognize the high probability that many stars of the Milky Way also have systems of planets revolving around them. Of the estimated 100 billion stars that compose the Milky Way galaxy, at least 100,000 stars are thought to be like our sun and may have created planets in their flattened disks.

## QUESTIONS

1. What is the difference in appearance of a planet and a star: (a) To the naked eye? (b) In a telescope?
2. Explain why planets sometimes appear to back up among the stars.
3. List the criteria for a good scientific model.
4. How did Galileo's discovery of the phases of Venus disprove the Ptolemaic theory?
5. What observation can astronomers make to prove that the earth revolves around the sun?
6. What is the shape of the orbits of the planets? How could you draw this shape accurately?
7. Kepler's second law may be abbreviated "equal area in equal time." What does it tell you about the motion of the planet?

8. True or false: The farther a planet is from the sun, the longer is takes to make one revolution.
9. Characterize the difference(s) between the terrestrial and the Jovian planets.
10. Why does Mercury show such a wide range in daytime and nighttime temperatures?
11. How many other planets have a nitrogen-oxygen atmosphere like that of the earth?
12. All the planets revolve about the sun in the same direction; however, two planets rotate backwards. Name these planets.
13. Why is Venus hotter on its surface than Mercury, even though it is farther from the sun?
14. List the reasons why man thought that Mars might support life.
15. True or false: We now have convincing evidence that life exists on Mars.
16. What is the primary constituent of the atmosphere of all the Jovian planets?
17. List the factors that make Jupiter distinctive from all other planets.
18. True or false: The rings of Saturn are a solid disk, each part of which orbits the planet in the same period.
19. The tilt of Uranus is specified as 98°. What can be said about any planet with a tilt more than 90°?
20. What observation (of Uranus) led to the discovery of Neptune?
21. State two peculiarities in the orbit of Pluto.
22. Pluto appears as only a speck in the largest telescopes. How may one judge its rate of rotation (spinning)?
23. True or false: There are so many asteroids that a spacecraft would certainly be hit if it passed through the asteroid belt.
24. Comets travel very rapidly near the sun but very slowly when far from the sun. What basic principle does this illustrate?
25. True or false: Comets possess a tail throughout their flight.
26. Typically meteor showers are associated with the orbital path of a comet. Can the times of these showers be predicted?
27. What evidence exists that would deny the idea that the sun simply captured its planets at random?
28. State in your own words how you think the solar system may have formed.

# 19

## stars and nebulae

The star that submits best to our close scrutiny is the sun. Of course the sun is an example of only one of many different types of stars, but it will serve as a basis for comparison. The sun is a huge, dynamic body. Its volume is over one million times that of the earth, and its mass is 300,000 times that of the earth. This means that the average density of the sun is only about one-fourth that of the earth, or 1.4 g/cm$^3$. Thus, we characterize the sun as more like a ball of gas. In the central core of the sun, where temperatures are very high (15 million° K) and pressures are measured in millions of kg/cm$^2$ due to the gravitational effects, atoms react with one another in ways that change their nuclei—a thermonuclear process. (See back end papers for explanation of ° K.)

## STAR ENERGY

Hydrogen atoms combine to produce a heavy form of hydrogen, deuterium, and a positron. Deuterium atoms combine with hydrogen atoms to form helium-3 atoms. A gamma ray is emitted in the process. Then helium-3 atoms combine to form helium-4 atoms and two hydrogen atoms that may participate in further reactions (see Figure 19.1).

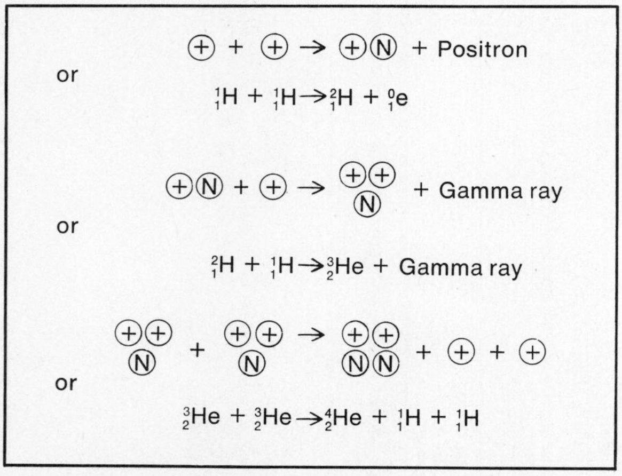

$$^1_1H + ^1_1H \rightarrow ^2_1H + ^0_1e + \text{Positron}$$

$$^2_1H + ^1_1H \rightarrow ^3_2He + \text{Gamma ray}$$

$$^3_2He + ^3_2He \rightarrow ^4_2He + ^1_1H + ^1_1H$$

**Figure 19.1.** The proton-proton cycle—a process whereby the sun converts part of its mass into energy every second.

The important thing to note is that the helium-4 atom formed has less mass than the four hydrogen atoms required to make it. What happened to the mass? It was converted into energy in the proportion denoted by $E = mc^2$, where $E =$ energy, $m =$ mass, and $c =$ speed of light. In fact, this is the source of energy of the stars: a thermonuclear reaction that converts mass to energy. Albert Einstein derived this relationship from theoretical considerations before any experiment had demonstrated it. Ultimately it was to lead to the hydrogen bomb. Yes, the stars shine by energy generated by the same basic process as the hydrogen bomb (thermonuclear fusion) but under a controlled situation. More properly, stars may be called *fusion reactors*.

It is interesting to see the magnitude of the constant of proportionality that relates energy and mass. $E = mc^2$ can be written $E/m = c^2$, and if $E$ is expressed in ergs and $m$ in grams, then $c = 3 \times 10^{10}$ cm/sec and $c^2 = 9 \times 10^{20}$ cm/sec (900,000,000,000,000,000,000). Hence in these units the ratio of energy to mass is $9 \times 10^{20}$ to 1. One might wonder how long the sun can last when one considers that 4.5 million metric tons of the sun's mass are converted into energy every second. Even at this rate, the sun's mass of hydrogen, if entirely used to produce helium, would last for more than 100 billion years. Other factors may control its destiny in a much shorter time however.

This great outpouring of energy expresses itself in several very dynamic ways. Sunspots appear and disappear on the visible surface of the sun periodically. These are relatively cool spots (see Figure 19.2). By contrast, extremely hot spots come and go, sometimes exploding as bright flares, releasing as much energy over a limited area as the remainder of the sun combined. X-ray observations reveal very energetic displays as well, and sometimes prominences erupt that send hydrogen gas hundreds of thousands of km out from the surface of the sun. All these activities are illustrated in Figure 19.3.

The sun also "wastes away" at the rate of 1 million metric tons/sec by expelling particles (primarily protons and electrons) in all directions. This is called

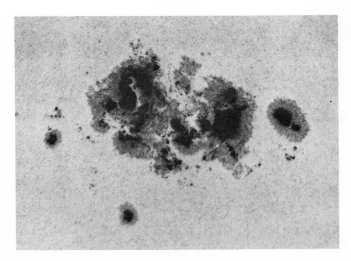

**Figure 19.2.** Sunspots are relatively cool spots on the visible surface of the sun. (Hale Observatories)

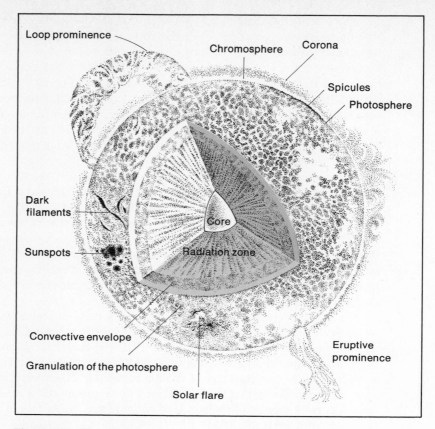

**Figure 19.3.** The active sun.

the *solar wind*, and it is this outflow of particles that drives the tail from the head of a comet.

Perhaps the most revealing device, in relation to the sun, is the spectrograph. When sunlight is collected and passed through the slit prism and focusing lens of the spectrograph, approximately 30,000 spectral lines appear (see Figure 19.4). Some of these lines are due to hydrogen (H) in the sun, others are due to sodium (Na), iron (Fe), calcium (Ca), magnesium (Mg), and argon (A). Over 70 elements have been identified on the sun as compared to the 92 naturally occurring elements on the earth. The spectrum reveals the composition of a region of the sun near its visible surface, not the core of the sun.

## STARS IN GENERAL

It would be accurate to say that stars come in all sizes and temperatures. If we think of our own solar system as extending out to the orbit of Mars (see Fig-

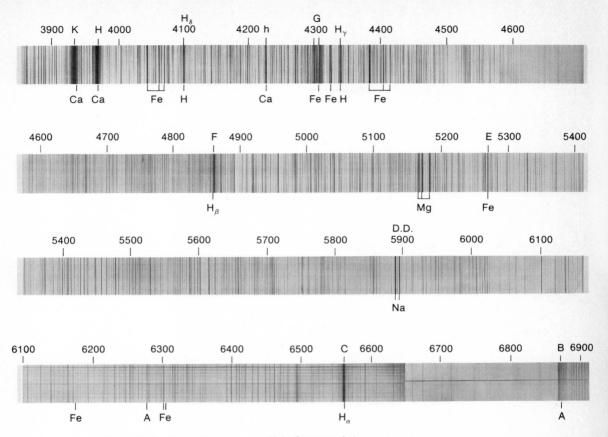

**Figure 19.4.** The solar spectrum. (Hale Observatories)

ure 19.5), it will assist us in comparing sizes of stars. Stars range in size down to as small as the earth (sometimes smaller) and up to as large a size as needed to fill the entire space defined by the orbit of Mars or slightly beyond. Yes, some very large stars have diameters of over 600,000,000 km.

The surface temperatures of stars also vary widely. Compared to the surface temperature of the sun (6000° K), we will find many stars in the range of 3000° K (or less) and many others with very high temperatures in the range of 50,000° K or higher. Within the last few years astronomers have extended their observations to record stellar energy in the infrared portion of the spectrum and as a result have discovered stars too cool to emit visible light—as cool as 300° K. Even the casual observer may detect the color of certain stars. For instance, Betelgeuse (in Orion), Aldebaran (in Taurus), and Antares (in Scorpius) are reddish in appearance, all about 3000° K. Sirius (in Canus Major), Rigel (in Orion), and Vega (in Lyra) are bluish in appearance, all between 25,000° K and 50,000° K.

Some stars appear much brighter than others. There are three basic factors that affect the apparent brightness of a star. First, you would probably suspect that

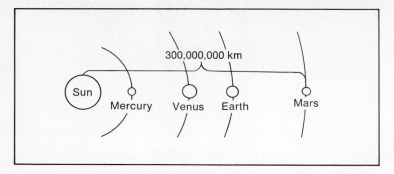

**Figure 19.5.** A few stars are large enough to fill the entire orbit of Mars.

the larger the star, the more surface it can radiate energy from, and therefore the brighter it would be. This is correct. A second direct relationship: The hotter the star, the more energy it radiates for each unit of surface area. In fact, a star of 12,000° K radiates 16 times as much energy per unit area as a star of 6000° K. Thus, these two factors (size and temperature) determine actual brightness. But because stars vary in their distance from us, similar stars may have different *apparent* brightnesses. For example, given two identical stars, if one is twice as far away, then it will appear one-quarter as bright. This element of distance, then, is the third factor that affects the apparent brightness of a star.

If we could place all the stars at the same distance, then only size and surface temperature would be factors, and we could compare star brightnesses as they really exist—called their *absolute magnitudes*. If a large number of stars are plotted onto a graph of absolute magnitude versus surface temperature, many points will fall along a diagonal line, labeled the "main sequence" in Figure 19.6. This shows a relationship one might expect: the hotter stars are typically the brighter stars; however, certain stars are bright in spite of being cool. The only possible explanation therefore involves their size. These cool but bright stars must be very large and so are called *red giants*. On the other hand, a few stars fall into the category of dim stars in spite of the fact that they are hot. These in turn must be very small and so are called *white dwarfs*.

## BINARY STARS

Many stars originate in pairs, and they continue to orbit a point somewhere between them due to their mutual gravitational effect (see Figure 19.7).

By observing their period of revolution, average separation, and the point about which they seem to orbit, the astronomer can compute the mass of each star. From such observations, the following relationship emerges. The more massive the star, the brighter it will be. Thus the mass of a star could be thought of as a fourth factor in determining its brightness (in addition to the factors of size, surface temperature, and distance mentioned in the previous section). The numbers that are shown alongside the Main Sequence in Figure 19.8 indicate the masses of stars

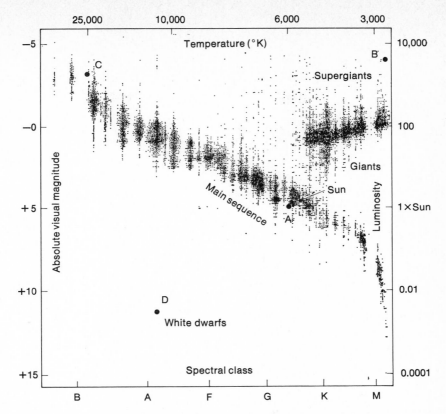

**Figure 19.6.** The H–R diagram, showing the surface temperature and absolute magnitude of approximately 6700 stars. (Yerkes Observatory)

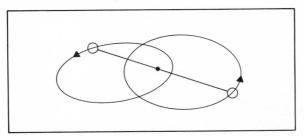

**Figure 19.7.** The orbital paths of stars in a binary system.

in solar masses (one solar mass equals the mass of the sun). We will see how the total amount of material that is gathered together in the creation of a star influences its life cycle.

## THE LIFE CYCLE OF A STAR

The great variety of star types visible in the universe is most likely due to the fact that we are viewing stars in different stages of their life cycle, some young and some older. Stars, like men, have a life cycle: they are born, have typical charac-

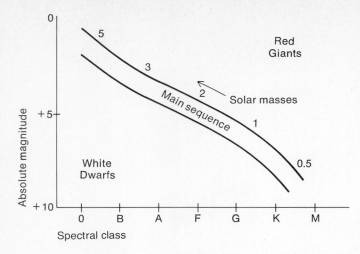

**Figure 19.8.** An H-R diagram showing the major subdivisions of stars.

teristics of youth, then settle down to become stable (Main Sequence stars), but in their old age may swell to become a red giant, then pass through a variable stage, perhaps becoming a nova, or supernova, and then shrivel up and die.

Figures 19.9 and 19.10 show places where astronomers think star formation is going on at the present time. In fact, Figure 19.11 represents direct evidence of changing condensations in the Orion nebula. Gravity is the force that creates stars out of gas and dust clouds (nebulae), and the dark granules shown in Figures 19.9 and 19.10 are thought to represent dusty regions where temperatures are low and molecular motion is slow. As gravity shrinks a huge section of a nebula, heat and pressure are generated, triggering the thermonuclear process described earlier in this chapter. Thus, a new force that resists gravity has been initiated, and the star begins to stabilize.

Young stars still in the process of stabilizing often vary rapidly in their light output and sometimes show *nebulosity* (gas and dust that are still in the process of condensing). The Pleiades in Figure 19.12 are an example of young stars. When the outflow of energy, created by the star, finally balances the infall tendency of gravity, the star becomes a Main Sequence star and remains as such most of its life.

**Figure 19.9.** The Lagoon nebula in Sagittarius. (Lick Observatory)

**Figure 19.10.** The nebula in Monoceros, birthplace of stars. (Hale Observatories)

A star like the sun has probably been a Main Sequence star for about 5 billion years and is expected to maintain that state for another 5 billion years. Then, when the supply of hydrogen fuel in the core of the sun begins to dwindle, the nuclear process will slow down, gravity will shrink the sun—heating it to even hotter levels, and the sun will react by throwing off a huge gas envelope that will then cool, turning the sun into a red giant. The sun will probably swallow up Mercury,

**Figure 19.11.** Herbig-Haro objects in Orion, showing growing knots of gas and dust. (Lick Observatory)

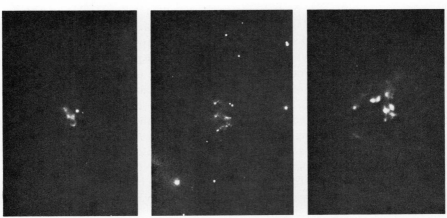

**Figure 19.12.** The Pleiades (Seven Sisters), showing nebulosity of gas and dust, which may still be condensing into these young stars. (Lick Observatory)

Venus, and Earth in their orbits. But this large envelope cannot be sustained, and the sun will finally shrink to become a white dwarf—a rather slow and unspectacular death.

More massive stars, however, go through their life cycle faster and come to a rather dramatic finale. They often explode as novae or supernovae, temporarily shining so brightly that they can be seen in the daytime, but then they fade to become *neutron stars,* stars in which the normal spaces that occur between protons, neutrons, and electrons are now occupied by these nuclear parts. Such a star is so dense that one single cubic centimeter of this material would weigh tons if brought to earth. Often such stars are rotating very rapidly, and they emit bursts of energy with each turn; hence they are called *pulsars.* One of the first pulsars to be discovered is located in the Crab nebula (see Figure 19.13); it is thought to be the remnant of a supernova that was recorded by the Chinese in A.D. 1054. Figure 19.14 shows that the Crab pulsar can be photographed "off" or "on" by means of a special rotating disc placed in front of the telescope. A possible mechanism whereby a star can appear to blink on and off 30 times per second is depicted in Figure 19.15—showing a condensed star rotating 30 times per second. The concentrated magnetic field may direct the star's remaining energy into two beams, each turning with the star like an airport beacon. With each rotation, an observer can witness a flash of light and a beep of radio energy.

Perhaps the nova and supernova represent one of the few examples of stellar evolution that has been recorded in the history of man. Typically one would not expect to see star changes because the overall life span of a star is usually measured in billions of years.

**Figure 19.13.** The Crab nebula, the results of a supernova explosion, first seen in A.D. 1054. (Hale Observatories)

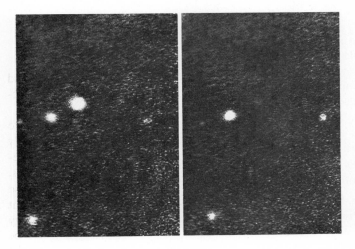

**Figure 19.14.** A pulsar, blinking on and off, caught in the act. (Lick Observatory)

## BLACK HOLES

One of the most interesting objects that has recently been discovered is the black hole. The black hole is thought to represent the way more massive stars die. When the nuclear "fires" of a very massive star dwindle, gravity tends to shrink the star with such force that even the nuclear structure must collapse; thus a star that was much larger than the sun and much more massive may now be collapsed to the size of an average city. The density of such a star is in the order of a thousand metric tons per $cm^3$ or more. In fact it may be true that the heart of the black hole can become infinitely dense. The surface gravity of such stars is so strong that light or radio energies cannot "get away," so they cannot be seen or detected. Then how

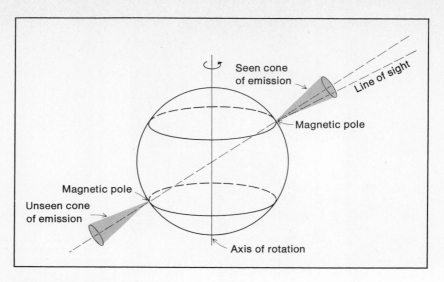

**Figure 19.15.** A model of a pulsar.

can we know that black holes actually exist? Perhaps it is because such a black hole exists in association with another star (a binary system), for example, a red giant (a star that has not evolved as far as the black hole). With the collapse of the stellar mass and the increase in surface gravity, the black hole may attract mass from its mate, the red giant. If material does fall into the black hole, astronomers believe that X-rays will be generated. A number of strong X-ray sources have recently been discovered and these sources are thought to be black holes.

The life cycle of a given star may be shown graphically by plotting successive stages in its evolution on a graph of absolute magnitude versus surface temperature. (see Figure 19.16). As one or both of these factors change, the location on the graph also changes.

## INTERSTELLAR MATERIAL

Not all of the material of the universe has condensed into stars, for as we have seen, huge glowing clouds of gas and dust exist between the stars (see Figure 19.17). These clouds glow (emit their own light) because hot stars are imbedded in them or exist nearby, and the ultraviolet radiation of those stars excite the atoms. Downward electron transitions in the atoms of these nebulae produce visible emissions; hence they are called *emission nebulae.* Other gas clouds show themselves by the absorption of certain wavelengths in starlight that passes through them. Even if a hydrogen gas cloud is not highly excited by nearby stars, it may be detected by its radio emissions. The hydrogen atom is known to emit a signal of 21 cm wave-

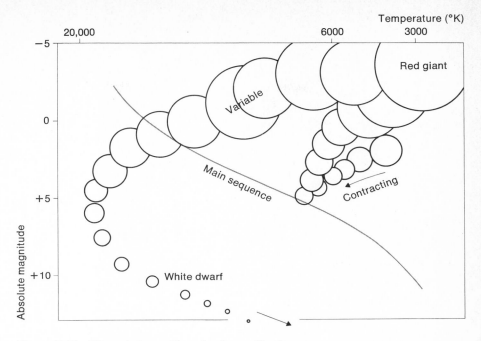

**Figure 19.16.** The evolutionary life-cycle of a star like the sun.

**Figure 19.17.** Glowing cloud of hydrogen gas; (a) the Orion nebula and (b) the Lagoon nebula. (Lick Observatory)

(a)  (b)

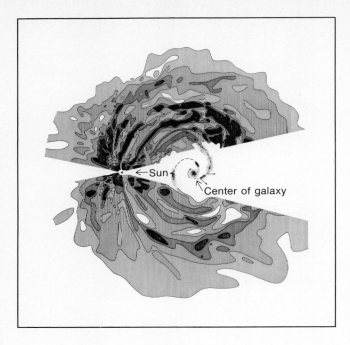

**Figure 19.18.** The 21-cm radio signal of hydrogen helped paint this picture of the Milky Way galaxy. (Leiden Observatory, Netherlands, and the Radio Physics Laboratory, Sydney, Australia)

length by merely a reversal in the spin of its electron. When the Milky Way galaxy is scanned on the 21-cm band, the distribution of hydrogen may be plotted (see Figure 19.18). The darker regions indicate higher concentrations of hydrogen.

## INTERSTELLAR MOLECULES

Astronomers have also detected more complex forms, namely, molecules, between the stars. A molecule typically vibrates, rotates, and translates (moves about), and in so doing generates radio signals that are unique to that molecule. By laboratory experimentation, the wavelength associated with each molecule may be determined. Then by directing a radio telescope that has been turned to the desired wavelength, the molecules can be identified (see Table 19.1).

## DUST NEBULAE

Typically, dust shows itself by obscuring the light from stars or emission nebulae beyond them. Such is the case in the Horsehead nebula (see Figure 19.19) and in the dark granules that are evident in certain nebulae [see Figure 19.17(b), p. 367]. We have already seen how these dark dust granules represent relatively cool places in the nebulae and may represent the centers of star formation. Dust may also exhibit itself as a reflection nebula surrounding young stars (see Figure 19.12,

**Table 19.1** Interstellar molecules

| MOLECULE | FORMULA | CHARACTERISTIC WAVELENGTH |
|---|---|---|
| Methylidyne | CH | 4300 Å |
| Methylidyne (ionized) | $CH^+$ | 3958 Å |
| Cyanogen radical | CN | 3875 Å, 2.6 cm |
| Hydroxyl radical | OH | 18.0, 6.3, 5.0, 3.7, 2.2 cm |
| Ammonia | $NH_3$ | 1.3, 1.2 cm |
| Water | $H_2O$ | 1.35 cm |
| Formaldehyde | $H_2CO$ | 6.6, 6.2, 2.1, 1.0, 0.2 cm |
| Hydrogen (gas) | $H_2$ | 1.060 Å |
| Carbon monoxide | CO | 2.7 cm |
| Methyl alcohol | $CH_3OH$ | 35.9 cm |
| Hydrogen cyanide | HCN | 3.4 mm |
| Cyanoacetylene | $HC_3N$ | 3.3 cm |
| Formic acid | HCOOH | 18.3 cm |
| Silicon monoxide | SiO | 2.3, 3.4 mm |
| Carbon monosulfide | CS | 2.0 mm |
| Formamide | $NH_2CHO$ | 6.5 cm |
| Carbonyl sulfide | OCS | 2.5 mm |
| Methyl cyanide | $CH_3CN$ | 2.7 mm |
| Isocyanic acid | HNCO | 1.36 cm, 3.4 mm |
| Methylacetylene | $CH_3CCH$ | 3.5 mm |
| Acetaldehyde | $CH_3CHO$ | 28.1 cm |
| Thioformaldehyde | $H_2CS$ | 9.5 cm |
| Methanimine | $CH_2NH$ | 5.8 cm |
| Hydrogen sulfide | $H_2S$ | 1.8 mm |

p. 364). The astronomer recognizes this type of nebula by the fact that it has the same spectrum as the nearby star (an absorption spectrum); whereas the emission nebula has its own typical spectrum (an emission spectrum).

## COSMIC RAYS

A cosmic ray is actually a charged atomic particle that has been accelerated to nearly the speed of light. It may consist of an electron, positron, proton (hydrogen nucleus), alpha particle (helium nucleus), or the nuclei of heavier elements. The accelerated particles carry very large amounts of kinetic energy (energy due to their motion). It is the motion of these particles, together with their mass, that makes

**Figure 19.19.** The Horsehead nebula in Orion, a dark cloud standing in front of a glowing nebula. (Hale Observatories)

cosmic rays out of otherwise normal atomic particles. Cosmic radiation is sensed, on the surface of the earth, from almost every direction; hence it is thought that these particles permeate the entire galaxy. Currently, many astronomers believe that the explosion of a supernova may be the source of such rays (particles).

QUESTIONS

1. What makes a star basically different from a planet?
2. In the core of the sun, hydrogen is continually converted to helium. What is the source of energy within this process?
3. The spectrum of the sun contains 30,000 dark lines. What have astronomers learned from this spectrum?
4. What evidence exists that both the sun and the earth may have formed from the same cloud of gas and dust?
5. What is the relationship between the surface temperature and the color of a star?
6. What factors determine the actual brightness of a star—that is, its absolute magnitude?

stars and nebulae

7. What factor, in addition to those stated in Problem 6, determines the apparent brightness of a star?
8. True or false: Stars are continually being "born" and are going through a life cycle.
9. What kind of star is thought to end its life as a black hole?
10. If light cannot escape a black hole, how can we detect the existence of such a hole?
11. Large amounts of neutral hydrogen have been found between the stars. How was this discovery made?
12. Molecules also exist between the stars. They generate radio energies. How can one molecule be distinguished from another?
13. Cosmic rays could be more properly called *cosmic particles*. How do they differ from interstellar dust?

# 20

the cosmos

We have been studying stars, nebulae, and cosmic particles. These are all members of a larger unit in the universe—the galaxy. The particular galaxy of which the sun and earth are members is called the *Milky Way galaxy*. It is a rather curious endeavor to try to decide what our galaxy looks like when we are a part of it and have no hope of seeing it from the "outside." At present rocket speeds, it would require at least 100,000,000 years to travel far enough to gain a view of the galaxy in the sense of seeing it as a whole from space. Therefore it may be helpful to look to other galaxies, of which there are many (perhaps 100 billion).

VARIETY IN GALAXIES

Galaxies show a great variety of sizes and shapes. Even a single telescopic view, shown in Figure 20.1, reveals this great variety. A closer look at certain galaxies shows that they may be classified as *ellipticals* if they are shaped like a ball (or football) (see Figure 20.2), *spirals* if they appear flattened like those in Figure 20.3, or *barred spirals* if they possess a barlike structure of brightness as shown in Figure 20.4.

Can you imagine how such galaxies would appear if you were embedded in each type? If we existed in an elliptical galaxy, the brightness of the night sky would be similar in all directions. On the other hand, if we existed in a flattened system, then a band of brightness corresponding to that disk should be apparent. On a clear, dark night we do see a fuzzy band across the sky that we call the Milky Way. A photograph through a telescope reveals that the fuzzy band is really a concentration of millions of stars, so the Milky Way band gives us a clue that the Milky Way galaxy is definitely a flattened system. In Figure 20.5 you see virtually the twin of the Milky Way—the Andromeda galaxy. The map of the Milky Way galaxy made by means of 21-cm radio observations, which we discussed in the last chapter (see Figure 20.6), clearly suggests the spiral arm structure of the galaxy. Astronomers have observed very dense clusters of stars that do not seem to be a part of the disk or nucleus of the galaxy but seem to form a "halo" around the nucleus (see Figure 20.7). These dense clusters are called *globular clusters*. From our point of view the globulars do not seem to be distributed evenly but appear in

**Figure 20.1.** A cluster of galaxies in Hercules, showing many different shapes of galaxies. (Hale Observatories)

**Figure 20.2.** A elliptical galaxy. (Hale Observatories)

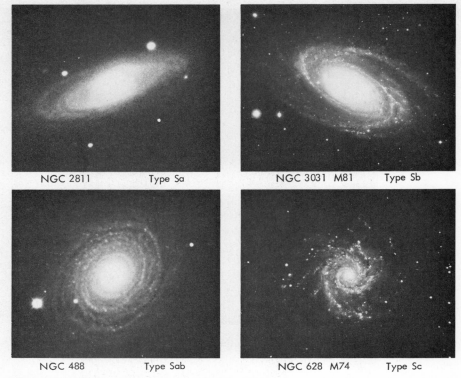

**Figure 20.3.** Spiral type galaxies. (Hale Observatories)

**Figure 20.4.** Barred-spiral type galaxies. (Hale Observatories)

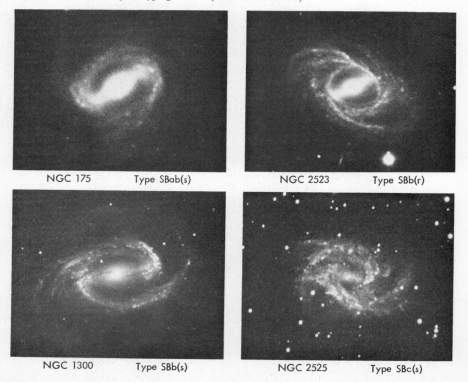

**Figure 20.5.** The great galaxy in Andromeda. (Hale Observatories)

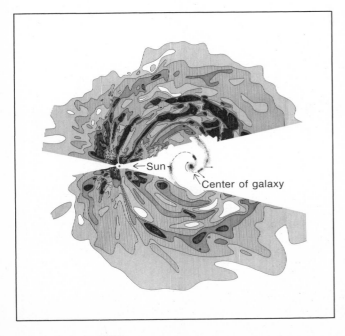

**Figure 20.6.** The 21-cm radio view of the Milky Way galaxy. (Leiden Observatory, Netherlands, and the Radio Physics Laboratory, Sydney, Australia)

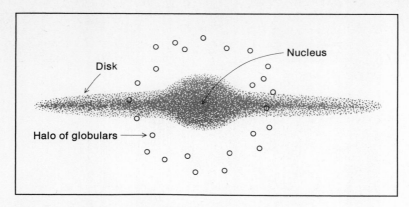

**Figure 20.7.** The halo of the Milky Way galaxy.

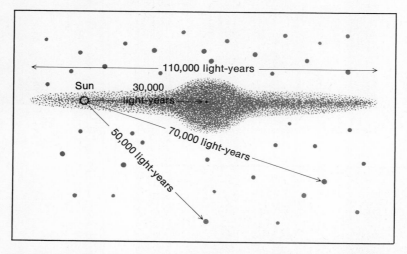

**Figure 20.8.** The distribution of globular cluster in relation to the Milky Way galaxy.

only one part of the sky. Hence, we conclude that we must be off center as shown in Figure 20.8.

The overall diameter of the Milky Way galaxy is estimated to be 110,000 light years, with the sun being 30,000 light years from the center. The sun is only one of 100 billion stars that compose this galaxy.

## UNUSUAL GALAXIES

A normal galaxy emits relatively small amounts of radio energy; however, there exist several classes of galaxies that emit radio energies in the order of one million times that of a normal galaxy. Cygnus A, shown in Figure 20.9, is such an example of such a galaxy. If we multiply the output of this radio galaxy by 100, we are led to an example of a peculiar radio galaxy, NGC 4486 (see Figure 20.10). The jetlike appendage suggests the possibility of a violent collapse or explosion as a source of strong radio noise.

**Figure 20.9.** Cygnus "A" a radio source. (Hale Observatories)

**Figure 20.10.** A peculiar galaxy, NGC 4486, showing jetlike appendige. (Lick Observatory)

# QUASARS

By the early 1960s radio astronomers were finding many new sources of radio emission, some of which appeared to be more like a star than a galaxy. These objects were called *quasistellar radio sources,* later contracted to *quasars.* Some of these objects have been photographed by optical telescopes as well, and it is interesting to note the jetlike appendage on an early discovery, 3C273 (see Figure 20.11). But what is more unusual is that when the spectrum of this quasar is recorded, it shows a Doppler shift that indicates a recessional velocity of 45,000 km/sec.

Today astronomers have spectra for well over 100 quasars, some of which have spectral red shifts that indicate velocities up to 90% of the speed of light.

**Figure 20.11.** A quasar, 3C273, showing a jetlike protrusion. (Hale Observatories)

## RED SHIFT LAW

Regardless of the type of galaxy we are viewing, a very interesting pattern in the motion of galaxies emerges. As you will recall from Chapter 8, when the source is moving away from the observer, its spectral lines are shifted toward the red end of the spectrum—red shifted. The spectrum of virtually every galaxy is red shifted, and the more distant galaxies have larger red shifts. Thus, not only do all the galaxies seem to be receding from us, but those that are farther away appear to be receding faster. Figure 20.12 shows the spectra of five galaxies, each more distant

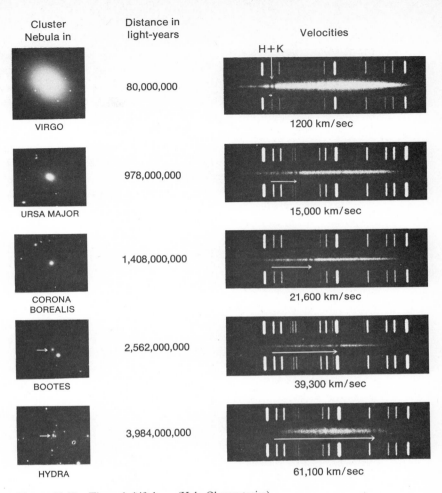

**Figure 20.12.** The red shift law. (Hale Observatories)

as you can see from their decreasing apparent brightness. The velocity of each of these galaxies is shown.

In 1929, the American astronomer Edwin P. Hubble, (1889–1953) showed that if you divide the velocity of each galaxy by its distance, you get almost the same result, about 15 km/sec for each million light years based on current observations. This is known as *Hubble's constant,* expressed as $H = v/r$. Because this pattern is so clearly established for galaxies over a billion light years away, most astronomers believe that it is safe to assume that the red shift can be used as a distance indicator. Certain of the quasars have unusually large red shifts in their spectra that correspond to velocities in the order of 270,000 km/sec (90% of the speed of light). If from this velocity we compute the distance according to Hubble's constant (15 km/sec per million light years), we get 18 billion light years.

Could any object at that great distance emit enough energy to make us aware of its existence? An ordinary galaxy could not be detected at so great a distance; hence the astronomers seek special explanations, such as new sources of energy or new processes. We should realize that if quasars are actually at these great distances, then we are looking backward in time billions of years. We are seeing objects as they existed nearer the beginning of the universe and perhaps some processes existed then that are not typical now.

There are some astronomers who would give a vastly different interpretation to the large red shift of quasars. They would explain that quasars are objects expelled from nearby galaxies at high velocity, and therefore no special explanation need be given for their apparent brightness. There are a number of objects with small red shifts that seem to be in proximity with galaxies having large red shifts.

It should be clear from the divergency of these two concepts of quasars that the matter is far from settled.

## COSMOLOGIES

When astronomers take the "big look" at the universe, they all agree that the red-shift pattern indicates an expanding universe—a universe that may have begun its expansion with a gigantic explosion called the *big bang*. Furthermore, most observers agree that the rate of expansion is slowing down—meaning that Hubble's constant tends to get smaller. The question is how fast is this rate changing. If the rate of expansion is slowing sufficiently, then the universe will end its expansion at sometime in the future and will begin to collapse under the force of gravity. Should the universe then begin expanding again with another big bang, we might call this a model of an *oscillating universe*.

It is possible that in spite of the slowing rate of expansion, the growth of the universe will never actually stop; hence this would indicate an open system. Many astronomers believe the universe to be an open system for another reason: If the total mass is below a certain level, then gravity will be too weak to halt its motion, and the universe will expand forever. Our present knowledge of that mass indicates that it falls far short of the critical level.

We have looked at the microcosm of the atom and have stretched our ability to visualize its structure. We have looked at the macrocosm of the universe and have tried to comprehend its size. Most of our lives are lived in a realm of experience somewhere between these two extremes.

## QUESTIONS

1. Which of the following might be contained within a galaxy: stars, solar systems, nebulae, star clusters, planets, atoms, the entire universe?
2. Our galaxy is called the Milky Way. This same title is given to a fuzzy band of light in the sky. What is the relationship between these two facts?

3. How do we know which basic form our galaxy takes when we live inside it and cannot see it from the same perspective as we see other galaxies?
4. How do we find our location within the Milky Way galaxy?
5. The term *quasar* is a contraction of what phrase?
6. List several facts about quasars—characteristics they have in common.
7. What seems to be indicated by the very large red shifts found in quasars?
8. What basic force tends to slow the expansion rate of the universe?
9. Is it possible for the universe to be slowing in its rate of expansion and still never stop expanding?
10. We see distant galaxies all receding from us. Does this mean that we are the center of the universe? Explain.

# 21

extraterrestrial life

It would never have occurred to an ancient observer to ask the question, "Am I alone in the universe?" for his universe consisted of only a flat earth with a canopy of stars overhead. Not until man realized that the sun is a star, and a rather ordinary star at that, did he begin to see the possibility for other planetary systems—planets going around other stars as their sun. But what is the probability that other systems exist? To answer this question we must look to theories concerning the origin of stars in general.

Astronomers believe that stars form as a result of condensations within vast clouds of gas and dust, largely due to gravity but perhaps also influenced by magnetic fields acting on plasma. As a huge gas and dust cloud condenses, it spins faster and faster (assuming at least a small rotational component in the original cloud) and flattens to form a disk of material surrounding the *protostar* (a star in the making). Then, depending on the mass of the protostar, which ultimately determines its surface temperature, planets may form in the disk as they did surrounding the sun. In the case of very massive stars, which are also very hot, materials may be blown away by the solar wind before planets can form. In the case of low-mass stars, the star may remain so cool that even if planets formed, life could not exist on them. So it seems more reasonable to assume that only stars about the mass and temperature of the sun may have planets circling them, some of which are suitable for life. Out of the 100 billion stars in the Milky Way galaxy, it seems reasonable to assume that at least 100,000 are similar to the sun in mass and surface temperature.

As in our own solar system, not all planets are habitable. Those closer to the sun than the earth are far too hot and those beyond Mars are far too cold, so any system of planets must have a limited habitable zone. This conclusion is based on the very narrow view that all life must be like our own. If we were to take a broader view of life, then much larger probabilities open up. Even within our present knowledge we should broaden our estimate a bit, for microorganisms exist in highly unlikely places—one species being able to live within the fuel tanks of jetliners, subsisting on kerosene. Others are known to survive a wide range of temperatures.

## WHAT IS LIFE?

All this really begs the question: What is life? What makes the living essentially different from the nonliving? We think of the ability to reproduce like kind as being unique among living organisms, yet certain nonliving compounds are able, in a sense, to reproduce themselves. Likewise, certain living organisms, like viruses, may lie dormant (appearing to have no reproductive process) for many years, then begin replication again. This is to suggest that the division between the living and the nonliving is not as sharp as one might think. We do know that in all earth forms that are clearly alive, replication is accomplished by means of a code within each cell that directs the replication process. The encoding molecule is called *deoxyribonucleic acid* (DNA). The question is, Should we impose this restriction on life elsewhere in the universe? Must life be as *we* know it—essentially based on organic molecules known as *carbon rings?* Perhaps living organisms in other solar systems sustain life on silicon molecules or other substances. Be this as it may, a further process seems to be characteristic of living organisms—that is, a process whereby certain materials and/or energies are utilized from the environment to sustain life. By this we mean the process of photosynthesis in plants or the various metabolic processes in animals.

## THE SEARCH FOR LIFE

In our search for life among the planets of our solar system or of others, we must utilize our knowledge of all such processes. For instance, the search for life on Mars (the Viking probes) consisted in part of a search for gases typically released as a product of metabolism. On the other hand, if our search for life were limited to those planets on which we could land space probes, we would have virtually exhausted the possibilities already. If we are to search for life on planets going around other stars, we must look to another aspect of some living creatures, namely, intelligence. Only if we can communicate with other intelligent beings, via some form of electromagnetic energy, can we traverse the great distances involved in our quest. In fact, for communication to be possible we must assume that intelligent beings on other planets have sufficiently advanced technology to make radio communications possible; that is, they must possess radio telescopes, transmitters, and receivers at least as adequate as our own. Our technology in this area has been developed only in the last 20 years. This is not to say that others haven't already advanced far beyond us in this respect and in many other respects. When we think of the history of radio technology in relation to the history of the earth, we realize that our technology has peaked up in only the last small fraction of earth time. Figure 21.1 suggests this peaking up.

## DIALOGUE OR MONOLOGUE?

If we hope to communicate with another civilization, is it essential that its technology "peak up" at the same time as ours? Perhaps not; in fact it may be more advantageous if it is otherwise. To understand such a statement, let's review the relationship of time and distance. Radio signals travel at the speed of light; hence if an intelligent community, with radio capabilities, were located at a distance of

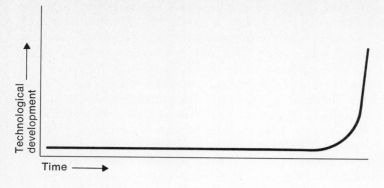

**Figure 21.1.** A time scale of technological development.

10,000 light years, it would require 10,000 years for their radio signal to reach us. If they had only recently developed radio capabilities and decided to beam a message in our direction, we would not receive it for another 10,000 years. Likewise our message, if sent today, would not reach them for 10,000 years. More advantageous would be the coincidence that the distant civilization perfected the radio 10,000 years ago and beamed a message then. We would just be receiving the information now. You can see how unlikely two-way communication would be unless the civilizations are physically very close. If you had only the possibility of one-way communication, within the span of several generations, what would you want to convey to an extraterrestrial civilization—or what would you hope they had communicated to you? What code or language would you use? On what wavelength would you transmit and/or listen?

Perhaps one could overcome the language barrier by transmitting pictures, as we have done from space probes or as we have been doing for the past 25 years via TV. Television and radio signals have been traveling outward from the earth for many years without any concerted effort to communicate with alien beings. Wouldn't it be a strange coincidence if the TV signals were the first to reach them? On the other hand, if we intentionally designed a visual message, it could include information as to our location in the universe, in the galaxy, and in the solar system; the nature of our being (man and woman); and the nature of our technology; but perhaps more vital than any of these would be how we solved our social problems in order to have survived so long. Perhaps a more advanced civilization might give us a clue as to how to solve some of our most perplexing problems!

The fundamental question that remains, assuming we believe it worthwhile to attempt communication, is a choice of one or more wavelengths on which to broadcast. Of the billions of possibilities, certain wavelengths may be good choices. The most prevalent element in the entire universe is hydrogen, and this atom emits a characteristic radio signal of 21-cm wavelength. Certainly any technically advanced civilization would be aware of this fact and would most likely be listening at that wavelength simply because so much can be learned about the universe in this way. Other atoms and molecules also have characteristic wavelengths that could be used with some degree of logic.

**Figure 21.2.** The Cyclops proposal for a radio telescope array. (NASA-Ames)

## OUR COMMITMENT

The question, during the past few years, centers about the commitment we are willing to make to the search for extraterrestrial life through communication. One proposal included a complex of radio antennae like that shown in Figure 21.2. The estimated cost of such a project is between 6 and 10 billion dollars. Perhaps it would be more reasonable to move in the direction of simply committing a certain percentage of existing radiotelescope time to this search. If such a commitment is made, perhaps we'll find that we are not alone in the universe.

## QUESTIONS

1. Why does it seem reasonable to assume that many other solar systems may exist in the Milky Way galaxy?
2. If the probability of stars having planets is one in a million, then what do you think may be the probability of life existing elsewhere in the universe? Substantiate your view.
3. What is the function of the DNA molecule in living cells?

4. Suppose intelligent beings existed on a planet 1000 light years away and they possessed a radio technology compatible with ours. What time delays would exist in trying to carry on a dialogue? How would you solve the problem of such a delay?
5. Why would the 21-cm radio wavelength be a natural for communication with extraterrestrial beings?
6. Devise a code or language with which one might communicate with extraterrestrial beings.

# appendixes

**Appendix 1** Orbital data of the planets

| PLANET | SYMBOL | SEMIMAJOR AXIS (A.U.) | SIDEREAL PERIOD | SYNODIC PERIOD | ECCENTRICITY OF ORBIT | INCLINATION OF ORBIT | AVERAGE ORBITAL SPEED (km/sec) |
|---|---|---|---|---|---|---|---|
| Mercury | ☿ | 0.387 | 87.97 days | 116 days | 0.2056 | 7.0° | 47.8 |
| Venus | ♀ | 0.723 | 224.7 days | 584 days | 0.0068 | 3.4° | 35.0 |
| Earth | ⊕ | 1.000 | 365.26 days | — | 0.0167 | 0.0° | 29.8 |
| Mars | ♂ | 1.524 | 687.0 days or 1.88 years | 780 days | 0.0934 | 1.8° | 24.2 |
| (Ceres[a]) | ① | 2.77 | 4.60 years | 467 days | 0.0765 | 10.6° | 17.9 |
| Jupiter | ♃ | 5.20 | 11.86 years | 399 days | 0.0484 | 1.3° | 13.1 |
| Saturn | ♄ | 9.54 | 29.46 years | 378 days | 0.0557 | 2.5° | 9.7 |
| Uranus | ♂ or ♅ | 19.18 | 84.01 years | 370 days | 0.0472 | 0.8° | 6.8 |
| Neptune | ♆ | 30.06 | 164.79 years | 367.5 days | 0.0086 | 1.8° | 5.4 |
| Pluto | ♇ | 39.44 | 248.40 years | 366.7 days | 0.2502 | 17.2° | 4.7 |

[a] An asteroid.

**Appendix 2** Physical and rotational data for the planets

| PLANET | DIAMETER km | DIAMETER $E^a = 1$ | MASS ($E^a = 1$) | DENSITY (WATER = 1) | PERIOD OF ROTATION | INCLINATION OF EQUATOR TO ECLIPTIC[b] | ALBEDO | SURFACE GRAVITY ($E^a = 1$) | VELOCITY OF ESCAPE (km/sec) |
|---|---|---|---|---|---|---|---|---|---|
| Mercury | 4,880 | 0.38 | 0.05 | 5.4 | 58$^d$ 15$^h$ | 7° | 0.07 | 0.39 | 4.3 |
| Venus | 12,108 | 0.95 | 0.82 | 5.3 | 243$^d$ 4$^h$ | 176° | 0.76 | 0.90 | 10.3 |
| Earth | 12,750 | 1.00 | 1.00 | 5.52 | 23$^h$ 56$^m$ | 23°27′ | 0.39 | 1.00 | 11.2 |
| Mars | 6,800 | 0.53 | 0.11 | 3.82 | 24$^h$ 37$^m$ | 25° | 0.18 | 0.38 | 5.1 |
| Jupiter | 142,800 | 11.20 | 317.9 | 1.33 | 9$^h$ 50$^m$ | 3° | 0.45(?) | 2.58 | 59.5 |
| Saturn | 120,000 | 9.41 | 95.2 | 0.69 | 10$^h$ 14$^m$ | 26°45′ | 0.61(?) | 1.11 | 35.6 |
| Uranus | 50,800 | 3.98 | 14.6 | 1.3 | 24$^h$ | 98° | 0.35(?) | 1.07 | 21.4 |
| Neptune | 49,500 | 3.88 | 17.2 | 1.7 | 22$^h$ | 29° | 0.62 | 1.40 | 23.6 |
| Pluto | 3,000 | 0.24 | 0.002 | 0.7 | 6$^d$ 9$^h$ 17$^m$ | ? | 0.50 | ? | ? |

[a] E is the earth.
[b] An inclination greater than 90° indicates retrograde rotation.

**Appendix 3**  Satellites of planets

| PLANET | SATELLITE | DISCOVERER | MEAN DISTANCE FROM PLANET (km) | SIDEREAL PERIOD (DAYS) | INCLINATION OF ORBIT TO PLANET'S EQUATOR | DIAMETER OF SATELLITE (km) | APPROXIMATE MAGNITUDE AT OPPOSITION |
|---|---|---|---|---|---|---|---|
| Earth | Moon | — | 384,405 | 27.322 | 23.5° | 3476 | −12.5 |
| Mars | Phobos | A. Hall (1877) | 9,380 | 0.319 | 1° | 27 | 11.5 |
|  | Deimos | A. Hall (1877) | 23,500 | 1.262 | 2° | 12 | 12.0 |
| Jupiter | V | Barnard (1892) | 180,500 | 0.498 | ∼0° | 160 | 13.0 |
|  | I Io | Galileo (1610) | 421,800 | 1.769 | ∼0° | 3658 | 5.5 |
|  | II Europa | Galileo (1610) | 671,400 | 3.551 | ∼0° | 3100 | 5.7 |
|  | III Ganymede | Galileo (1610) | 1,070,000 | 7.155 | ∼0° | 5300 | 5.0 |
|  | IV Callisto | Galileo (1610) | 1,884,000 | 16.689 | ∼0° | 5000 | 6.3 |
|  | XIII | Kowal (1974) | 11,100,000 | 239 | 26.7° | 10 | 20 |
|  | VI | Perrine (1904) | 11,470,000 | 250.57 | 27.6° | 120 | 14.0 |
|  | VII | Perrine (1905) | 11,800,000 | 260.10 | 24.8° | 40 | 17.5 |
|  | X | Nicholson (1938) | 11,850,000 | 263.55 | 29.0° | 20 | 19.0 |
|  | XII | Nicholson (1951) | 21,200,000 | 617.0r[a] | 147.0° | 20 | 18.5 |
|  | XI | Nicholson (1938) | 22,600,000 | 692.5r | 164.0° | 24 | 19.0 |
|  | VIII | Melotte (1908) | 23,500,000 | 735.0r | 145.0° | 20 | 17.5 |
|  | IX | Nicholson (1914) | 23,700,000 | 758.0r | 153.0° | 22 | 19.0 |
|  | XIV | Kowal (1975) | — | — | — | — | 20 |
| Saturn | Janus | A. Dollfus (1966) | 168,700 | 0.749 | ∼0° | 350 | 14.0 |
|  | Mimas | W. Herschel (1789) | 185,800 | 0.942 | ∼0° | 520 | 12.0 |
|  | Enceladus | W. Herschel (1789) | 238,300 | 1.370 | ∼0° | 600 | 12.0 |
|  | Tethys | Cassini (1684) | 294,900 | 1.888 | ∼0° | 1200 | 10.5 |
|  | Dione | Cassini (1684) | 377,900 | 2.737 | ∼0° | 800 | 11.0 |
|  | Rhea | Cassini (1672) | 527,600 | 4.518 | ∼0° | 1300 | 10.0 |
|  | Titan | Huygens (1655) | 1,222,600 | 15.945 | ∼0° | 4800 | 8.3 |
|  | Hyperion | Bond (1848) | 1,484,100 | 21.277 | ∼0° | 400 | 13.0 |
|  | Iapetus | Cassini (1671) | 3,562,900 | 79.331 | 14.7° | 1300 | 11.0 |
|  | Phoebe | W. Pickering (1898) | 12,960,000 | 550.45r | 150.1° | 300 | 14.0 |
| Uranus | Miranda | Kuiper (1948) | 130,000 | 1.414r | 0° | — | 19.0 |
|  | Ariel | Lassell (1851) | 191,000 | 2.520r | 0° | 600 | 15.0 |
|  | Umbriel | Lassell (1851) | 266,000 | 4.144r | 0° | 400 | 16.0 |
|  | Titania | W. Herschel (1787) | 436,000 | 8.706r | 0° | 1000 | 14.0 |
|  | Oberon | W. Herschel (1787) | 583,400 | 13.463r | 0° | 800 | 14.0 |
| Neptune | Triton | Lassell (1846) | 355,500 | 5.877r | 160° | 4000 | 14.0 |
|  | Nereid | Kuiper (1949) | 5,567,000 | 359.881 | 27.7° | 320 | 19.0 |
| Pluto | Charon | Christy (1978) | 20,000 | 6.39 | — | 1000 | 20.0 |

[a] The notation "r," when shown with the sidereal period of satellite, indicates that the satellite orbits the planet in retrograde motion.

# Appendix 4  The twenty brightest stars

| STAR | RIGHT ASCENSION (1950) | DECLI-NATION (1950) | DIS-TANCE (PARSECS) | PROPER MOTION | SPECTRA OF COMPONENTS[a,c] | | | VISUAL MAGNITUDES OF COMPONENTS[b,c] | | | ABSOLUTE VISUAL MAGNITUDES OF COMPONENTS[c] | | |
|---|---|---|---|---|---|---|---|---|---|---|---|---|---|
| | | | | | A | B | C | A | B | C | A | B | C |
| | (h) (m) | (°) (') | (pc) | (") | | | | | | | | | |
| Sirius | 6 42.9 | −16 39 | 2.7 | 1.32 | A1V | wd | — | −1.47 | +7.1 | — | +1.4 | +10.5 | — |
| Canopus | 6 22.8 | −52 40 | 30 | 0.03 | F0Ib | — | — | −0.72 | — | — | −3.1 | — | — |
| α Centauri | 14 36.2 | −60 38 | 1.3 | 3.68 | G2V | K5V | M5V | −0.01 | +1.5 | +10.7 | +4.4 | +5.8 | +15 |
| Arcturus | 14 13.4 | +19 27 | 11 | 2.28 | K2III | — | — | −0.06 | — | — | −0.3 | — | — |
| Vega | 18 35.2 | +38 44 | 8.0 | 0.34 | A0V | — | — | +0.04 | — | — | +0.5 | — | — |
| Capella | 5 13.0 | +45 57 | 14 | 0.44 | G0II | M1V | M5V | +0.09 | +10.2 | +13.7 | −0.7 | +9.5 | +13 |
| Rigel | 5 12.1 | −8 15 | 250 | 0.00 | B8Ia | B9 | — | +0.10 | +6.6 | — | −6.8 | −0.4 | — |
| Procyon | 7 36.7 | +5 21 | 3.5 | 1.25 | F5IV-V | wd | — | +0.38 | +10.7 | — | +2.7 | +13.1 | — |
| Betelgeuse | 5 52.5 | +7 24 | 200 | 0.03 | M2Iab | — | — | +0.41v | — | — | −5.5 | — | — |
| Achernar | 1 35.9 | −57 29 | 20 | 0.10 | B5V | — | — | +0.47 | — | — | −1.6 | — | — |
| β Centauri | 14 00.3 | −60 08 | 90 | 0.04 | B1III | — | — | +0.63 | — | — | −4.1 | — | — |
| Altair | 19 48.3 | +8 44 | 5.1 | 0.66 | A7IV,V | — | — | +0.77 | — | — | +2.2 | — | — |
| α Crucis | 12 23.8 | −62 49 | 120 | 0.04 | B1IV | B3 | — | +1.39 | +1.9 | — | −4.0 | −3.5 | — |
| Aldebaran | 4 33.0 | +16 25 | 16 | 0.20 | K5III | M2V | — | +0.86v | +13 | — | −0.2 | +12 | — |
| Spica | 13 22.6 | −10 54 | 70 | 0.05 | B1V | — | — | +0.91 | — | — | −3.6 | — | — |
| Antares | 16 26.3 | −26 19 | 120 | 0.03 | M1Ib | B4V | — | +0.92v | +5.1 | — | −4.5 | −0.3 | — |
| Pollux | 7 42.3 | +28 09 | 12 | 0.62 | K0III | — | — | +1.16 | — | — | +0.8 | — | — |
| Fomalhaut | 22 54.9 | −29 53 | 7.0 | 0.37 | A3V | K4V | — | +1.19 | +6.5 | — | +2.0 | +7.3 | — |
| Deneb | 20 39.7 | +45 06 | 430 | 0.00 | A2Ia | — | — | +1.26 | — | — | −6.9 | — | — |
| β Crucis | 12 44.8 | −59 24 | 150 | 0.05 | B0.5IV | — | — | +1.28 | — | — | −4.6 | — | — |

[a] The Roman numerals after the spectral classifications have the following meanings: Ia or Ib, *supergiant;* II or III, *giant;* IV, *subgiant;* V, main-sequence star. The notation "wd" indicates *white dwarf.*
[b] The notation "v" following the magnitude indicates a *variable star.*
[c] When entries are shown in both A and B columns, the star is known to be a binary system. When an entry is also shown in the C column, the system is known to have three components.

# index

## A

Absolute humidity, 276
Absolute magnitude, 360
Absolute zero, 70
Absorption spectrum, 136
Acceleration, 37
  due to gravity, 51
Acetylene, 234
Acid-base reaction, 229
Acids, 227
Action-reaction, 40
Active metals, 195
Adams, John, 345
Advancing polar front, 292
Age of Aquarius, 308
Age of earth rocks, 265
Age of Pisces, 309
Alchemy, 16
Aldebaran, 359
Alexander the Great, 13
Alexandria, 13
Al-Hazen, 24
Alpha particle (helium nucleus), 265, 369
Alphonso X, 22
Alternating current generator, 116
Amino acids, 241
Amplitude of a wave, 81
Anaximenes, 9, 160
Aneroid barometer, 275
Angular momentum, 42
Antares, 359
Antiacid, 229
Antinode, 91
Apparent brightness of stars, 360
Aquinas, St. Thomas, 23
Arabic influence, 20
Archimedes, 14

Aristarchus, 13
Aristotle, 11, 23
Asteroids, 348
Astrology, 6
Atmosphere, 272, 276
Atmospheric pressure, 274
Atom, 159
  Bohr model, 134
  early concept of, 10
  $p$-orbitals, 200
  radioactive, 161
Atomic energy, 176
  levels of, 190
Atomic model of a charged body, 100
Atomic theory, 160
Atomic weight, 171, 185
Atomos, 160
Autumn equinox, 308
Average velocity, 36
Avogadro, Amadeo, 226
Avogadro's number, 226

## B

Babylonia, 6
Bacon, Francis, 29
Bacon, Roger, 24
Bagdad, 20
Balancing equations, 225
Barometer, 275
Barred spiral galaxies, 374
Barrel effect, 54
Barycenter, 311
Bases, 227
Battery, 106
Becquerel, A. H., 161
Benedictine Order, 23
Bernoulli, Daniel, 160

Beta particles, 265
Betelgeuse, 359
Big Bang, 382
Binary stars, 360
Binary system, 301, 310
Binocular, 142
Biodegradable, 241
Black holes, 365
Blink Microscope, 346
Bode, Johann E., 348
Bohr, Niels, 134
Bohr model of the atom, 134
Boyle, Robert, 29
Boyle's law of gases, 207
Brahe, Tycho, 323
Brittany (France), 8
Bronze age, 3
Bruno, Giordono, 29

C

Calendars:
    Egyptian, 5
    Julian, 6
Callisto, moon of Jupiter, 341
Calorie, 71
Camera, 141
Carbohydrates, 236
Carbon, 232
Carbon-14, 266
Carbon monoxide, 295
Carbon rings, 235, 387
Cavendish, Lord, 49
Cellulose, 237
Celsius, 69
Cenezoic era, 267
Centigrade, 69
Centimeter, 32
Centrifugal force, 53
Centripetal force, 53, 56
Change of state, 217
Charging an electroscope, 101
Charging by induction, 101
Charles, Jacques A., 208
Chemical bonds, 194
Chemical energy, 221
Chemical equations, 225
Chemical reaction, 184
Chemistry of living organisms, 231
Chinese science, 20

Cholesterol, 239
Church dogma, 23, 28
Circular motion, 53
Clouds:
    forms, 291
    seeding, 289
Colliding plates, 253
Coma, 350
Comet, 349
Compounds, 185
Condensations in sound waves, 87
Conductor, 106
Conservation of energy, 68
Constantinople, 22
Constructive interference, 84
Copernican model, 26
Copernicus, Nicholaus, 25
Cordova, Spain, 21
Coriolis effect, 280
Cosmic rays, 369
Cosmologies, 382
Cosmos, 373
Coulomb, a quantity of electrical charge, 104
Coulomb, Charles, 103
Coulomb's law, 103
Covalent bonding, 196
Crab pulsar, 364
Crystalline form, 215
Crystals:
    sphere, 11
    structure, 261
Curie, Marie S., 161
Cyclone, 293
Cygnus A, 379

D

Dalton, John, 160
Dark ages, 20
Dating organic material, 266
Daughter product, 265
da Vinci, Leonardo, 24
de Broglie, Louis V., 168
Decay, radioactive, 174
Deferent, 17
Democritus, 10, 160
Density, 34
Destructive interference, 84
Detergents, 239

Deuterium, 356
Dialogue, extraterrestrial, 387
Diamond, 213
Diffraction, 85
Direct current generator, 122
Dispersion, 133
DNA (deoxyribonucleic acid), 242, 387
Doppler Effect, 137
Doppler shift, 380
Dust nebulae, 368
Dynamic earth, 245

## E

Ear, human, 90
Early humans, 2
Earth:
   history of, 258
   size of, 13
Earth-moon system, 301
Earthquakes, 246
Eclipses, 313
Ecliptic, 305
Efficiency, 61
Egypt, 4
Einstein, Albert, 154
Electrical energy, transmission of, 120
Electrical force, 98, 104
Electric circuits, 107
Electricity, 97
Electrolyte, 106
Electromagnetic spectrum, 125
Electromotive force, 106
Electron, 134, 163, 369
Electronegativity, 201
Electron microscope, 145
Electron orbitals, 191
Electroscope, 100
Elements, 185
   family groups of, 185
Elliptical galaxies, 374
Emission nebulae, 366
Endothermic reaction, 222
Energy, 59
   atomic, 176
   conservation of, 68
   gravitational, 66
   kinetic, 67
   potential, 65
England, Stonehenge, 8

Entropy, 74
Enzymes, 242
Epicycle, 16
Eratosthenes, 13
Erosion, 257
Ethene, 234
Euphrates river, 4
Europa, moon of Jupiter, 341
Evening star, 331
Exothermic reaction, 222
Extraterrestrial life, 385
Eye, human, 140

## F

Family groups of elements, 185
Fitzgerald, George F., 154
Flint tools, 3
Fluorescent tube, 218
Foods, 236
Foot, 32
Force, 37
   centrifugal, 53, 56
   centripetal, 53, 56
Force field, 104
Foucault, Jean, 302
Foucault pendulum, 302
Fructose sugar, 236
Fusion, thermonuclear, 179
Fusion reactor, 357

## G

Galaxies:
   unusual, 378
   variety in, 374
Galileo Galilei, 26, 51
Gamma rays, 162, 265
Ganymede, moon of Jupiter, 341
Gases:
   laws, 207
   properties of, 206
Generator, 119
   alternating current, 116
   direct current, 122
Geological eras, 267
Geology of the ocean floor, 287
Geuricke, Otto von, 29
Gilbert, William, 29
Globular clusters, 374

Glucose sugar, 236
Glutamic acid, 241
Goldstein, Richard, 342
Gram-formula-weight, 226
Grand Canyon, 267
Graphite, 214
Gravitation, concept of, 46
Gravitational energy, 66
Gravitational force, 56, 104
Gravity, 45, 51
    universal nature of, 49
Great Red Spot, Jupiter, 340
Greece, 9
Greenhouse effect, 276
Ground state of energy, 166

## H

Half-life of a radioactive element, 265
Halley's comet, 350
Halogens, 195
Harmonics, 94
Heat, 69, 71
    of fusion, 217
    mechanical equivalent, 72
    of vaporization, 217
Heraclides, 13
Heraclitus, 9, 160
Hero, 15
Hertz, Heinrich, 167
High fidelity in music systems, 94
Hipparchus, 16
Horizontality, principle of, 268
Horsepower, 65
Hubble, Edwin P., 381
Hubble's constant, 381
Humidity, 276
Hurricane, 293
Hydrocarbons, 232, 295
Hydrogenation, 238
Hydronium ion, 227
Hydroxide ion, 228

## I

Igneous rock, 258
Impulse, 41
Inch, 32
Inductive method, 29

Inertia, 37, 46
Instantaneous velocity, 36
Insulator, 106
Interference, 84
Interstellar material, 366
Interstellar molecules, 368
Io, moon of Jupiter, 341
Ionic pairs, 185, 195
Isomer, 234
Isotope, 171, 263

## J

Joule, James, 72
Jovian planets, 327
Jupiter, 339
    moons of, 341

## K

Kelvin, 70
Kepler, Johannes, 323
Kepler's laws, 324
Kilocalorie, 72
Kinetic energy, 67
Kinetic-molecular theory, 207

## L

Leucippus, 10
Lever, 62
    law of, 14
Leverrier, Joseph, 345
Life, 386
Life cycle of a star, 361
Light:
    speed of, 127
    nature of, 129
Lipids (fatty acids), 237
Liquid state, 209
London-type smog, 297
Longitudinal waves, 85
Lorentz, Hendrick A., 154
Lorentz contraction, 154
Lowell, Percival, 332, 346
Lucretius, 10
Lunar eclipse, 316
Luther, Martin, 25

## M

Magnetic declination, 114
Magnetic north pole, 113
Magnetism, 97, 109
  atomic model of, 111
Main sequence of stars, 360
Mariner 9, 332
Mariner 10, 329
Mars, 331
Marsh gas, 296
Mass, 34
Mass and energy equivalency, 156
Matter:
  states of, 205
  units of, 184
Mechanical advantage, 61, 63
Mendeleev, Dmitri, 185
Mercury, the planet, 328
Mesopotamia, 4
Mesosphere, 273
Mesozoic era, 266
Metals, 189
Metamorphic rock, 258
Meteor, 351
Meteorite, 352
Meteoroid, 352
Meter, 32
Metric system, 32
Michelson, Albert A., 151
Michelson-Morley experiment, 152
Microscope, optical and electron, 144
Mid-Atlantic Ridge, 266
Milky Way galaxy, 374
Millimeter, 32
Minerals, 261
Mixture, 185
Mohammed, 23
Mohole Project, 250
Mole, 226
Molecular kinetic energy, 69
Molecular motion in solids, 216
Momentum, 40
Moon:
  age of, 318
  trip to the, 317
  history of, 318
Moon of Pluto, 348
Moorish influence, 21
Morley, Edward W., 151
Moslems, 21
Motion, 31
  circular, 53
Musical sounds, 90

## N

Neap tides, 284
Nebulae, 355
  dust, 368
Nebulosity, 362
Negative ion, 106
Neolithic period, 3
Neptune, 345
Neutral solution, 228
Neutron stars, 364
Newton, a unit of force, 53
Newton, Sir Isaac, 37
Newton's First Law, inertia, 37
Newton's Second Law, $F = ma$, 38, 52
Newton's Third Law, action-reaction, 39, 53
Nile, 94
Nitric oxides, 296
Nitrogen, 276
Node, 91
Nomad, 3
Nonmetals, 189
Northern lights, 115
Nuclear reaction, 184
Nucleus, 165
  forces in, 170

## O

Ocean currents, 281
Ocean of air, 271
Ocean of water, 271
Ockham, 24
Octane, 233
Oersted, H. C., 110
Ohm's law, 107
Open universe, 382
Orbiting satellites, 56
Oresme, Nicolas, 24
Organic molecules, 232
Organic succession, principle of, 268
Organ pipe, 92

Origin of the solar system, 353
Orion nebula, 362
Oscillating charged particle, 126
Oscillating universe, 382
Outer core of earth, 250
Overtones (harmonics), 94
Oxidation reaction, 224
Oxidation-reduction reaction, 224
Oxygen atom, 199

## P

Paleozoic era, 267
Palomar, Mt., 142
PAN, a pollutant, 297
Parallel circuits, 108
Particulate matter, 297
Path of totality, 313
Pauli, Wolfgang, 192
Pendulum, 68
Periodic nature of elements, 183
Periodic table, 193
Period of a wave, 81
Phases of the moon, 311
pH of a solution, 228
Photochemical smog, 296
Photosynthesis, 236
Pioneer 11, 342
Pisa, Leaning Tower of, 51
Planets, physical properties, 326
Plasmas, 218
Plastic wrap, 198
Plate tectonics, 251
Plato, 11, 23
Platonic solids, 12, 213
Pluto, 346
　moon of, 348
Plutonic rock, 258
Polar covalent bonds, 197
Pollution:
　of the atmosphere, 295
　solution to the, 298
Polyunsaturated fat, 238
$p$-orbitals in the atom, 200
Positive ion, 106
Positron, 369
Potential energy, 65
Power, 64
Power loss, 120

Precambrian era, 267
Precession of the equinoxes, 308
Proper length, 155
Proper time, 155
Protein, 241
Proton, 134, 369
Ptolemaic model, 17
Ptolemy, 17, 26
Pulsars, 364
P waves (earthquakes), 248
Pythagoras, 9

## Q

Quantum mechanical model, 191
Quantum theory of radiation, 166
Quartz, 261
Quasars (quasi-stellar radio sources), 380

## R

Ra, sun god, 5
Radioactive atoms, 161
Radioactive elements, 263
Radioactivity, 171
Raisin pudding model of the atom, 165
Rarefactions in sound waves, 87
Rebirth of science, 19
Red giant star, 360
Red shift law, 380
Reduction reaction, 225
Reflection, 83
　law of, 131
Reformation, 25
Refraction, 83, 131
Refractor telescope, 131
Relative humidity, 277
Relative motion, 150
Relativity, 149
Resonant wavelengths, 91
Rest mass, 34, 156
Revolution of the earth, 302
Rigel, a star, 359
Rings of Uranus, 345
RNA (ribonucleic acid), 242
Rocks, age of, 265
Rotation of the earth, 302
Rutherford, Ernest, 162, 165

## S

Salt, 195, 229
Saturn, 342
Schiaparelli, Giovanni, 331
Search for extraterrestrial life, 387
Second, a unit of time, 35
Sedimentary rock, 258
Seismograph, 248
Seismologist, 248
Series circuits, 108
Signs of the zodiac, 7, 305, 322
Simple harmonic motion, 80
Sine wave, 117
Sirius, 359
Sixty-cycle current, 118
Soap, 239
Solar heating, 277
Solar system, 321
  origin of, 353
Solar wind, 358
Solenoid, 112
Solid state, 211
Sonar, 89
Sonic boom, 88
Sound, 77, 86
  musical, 90
  quality of, 94
  speed of, 88
Spectrum, 133
  of the sun, 136
Speed of light, 127
Spiral galaxies, 374
Spring tides, 284
Square meter, 33
Standing waves, 90
Stars, 355
  binary, 360
  black holes, 365
  energy of, 356
  in general, 358
  life cycle of, 361
States of matter, 205
Steam engines, 73
Steam turbine of Hero, 16
Steroids, 241
Stone age, 2
Stonehenge, 8
Stratosphere, 272
Strong acid, 228
Subduction, 253
Sublimation, 218
Sucrose sugar, 237
Sun's spectrum, 136
Superposition, principle of, 268
Supersonic transport, 89
Superstition, 7
S waves (earthquakes), 248
Synchronous rotation and revolution, 319
System Internationale, 33

## T

Telescope, 141
  refractor, 131
Temperature, 69
  function of, 274
Terrestrial planets, 326
Thales, 9, 160
Thermal inversion, 295
Thermodynamics, law of, 74
Thermonuclear fusion, 179
Thermosphere, 273
Thompson, Joseph J., 163
Tides, 281
  effect of, 285
Tigris river, 4
Time, 35
Toledo, Spain, 21
Tombaugh, Clyde, 346
Torque, in a magnetic field, 115
Torricelli, Evangelista, 29, 275
Transformer, 120
Transmission of electrical energy, 120
Transverse waves, 78
Triton, moon of Neptune, 346
Troposphere, 272
21-cm radio signal, 374, 388
Typhoons, 293

## U

Uniformitarianism, principle of, 268
Units of matter, 184
Unusual galaxies, 378
Uranium, 265

Uranus, 343
   rings of, 345
Urea, 232
U-shaped valley, 257

## V

Valence, 193
Van Allen Belts, 114
Vega, a star, 359
Velocity, 35
   instantaneous, 36
Venus, 330
   phases of, 308
Vernal equinox, 308
Vibrations in sound, 95
Viking 1 and 2, to Mars, 334, 387
Volcanic rock, 258
Volcanos, 250
Voltaic cell, 106
Voyager 2, spacecraft, 345
V-shaped valleys, 257

## W

Wanderers in the sky, 322
Water molecule, 240
Wavelength, 81, 128
Waves, 77
Weak base, 229
Weather:
   fronts, 292
   maps, 293
   principle of prediction, 287
Weathering, 257
Weight, 53
White dwarf star, 360
Wind, currents of air, 278
Wohler, Friedrich, 232
Work, 60

## X

X-ray sources, 366

## Y

Yard, a unit of length, 32
Year, a unit of time, 4

## Z

Zeeman, Peter, 191
Zeeman effect, 191
Zodiac, 7

Powers-of-ten notation

In writing very large or very small numbers, it is convenient to use the following system of notation:

$10^1 = 10$
$10^2 = 10 \times 10 = 100$
$10^3 = 10 \times 10 \times 10 = 1000$
$10^4 = 10 \times 10 \times 10 \times 10 = 10,000$

Following this pattern:
$10^{12} = 1,000,000,000,000$

A light-year is approximately equivalent to 6,000,000,000,000 miles, which could be written $6 \times 1,000,000,000,000$ miles, or $6 \times 10^{12}$ miles—a much simpler notation.

In a very similar way:
$10^{-1} = 0.1 = 1/10$
$10^{-2} = 0.01 = 1/100$
$10^{-3} = 0.001 = 1/1000$

Following this pattern:
$10^{-7} = 0.0000001$

The wavelength of blue light is approximately 0.0000005 m, but this number is equal to $5 \times 0.0000001$, therefore it may be written $5 \times 10^{-7}$ m.

**Summary:**
If given $7 \times 10^9$, move the decimal nine places to the right, which produces 7,000,000,000.

If given $7 \times 10^{-9}$, move the decimal nine places to the left, which produces 0.000000007.

Constants with useful approximations[a]

Pi $(\pi) = 3.14159$
$\cong 22/7$
Velocity of light, $c = 2.99793 \times 10^{10}$ cm/sec
$\cong 300,000$ km/sec
$\cong 186,000$ mi/sec
Constant of gravitation,
$G = 6.67 \times 10^{-8}$ dyne-cm$^2$/g$^2$
Angstrom unit, Å $= 10^{-10}$ m
Astronomical unit, A.U. $= 1.49598 \times 10^{11}$ m
$\cong 150,000,000$ km
$\cong 93,000,000$ mi
Parsec $= 206,265$ A.U.
$= 3.262$ light-years
Light-year $= 9.4605 \times 10^{15}$ m
$\cong 9.5 \times 10^{12}$ km
$\cong 6 \times 10^{12}$ mi
Mass of the sun, $m_s = 1.991 \times 10^{33}$ g
Mass of the earth, $m_e = 5.98 \times 10^{27}$ g
Mass of the proton, $m_p = 1.672 \times 10^{-24}$ g
Mass of the neutron $= 1.674 \times 10^{-24}$ g
Mass of the electron $= 9.108 \times 10^{-28}$ g
Charge of the electron
$= 1.60 \times 10^{-19}$ coulomb
Avogadro's number
$= 6.025 \times 10^{23}$ molecules/mole
Planck's constant $= 6.625 \times 10^{-34}$ joule-sec
Acceleration due to gravity $= 980$ cm/sec$^2$
$= 32$ ft/sec$^2$

[a] Round-number approximations are indicated by $\cong$.